Palgrave Studies in Sustainable Business In Association with Future Earth

Series Editors

Paul Shrivastava, Pennsylvania State University, University Park, PA, USA

László Zsolnai, Corvinus University of Budapest, Budapest, Hungary

Sustainability in Business is increasingly becoming the forefront issue for researchers, practitioners and companies the world over. Engaging with this immense challenge, Future Earth is a major international research platform from a range of disciplines, with a common goal to support and achieve global sustainability. This series will define a clear space for the work of Future Earth Finance and Economics Knowledge-Action Network. Publishing key research with a holistic and trans-disciplinary approach, it intends to help reinvent business and economic models for the Anthropocene, geared towards engendering sustainability and creating ecologically conscious organizations.

More information about this series at
http://www.palgrave.com/gp/series/15667

Thomas Walker · Dieter Gramlich · Kalima Vico · Adele Dumont-Bergeron
Editors

Water Risk and Its Impact on the Financial Markets and Society

New Developments in Risk Assessment and Management

futurearth
research for global sustainability

palgrave macmillan

Editors
Thomas Walker
Emerging Risks Information Centre (ERIC)
John Molson School of Business
Concordia University
Montreal, QC, Canada

Kalima Vico
Emerging Risks Information Centre (ERIC)
John Molson School of Business
Concordia University
Montreal, QC, Canada

Dieter Gramlich
Department of Banking
DHBW - Duale Hochschule
Baden-Württemberg
Heidenheim an der Brenz, Germany

Adele Dumont-Bergeron
Department of English
Concordia University
Montreal, QC, Canada

ISSN 2662-1320 ISSN 2662-1339 (electronic)
Palgrave Studies in Sustainable Business In Association with Future Earth
ISBN 978-3-030-77649-7 ISBN 978-3-030-77650-3 (eBook)
https://doi.org/10.1007/978-3-030-77650-3

This Palgrave Macmillan imprint is published by the registered company Springer Nature Switzerland AG
The registered company address is: Gewerbestrasse 11, 6330 Cham, Switzerland

Acknowledgements

We acknowledge the financial support provided through the Global Risk Institute, the Mitacs Research Training Award Program, and the Higher Education Institutional Excellence Program 2020 of the Ministry of Innovation and Technology (TKP2020-IKA-02) at Corvinus University of Budapest. In addition, we appreciate the excellent copy-editing and editorial assistance we received from Tyler Schwartz, Sherif Goubran, and Gabrielle Machnik-Kekesi. Finally, we feel greatly indebted to Arlene Segal, Joseph Capano, Anne-Marie Croteau, and Norma Paradis (all at Concordia University) who in various ways supported this project.

Contents

Notes on Contributors

Luc Bonnafous is a Disaster Risk Management Specialist at the World Bank. He focuses mostly on climate risk, both from a scientific and modeling perspective and from a project implementation standpoint. Prior to this, he was a climate researcher at Beyond Ratings, and a Staff Associate at the Columbia Water Center.

Giulia Christianson is a Senior Associate at WRI's Finance Center, where she leads the Sustainable Investing Initiative. The initiative develops research and tools designed to encourage and empower mainstream investors to pursue sustainable investment strategies—re-directing private capital flows toward a sustainable, inclusive, and low-carbon future. Giulia contributed to the inception of WRI's Sustainable Investing Initiative in 2015. Since joining WRI in 2011, her work has also spanned other finance topics, including mainstreaming climate change within financial institutions, improving the effectiveness of public finance in mobilizing private sector climate investment, and scaling environmental entrepreneurship in emerging markets.

Paulina Concha Larrauri is a Researcher at the Columbia Water Center at Columbia University. She holds a B.S. in Chemistry from the ITESM in Monterrey, Mexico, and an M.S. degree in Environmental Engineering from Columbia University. Prior to starting a career in water, she worked for five years in engineering at Procter & Gamble. Her research focuses on risk assessments of tailings dams, rainwater harvesting, developing a risk-based approach for the valuation of water in the mining industry, developing strategies for water management and climate risk in agricultural supply chains, and evaluating the feasibility of implementing decentralized water systems, among others.

Erica Coulombe is the Vice-President of Millani Inc., a consulting firm based in Montreal, and has more than seven years of consulting experience in sustainability and responsible investment in both Canada and Europe. After an internship with the World Resources Institute in Washington, DC, Erica joined Deloitte's Sustainability Group before moving to France where she worked for GreenFlex, a leading sustainability consulting firm based in Paris. In March 2017, Erica joined Millani where she provides advisory services in ESG integration for both investors and companies. Having worked and supported companies of all sizes and sectors, from SMEs to institutional investors, Erica has developed a deep understanding of sustainable development and can bridge the gap between companies and investors on environmental, social, and governance issues.

Milla Craig is the Founder and President of Millani Inc., an independent advisory firm helping investors to integrate ESG issues into their investment decisions, companies to communicate their material ESG issues to investors, boards of directors to understand their ESG responsibilities, and capital markets participants to create their proprietary ESG strategies. Prior to founding Millani, Milla worked for more than 15 years in the institutional equity markets for major Canadian financial institutions such as RBC Dominion Securities and Scotia Capital. Throughout her career, she has worked closely with senior management and investor relations teams of Canadian publicly listed companies, as well as financial analysts and investment managers of several Quebec pension and investment funds.

Adele Dumont-Bergeron is a Graduate Student in English Literature and Creative Writing at Concordia University, Montreal. She currently serves as a Research Associate in the Department of Finance. Adele has copyedited over a hundred chapters and articles by professionals in finance for publication. She recently completed research projects about plastic taxes and social impact bonds, which have resulted in co-authored articles. Her research now focuses on water risks and their impacts on the financial system and society.

Nourhane ElGarhy is an ESG Consultant at Millani Inc. An economist by training, Nourhane has extensive experience in finance, with an emphasis on impact and responsible investing. After working for several years in corporate and investment banking where she supported organizations across various sectors, she joined the Trottier Family Foundation where she led the design and implementation of sustainable finance and impact investing strategies and frameworks. Nourhane leverages her expertise in finance and her passion for sustainability to support organizations in navigating the new business paradigm and create lasting value for all stakeholders through better integration of sustainability principles and management of environmental, social, and governance risks and opportunities.

John Gibson has worked in different areas of the water industry for more than twenty years. His main focus has been research and municipal infrastructure, including both wastewater and drinking water. He spent ten years at the Wastewater Technology Centre, part of the Canada Centre for Inland Waters. More recently John was awarded a Ph.D. in civil engineering from the University of Toronto. John's research focuses on drinking water distribution systems, fire protection, and water quality.

Dieter Gramlich is a Full Professor of Banking & Finance at DHBW—Baden-Württemberg Cooperative State University, Heidenheim, where he serves as head of the Banking Department. He received his Ph.D. from the University of Mannheim and his Habilitation degree from the University of Halle. His research focuses on financial risk and return management, systemic financial stability, and sustainable finance. He has published widely in these areas and recently published a book on the topic of emerging risk management.

Karsten Haustein is a Trained Meteorologist and Climate Scientist with expertise in extreme weather event attribution. He worked in the World Weather Attribution team at the Environmental Change Institute of the University of Oxford, where he also carried out the work on the Global Warming Index. He has recently joined the German Institute for Climate Services (GERICS) in Hamburg, Germany, where he is now working as a scientist in the ClimXtreme project on extreme weather events, an endeavor which aims at bridging the gap between stakeholders and scientific data analysts.

Laureline Josset is an Associate Research Scientist at the Columbia Water Center, where she works on the evaluation of water stresses through integrated assessments, the optimization of management strategies for water quantity and quality, and transitions in the food-energy-water nexus. Collaborating with actors from government and civil society, Laureline focuses on the quantification of risks arising from climate uncertainty and the provision of data to inform decisions. Before joining Columbia, she obtained a Bachelor's and Master's degree in Physics from the École polytechnique fédérale de Lausanne, Switzerland, and a Ph.D. in Earth Sciences from the University of Lausanne. Laureline teaches classes on water system analysis and groundwater management for Columbia's Sustainable Management Program, with a particular emphasis on conceptual modeling and system thinking.

Bryan Karney is a Professor of Civil Engineering at the University of Toronto and has been the Associate Dean of the university's Cross-Disciplinary Programs Office since 2009. He is a Principal of HydraTek with more than thirty years of experience in providing hydraulic and hydraulic transient consulting services on a wide range of fluid systems. Bryan has spoken and written widely on subjects related to water resource systems, energy issues, hydrology, climate change, optimization, engineering education, and engineering ethics. He has been Associate Editor for the ASCE's Journal of Hydraulic Engineering and is now Associate Editor for IAHR's Journal of Hydraulic Research. Bryan has been the recipient of a number of teaching and research awards.

Deepak Krishnan is the Associate Director for the World Resources Institute (WRI) India's Energy Program and leads work on clean energy initiatives (renewable energy & energy efficiency) including in different consumer categories, Clean Energy Transitions, and Water-Energy nexus. Prior to this, he worked with the energy consulting practices of PricewaterhouseCoopers and Deloitte. He is a certified Energy Risk Professional. He is an electrical engineering graduate and holds a master's degree in electric power systems.

Upmanu Lall is the Director of the Columbia Water Center and the Alan and Carol Silberstein Professor of Engineering at Columbia University. He is a world-renowned expert in statistical and numerical modeling of hydrologic and climatic systems and water resource systems planning and management. He has pioneered statistical methods and their application to the prediction of hydrologic and climate conditions, and advanced tools for decision analysis and risk management.

Charlotte Lombardi is a Senior ESG Consultant at Millani Inc. She has extensive experience in non-financial corporate disclosures, allowing her to advise clients on how to best communicate ESG issues. After working in strategy consulting for professional service firms for several years, she commenced work at RY, a leading communications agency in London, UK. There, she specialized in stakeholder engagement and advised on non-financial disclosures and sustainability communications for British and Continental European listed companies.

Tianyi Luo is the Acting Director for the Aqueduct Project at the Global Water Program. He directs Aqueduct data and tool development and oversees advanced geospatial and statistical analysis for Aqueduct applications. Tianyi has worked directly with more than a dozen multi-national companies on quantifying water risks and associated financial implications in value chains, as well as international development banks and aid agencies on assessing regional water security and physical climate risks. Besides engaging in corporate and national water risk assessment, Tianyi also leads the Water Program's efforts on the water-energy nexus, particularly for power generation and unconventional gas sectors, and works across WRI's Climate, Energy, Finance, and Economics programs and centers.

Paras Mahajan is pursuing a Master's Degree in Commerce from Hindu College at the University of Delhi. He has prior experience in business development and portfolio management at the National Skill Development Corporation (NSDC). He pursued a Bachelor of Business Administration (Finance and Investment Analysis) from Shaheed Sukhdev College of Business Studies. His earlier research work on embedded water derivatives has been published in the International Journal of Management Studies. He has a growing research interest in finance, financial product development, and sustainable finance.

Jack McClamrock was a Research Analyst at the World Resources Institute.

Dharisha Mirando leads CWR's investor engagement and water risk valuation work. She is the Lead Author of the recent series of reports on coastal threats and developed the CWR APACCT 20 Index to benchmark cities against coastal threats. She has also published reports with

Manulife Asset Management, the Asia Investor Group on Climate Change (AIGCC), and CLSA and is a frequent commentator on water and climate risks and the need for adaptation on Bloomberg. She hails from the finance industry—prior to joining CWR, she worked in the investment team of a long-only public equities fund and in the impact investment space in London and Singapore.

Hamendra Kumar Porwal has written numerous articles, which have been published in reputed national and international journals. He has chaired and presented various papers in national and international seminars. Hamendra has teaching experience (both at national and international institutions) of more than three decades.

Quintin Rayer has worked for actuarial and investment consultancy firms and a multi-national European bank, including broad experience in quantitative fund and risk analysis. He is a Fellow of the Institute of Physics, a Chartered Fellow of the CISI, and a Chartered Wealth Manager. Quintin has applied skills gained from his Oxford University Physics Doctorate and while working in engineering to finance. He is the second UK graduate from the Sustainable Investment Professional Certification (SIPC) program. In January 2017 Quintin joined P1 Investment Management, founding their ethical and sustainable investing proposition.

Guneet Sandhu is a Ph.D Student in the Sustainability Management Program at the University of Waterloo, Canada. Her research focuses on developing a locally attuned interdisciplinary water risk management framework and decision-support tools to foster sustainable water management in the corporate and financial sector. She holds a Master of Environmental Studies in Sustainability Management (Water) from the University of Waterloo along with a Master of Engineering (Chemical) from Cornell University, USA, and a Bachelor of Engineering (Chemical) with Honors from Panjab University, India. She has previously worked as a project engineer for the non-profit AguaClara Reach, where she implemented sustainable drinking water treatment technologies in under-served communities.

Jason Siegel is a Ph.D. Candidate in the Department of Earth & Environmental Engineering at Columbia University's Henry Krumb School of Mines.

Debra Tan heads China Water Risk (CWR), a think tank that aims to embed water and climate risks in business and finance. Working with financial and government-related institutions and corporates, she built CWR from an idea into a 'go-to' resource on water risks. The reports ex-banker Debra has written and curated on the financial implications of sectoral water risks are considered groundbreaking and instrumental in understanding Asia's water challenges. Debra has published in influential policy journals in China, has contributed to UN publications, and sits on global advisory and technical councils for corporations as well as the Climate Disclosure Standards Board (CDSB) and the Carbon Disclosure Project (CDP). She spends her spare time exploring Himalayan glaciers.

Kalima Vico is a Research Associate at the John Molson School of Business at Concordia University, Montreal. She has previously served with Concordia's David O'Brien Centre for Sustainable Enterprise. She holds a Bachelor of Commerce in finance with a concentration in economics from Concordia University. She has participated in and worked on well over 30 research papers and helped launch three books within the Finance Department.

Thomas Walker is a Full Professor of Finance at Concordia University in Montreal, Canada. He previously served as an Associate Dean, Department Chair, and Director of Concordia's David O'Brien Centre for Sustainable Enterprise. Prior to his academic career, he worked for firms such as Mercedes Benz, KPMG, and Utility Consultants International. He has published over 70 journal articles and books.

Pete Walton is a Knowledge Exchange Research Fellow and leads academic at UKCIP, University of Oxford, where he works with a range of stakeholders in the UK and abroad, supporting them in understanding how to build resilience to a changing climate. Pete's academic and professional career combines both climate change and education, providing him with the expertise in adaptation and climate change impacts and the skills to communicate the practical implications of such impacts to a wide range of audiences.

Olaf Weber is a Professor at the University of Waterloo's School of Environment, Enterprise, and Development and holds the position as the university's Research Chair in Sustainable Finance. His research and teaching interests address the connection between financial sector players, such as banks and sustainable development, and the link between sustainability and financial performance of enterprises. His research focuses on the impacts of the financial industry on sustainable development, the role of voluntary and regulatory mechanisms for the financial sector to become more sustainable, social banking and impact investing, the materiality of sustainability risks, and opportunities for investors and artificial intelligence as a tool to analyze environmental, social, and governance (ESG) performance.

Michael O. Wood is the Associate Director of Undergraduate Studies and a Continuing Lecturer at the School of Environment, Enterprise, and Development (SEED) at the University of Waterloo. He holds a Ph.D. in strategy and sustainability from the Ivey Business School. His research examines organizational perceptions and responses to sustainability issues through the lens of space, time, scale, and social license to operate within the contexts of the insurance industry, mining, carbon management, the Blue Economy, and climate change and global security. His research has most recently been published in the Canadian Water Resources Journal, the Academy of Management Review, the British Journal of Management, and Resources.

Yuanchao Xu specializes in sectoral/regional water and climate risk assessment as well as the regulatory risk interpretation in China for CWR. His contributions to CWR's water risk valuation work include the writing of the water risk chapter of the "Green Finance Series – Case Studies on Environmental Risk Analysis of Financial Institutions" led by the Chair of the Green Finance Committee of China as well as presenting key valuation methodologies at China's central bank. As a hydrologist, Yuanchao also leads CWR's collaboration with the China-Europe Water Platform (CEWP) and writes extensively on China's transformation into an ecological civilization; his views have been quoted by Bloomberg.

Lihuan Zhou is an Associate at WRI's Sustainable Finance Center, where he mainly works on the Sustainable Investing Initiative and conducts research to advance sustainable investment practices in the private sector. He also contributes to the center's work on cross-border investments from China. Lihuan holds a Master of Science in Applied Economics from Johns Hopkins University and dual bachelor degrees in Environmental Sciences and Economics from Peking University. He is a CFA Charterholder and a member of the CFA Society in Washington, DC.

List of Figures

Introducing Water Risk: A Framework for (Integrated) Water Risk Assessment and Management

Data for Water Risks: Current Trends in Reporting Frameworks, Shortcomings, and the Way Forward

The Developing Field of Water Risk Valuation for the Financial Industry

Financial Implications of Parched Power: Insights from an Analysis of Indian Thermal Power Companies

Water Insecurity and Climate Risk: Investment Impact of Floods and Droughts

Chronic Coastal Water Threats Warrant a Valuation Re-Think

Multi-dimensional and Interacting Water and Climate Risks and Pricing Them in the Industry Context

List of Tables

Water Insecurity and Climate Risk: Investment Impact of Floods and Droughts

Forward Pricing of Embedded Water: A Step Toward Sustainable Development in Agriculture

Multi-dimensional and Interacting Water and Climate Risks and Pricing Them in the Industry Context

Introducing Water Risk: A Framework for (Integrated) Water Risk Assessment and Management

Adele Dumont-Bergeron and Dieter Gramlich

1 The Water Challenge

It is well known that water is necessary to life. Without water, organisms cannot grow and thrive, yet alone survive. Despite this knowledge, water is largely taken for granted, at the national and international levels. Water is a finite resource; it is not something that can be grown or scientifically engineered. The current amount of water on the planet represents what we must work with. Since 97% of the total water is salt water, only 2.5–3% of water is freshwater that can be used for consumption, a number that decreases as the quality of freshwater worsens (National Oceanic and Atmospheric Administration, 2020). Glaciers, lakes, reservoirs, ponds, rivers, streams, wetlands, and groundwater all make up

A. Dumont-Bergeron (✉)
Department of English, Concordia University, Montreal, Canada

D. Gramlich
Department of Banking, DHBW - Duale Hochschule Baden-Württemberg, Heidenheim an der Brenz, Germany
e-mail: dieter.gramlich@dhbw-heidenheim.de

T. Walker et al. (eds.), *Water Risk and Its Impact on the Financial Markets and Society*, Palgrave Studies in Sustainable Business In Association with Future Earth,
https://doi.org/10.1007/978-3-030-77650-3_1

freshwater. This amount of freshwater is globally shared among 7.7 billion people, ecosystems, and industries (Boretti & Rosa, 2019). Many sectors are not only water-dependent, but also voracious when it comes to freshwater consumption. These sectors include agriculture, beverage, textile, mining, energy, and transport, with agriculture accounting for about 70% of human water use because of such practices as crop irrigation and livestock rearing (WWF, 2011; WWF Germany, 2019). Boretti and Rosa (2019) report that, over the last century, the global water demand has seen an increase of 600%, a number which will continue to rise as population is expected to grow between 9.4 and 10.2 billion by 2050. More people on earth necessarily means more pressure on the limited quantity of water in order to grow food and feed everybody (Boretti & Rosa, 2019).

While the quantity of water is a global problem, it is by no means a fairly distributed one. Indeed, already in 2020, some regions suffer more from water scarcity than others, among them are Central-South America, the Middle East region, Eastern Europe, sub-Saharan Africa, and areas of Central-South Asia, including China (WWF, 2011). Therefore, while it is a global problem, it is simultaneously a local problem that requires local solutions. Naturally, droughts in the United States (US) and droughts in India affect differently their respective population because of the countries' socioeconomic situation as well as geographic location. Consequently, the remedial actions—and the available solutions—will differ.

The limited amount of freshwater is part of the conversation on **water scarcity**, which is defined as "the volumetric abundance, or lack thereof, of freshwater resources" (PRI & WWF, 2018). Water scarcity, at its most simplistic, refers to quantity. It thus relates to the physicality of water and its material risk, which could highly affect the gross domestic product (GDP) of some regions, costing them up to 6% of their GDP (UNESCO, 2020).

This unequal division of water resources has notably led to the adoption of the Sustainable Development Goal (SDG) number 6 by the United Nations (UN) in 2015: To "ensure availability and sustainable management of water and sanitation for all by 2030" (UN, 2020). In 2017, 2.2 billion lacked access to safely managed drinking water while 4.2 billion had the same problem with sanitation. During the COVID-19 pandemic of 2020, a lack of access to safe water and sanitation signified that three billion could not prevent the spread of the disease through basic handwashing (UN, 2020). These numbers shared by the UN illustrate the significant role of water in preventing and curing diseases, which can be forgotten when discussing water scarcity.

Globally, it is estimated that four billion people already suffer from severe drought at least one month per year (UNESCO, 2020). Droughts generate consequences that reach far beyond the direct impacts (water scarcity, crop losses, local food shortages, and forest fires), such as migration, unemployment, and social unrest (WWF, 2019a). Although droughts tend to recur seasonally, their frequency as well as magnitude is further exacerbated by climate change. Not only are droughts extreme weather events to monitor, but so are floods, extreme precipitation, hurricanes, heatwaves, and so on, which all have an impact on water resources, whether in terms of overflow or dryness (IPCC, 2008, 2018). For instance, heavy precipitation leads to soil erosion, adds pressure on infrastructure, and affects the quality of groundwater (UNECE, 2009).

Due to climate change, these events are happening at a faster rate, occasioning short-term and long-term damages to ecosystems, as well as the economy (UNESCO, 2020; WWF & ABInBev, 2019). For example, droughts between 1991 and 2013 in sub-Saharan Africa have incurred agricultural production losses amounting to US$31 billion. While cases from developing countries are perhaps too historically familiar, droughts affect various regions of developed countries, increasingly so. The US is one of them. Between 2011 and 2013, three drought and heatwave events led to a US$60 billion loss because of their impacts on various sectors, such as agriculture and energy (Alizadeh et al., 2020).

In addition, quality is part of the water problem too. The pollution of water bodies represents a human challenge, as it is associated with human actions driven by population and economic growth among other factors (Boretti & Rosa, 2019). Indeed, industries have their fair share of blame as they annually release 300–400 megatons of waste in the water (Boretti & Rosa, 2019). In terms of sanitation, untreated sewage spills into water in 90% of developing countries. This percentage signifies that "730 million tons of sewage and other effluents are discharged into the water" every year (Boretti & Rosa, 2019, p. 2), thus greatly affecting the availability of clean freshwater. When water quality is affected, the effects are multifold: Biodiversity is threatened, consumption is impossible for both personal and economical gains, diseases proliferate, and nutrient and chemical overloads become dangerous and pathogenic (Boretti & Rosa,

2019; UNESCO, 2020). By 2050, about 3.9 billion (around 40%) of the world population will suffer from water scarcity and will live under severe water stress (UNESCO, 2020). Therefore, quality of freshwater, which is part of water stress, is just as worrying as water scarcity.

In short, **water stress** is defined as "the ability, or lack thereof, to meet human and ecological demand for freshwater" (PRI & WWF, 2018). A broader category than water scarcity, factors of water stress include quality, accessibility, and availability. Factors that strain the quantity and quality of freshwater are usually part of the conversation on water stress. Population growth, for example, represents a driver of water stress, as it affects the quality, the availability, and sometimes the accessibility of freshwater.

Water, therefore, should be of personal concerns for individuals and industries, as the problems its absence would create are phenomenal. At the economic level, such consequences would be in the billions of dollars. To be more precise, "$415 billion in revenue may be at risk from lack of water availability for irrigation or animal consumption" (Ceres, 2019, p. 2). If the agriculture sector takes a hit of US$415 billion in revenue, we can only imagine how the other sectors would be impacted. Yet, it is important to note that this number encompasses the negative effects that water scarcity would have on the supply chain; it is a cost that farmers, transporters, and retailers (i.e., the supply chain) would absorb, thus having widespread socioeconomic impacts at the national and international level. A report by Water Aid et al. (2017) reveals that one in five individuals is estimated to work in a globalized supply chain. Such a loss also implies that production has been disrupted, either in the short term or the long term, resulting in a decrease of goods. For the agriculture sector, it represents less food for consumers, thus contributing to inequalities. At its core, water is a basic human right on which prosperity and well-being rely.

2 Defining Water Risk and Discussing Opportunities

PRI and WWF (2018, p. 7) provides a helpful definition in thinking about **water risk**: "The possibility of an entity experiencing a water-related challenge (e.g., water scarcity, water stress, flooding, infrastructure decay, drought)." First, as this definition mentions, water scarcity and water stress represent water-related challenges. They are therefore part of the broader category of water risk, which also include climate change events. Secondly, by definition the word "risk" implies the possibility (or probability) of a threat to occur, hence relating uncertainty to risk. Part of the goals of academics and organizations alike is not necessarily to predict the materiality of these risks (i.e., when risks become events), but rather to find adequate ways to prevent and mitigate the consequences of these risks. In other words, the aim is to understand the potential scenarios in order to act as best as possible under uncertainty (Garnier et al., 2015; United Nations Global Compact et al., 2020).

Water risk presents a challenge; it is not yet a lost cause. Challenges should be stimulating because they offer problems that society must solve. In that sense, they motivate the collective minds, which lead to an important aspect: opportunities. While water risk represents national and international challenges, it also represents opportunities to create multidisciplinary knowledge, improve or generate tools, innovate technology and infrastructures, teach about water, risks, and water efficiency, and most importantly for this book, help investors and companies toward sustainable decision-making. Some of these opportunities relate to what has been deemed "unconventional" water resources (UNESCO, 2020). These resources include:

- Wastewater reuse (also known as reclaimed water), including gray water reuse (from sinks, showers, washing machines);
- Water recycling;
- Rainwater or stormwater harvesting;
- Desalination of salt water.

In the cases of water reuse and water recycling, most times these processes require water treatment. In some cases, depending on the treatments, water reuse can become potable water, while a lesser treatment might be enough for irrigational or industrial purposes (Nicholson & Vespa, 2010). Similarly, rainwater catchment also requires the necessary installations: A roof or a drain is necessary, as well as some storage equipment, to capture non potable water (Garnier et al., 2015). Likewise, treatment of harvested water can be done to improve its quality. Another low-cost derivative of water harvesting is the catchment of atmospheric moisture (fog) in highly foggy regions (UNESCO, 2020). Finally, the desalination of water also necessitates a specific technology, and one that warrants a massive input of energy. Some regions, such as the Gulf Cooperation Council countries (GCC), already rely on desalination to meet their water requirements (Darwish & Zubari, 2020; UNESCO, 2020). Since these unconventional methods are emerging, there are opportunities to invest in these technologies, as well as to do further research.

Furthermore, companies face a wide array of challenges that are deemed material. Out of these challenges emerge direct impacts for businesses, including water price and availability, as well as the disruption of production, or even losses, such as with crop irrigation. Additionally, these risks engender indirect impacts that may include surges "in electricity prices, macroeconomic decreases in consumer spending, stranding of corporate infrastructure, or loss of access to markets or growth opportunities as a result of water shortages" (Vivid Economics, 2020, p. 9), only to name a few.

The term water risk hence refers to a comprehensive, multifaceted context. Water risk expresses the possible affectedness of individuals and institutions from water-related challenges and their potentially negative or positive impact. It materializes in different forms whereby the types of water risks differ in the way they affect people and organizations. Many of them emerge in relation to the following risks:

- Physical water risks;
- Regulatory water risks (transition water risks);
- Reputational water risks.

Physical water risks relate to materiality issues because of water-related events. They result from problems in water quantity and quality. For example, water scarcity can disrupt business operations, as water may be used to produce goods, irrigate, process, cool, and clean (Barton, 2010), and consequently represents a **physical water risk** for companies (CDP, 2015; WWF, 2019b). Without water, many parts of the supply chain are threatened, often unbeknownst to corporations. Other types of physical risks include polluted water and flooding (WWF, 2011). Naturally, sectors use different quantities according to their needs, but an insufficiency of water at some point in a water-intensive supply chain can cause important damages in revenue to companies. For example, droughts and heatwaves can significantly impact the energy sector, especially when installations rely on hydropower, or use cooling water with coal or nuclear energy. Before 2013, California produced 18% of its electricity using hydropower. When droughts started in 2013 and continued through 2016, the percentage decreased to 10.5%, even dropping to 7% in 2015. As a result, California turned to less sustainable and more expensive sources of power, which increased electricity costs at US$2.45 billion and generated pollution (WWF, 2019a). Experts have forecast, under a moderate climate change scenario, that California could lose 10–20% of its total hydropower, amounting to an annual US$440–880 million loss (Barton, 2010). Moreover, in 2016, El Niño led to a water-level drop of 13% in Zambia. For a country where 95% of its power comes from hydropower, such a decrease put the country's electricity security in jeopardy (WWF & ABInBev, 2019). A low-level river due to droughts, such as the Rhine river in Germany in 2018, can also prevent boats and cargos from reaching their destination. In the case of the Rhine, they had to lighten their loads to pass (WWF Germany, 2019).

Regulatory risk, also referred to as "transition risk," appears in the form of laws, policies, and regulations imposed by the government as means to control freshwater use and wastewater discharge (WWF, 2011; WWF Germany, 2019). Licenses and sanctions, as well as water and wastewater pricing, materialize as additional costs for a company. Conversely, a lack of stringent regulatory rules can have an adverse effect on the company due to other companies' use and may lead to physical and reputational risks (WWF Germany, 2019). Unanticipated, these changes in regulation can be quite costly. Yet, they often spark from physical (e.g., scarcity risk) and reputational (e.g., worried citizens) concerns that lead to additional pressure to the governments to make changes (Barton, 2010). Similarly, responses from governments such as the implementation of "net zero transition policies" also represent risks because of the changes in the business-as-usual design. Nonetheless, these objectives to attain net zero emissions strive to encourage actions and opportunities toward innovation, rather than being detrimental to entities (Vivid Economics, 2020).

When a company and/or its supply chain perpetrates actions that negatively affect the environment, it may suffer from reputational damage. More precisely, **reputational water risk** occurs when a company poorly manages water resources, thus affecting communities and ecosystems (PRI & WWF, 2018). Consequently, a company's reputation might be harmed as the result of tensions between the company's actions and local communities regarding local resources. In developing countries where water is perceived as a scarce resource, the population tends to raise public awareness and boycott the product as a way to oppose the industrial use of water (Barton, 2010). These kinds of campaign can hurt the reputation of the company and its brand, usually costing sales, and sometimes even costing its license to operate (Barton, 2010). In 2017, PepsiCo was the subject of such action as it kept producing soft drinks in Tamil Nadu, India, despite the ongoing drought. As a result, boycotts from retailers and consumers led to a decrease in purchase of the product (WWF Germany, 2019). These damages can emerge quickly and spread from the local community to the global scandal, thus blackening a company's good name (WWF, 2011).

What is important to note here is that physical, regulatory, and reputational risks are intertwined: One may lead to another as drivers or emerge as consequences. In a way, these risks are branches of water risks, constantly interacting with one another (see Fig. 1). Regardless of how these risks materialize, they offer opportunities for a company to improve their risk management as well as for governments to establish more stringent measures and frameworks. In that sense, these risks create behavioral opportunities toward a collective accountability, where each party attempts to do what is best for the common good.

The goal in discussing these consequences is to illustrate that the effects of the current drivers of water risk, such as population and economic growth, can be addressed, and potentially mitigated through actions that aim to diminish the physical, regulatory, and reputational risk of a company. Yet, the projected mid- to long-term consequences must be addressed *now*. While it is important that corporations have focused in the last decades on decreasing their greenhouse gas (GHG) emissions, air pollution, and more recently the global COVID-19 pandemic, they are not the only threats for risk managers to focus on (Ceres, 2015).

As demonstrated in the previous sections, water risk presents a great challenge, with far-reaching impacts in various sectors. Even though investors and companies are becoming increasingly aware of the ramifications of water risk on their business operations, awareness and actions are still lagging, in part because frameworks to assess, record, and report are neither globally adopted and widespread, nor locally suited (Ceres, 2015; Christ & Burritt, 2017). In addition, these frameworks need qualitative and quantitative information about the direct and indirect relationships

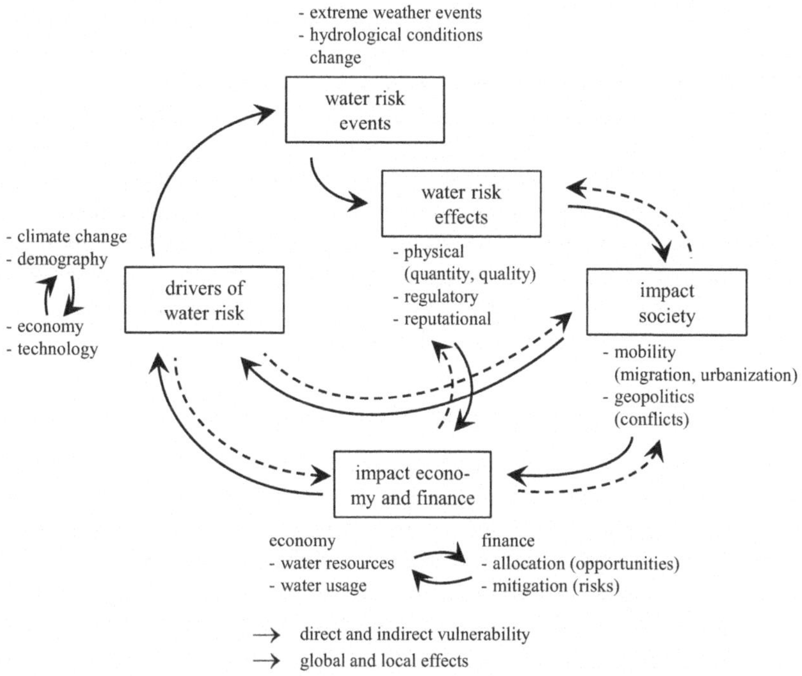

Fig. 1 A framework for water risk dimensions and impacts

between water, economy, and society, particularly, as regards the impact of water risk on supply chains, production infrastructure, and distribution systems. However, as the following chapter from Concha Larrauri and Josset shows, the formats of requested data as well as the readiness of stakeholders to report them must be further developed.

It is well known that investors and the financial system have a key role to play in addressing this challenge (Hogeboom et al., 2018). Water risk may hit the financial system directly. As the contribution to this book from Mirando and Tan on coastal water threats shows, major financial centers in Asia including Hong Kong and Singapore are exposed to a rising sea level and extreme weather events. More comprehensively, the financial system is jeopardized through its large investments in water-sensitive assets. Water risk increases the default risk from loans and securities held on companies, countries, and households in water-sensitive areas. The chapter from Xu and Tan demonstrates the increasing interest of financial investors to incorporate water risk into risk evaluation, here based on evidence from power companies in China. Conversely, the financial sector benefits from funding opportunities of water-related technology. In their chapter, Craig, Coulombe, Lombardi, and ElGarhy present possibilities to benefit from water innovations and to enact positive changes. Notably, the European Union (EU) has adopted a plan to move forward toward a circular and sustainable economy including the reuse of water (Green Deal). The EU assigns financial institutions a leading role in this process through the allocation of green money and puts "sustainable finance at the heart of the financial system" (European Commission, 2020, p. 10).

3 Frameworks for Modeling Water Risk

Water risk is a new concept that has emerged in the last decades. As a result, the tools to assess, report, and mitigate this risk are still emerging along the frameworks to understand the various effects of this phenomenon, with the goal of providing a blueprint on how to act. This book presents additional approaches on how to cope with water-related challenges. However, although the suggested approaches are very sophisticated and almost operational, a better understanding is achieved when positioning them into a more comprehensive and overarching context. Taking into account the background of the model highlights the multiple dependencies the model must comply with and can increase the effectiveness in model application. Overall, an integrated approach respecting the multiple connectivity between sources, forms, and effects of water risk along with the exposedness of stakeholders as well as their potential actions and feedbacks from them must be applied. To this extent, we suggest the framework represented in Fig. 1.

The figure addresses the characteristics of water risk, incorporates the ways water challenges affect different stakeholders in the water context, and points to the stakeholders' potential actions. The links between the different elements represent their interconnectivity and joint impact on the overall setting of water risk. We conceive the drivers of water risk (left) to be at the origin of the overall transmission process. They manifest as specific water-related events that then have an impact on the economy and society. Responses from the real economy and the financial markets as well as from society in general (e.g., way of life) interact and feedback on the drivers of water risk (e.g., via growth, urbanization, etc.). Overall, a balance is needed between the available quantity and quality of water on the one hand and its multiple uses in economy and society on the other. Models for the assessment and management of water risk ideally have to comply with this comprehensive and interactive perspective.

Schneider (1997) describes integrated assessment modeling (IAM) as a multidisciplinary tool that helps decision-making because it offers insights as to the physical, ecological, and social dimensions of a global climate change problem. In that sense, Schneider highlights how integrated assessment modeling plays a key role in guiding toward "more rational" decision-making:

> The role of any analytic method, IAMs included, is to help elucidate how various policy choices could alter the likelihood or costs of various options and/or consequences. For this reason, the discipline of economics, since it has the best developed formalism and empiricism for cost/benefit analyses, is in a particularly advantaged position to contribute to IAMs. (Schneider, 1997, p. 230)

Nowadays, economists continue to use IAMs to weigh different climate policy options by assessing the cost and benefits of each scenario (Ackerman et al., 2009), just as scholars in hydrology use modeling for water resources planning (O'Connell, 2017). Because the threats and effects of water risk are manifold and long-lasting, integrated and comprehensive frameworks are necessary for informed decision-making, consequently meeting the needs of all parties involved. In the case of water risk, these frameworks and methodologies can be used in risk management to protect a river's clean water supply, for instance (Hester & Harrison, 1998). It is important to note that multiple modeling frameworks exist, one example being the integrated water resources management (IWRM), which attempts to encompass various levels of water management (technical, management, integrative) to meet the needs of various users (Grigg, 2016).

So far in public reports targeting investors, discussions of water risk in the financial sector have predominantly revolved around raising awareness related to the physical risk of water scarcity on the supply chain or presenting investment opportunities (see PRI & WWF, 2018; WWF, 2019b; United Nations Global Compact et al., 2020; Vivid Economics, 2020). Yet, because of the various interconnected drivers of water risk (e.g., water resources and ecosystems; climate change; governance (institutions); technology; economy and security; agriculture; infrastructure; demography; ethics, culture and society; and politics (Diggle, 2013)), the ecological, societal, technological, and economical dimensions of water risks must be increasingly considered so that companies can rapidly adapt to become resilient in the face of global climate change (United Nations Global Compact et al., 2020).

Because it is difficult to assess and report the extent of damages (e.g., water footprint and lifecycle assessments) caused by climate change events, the full picture remains blurry (WWF, 2013, 2019b). The human aspect (e.g., the complexity of human choices regarding water use and water risk) has been a particularly challenging factor to include in modeling (O'Connell, 2017). As a result, there is a need to develop tools and methodologies that will enable a more holistic understanding of the ways drivers of water risk interact with one another, which will help break down barriers that come from a lack of understanding, planning, and financing (United Nations Global Compact et al., 2020). Indeed, "new approaches that better account for an array of basin water risks, operational water risks, and a more nuanced view of responses would provide a much more accurate picture of the performance, risk, and opportunity of investments" (WWF, 2019b, p. 19). For example, how do population and economic growth affect water scarcity, thus impacting social and industrial water use and discharge, after extreme weather events? Does heavier precipitation stress enough water infrastructure in urban centers to threaten water quality, thus having repercussion on health, industries, and individual use? Do these problems present attractive opportunities for short-term and long-term investments to increase corporate performance?

As these questions imply, the drivers of water risk may overlap. For example, in the case of loose regulation that leads to water stress, which climate change then amplifies, the various drivers coexist and may exponentially jeopardize the resource. In turn, companies located in that region are sure to face a physical risk (e.g., a lack of freshwater availability for operations), a regulatory risk (e.g., new policies from the government), and a reputational risk (e.g., pressure from local communities and public campaigns if companies continue to deplete the resource despite water scarcity and stress issues). There is therefore a strong need to assess the interlinkages and interdependencies of water risk drivers and sectors, i.e., this nexus between water, energy, food, land, and climate that hinders the progress toward sustainable societies and institutions (Laspidou et al., 2020; Michel, 2020).

It is also important to underline that looking at a water risk—and a potential solution—through a focalized lens would be misleading. To take the example of desalination in a water scarce country, such a process appears, theoretically, as a viable solution to water scarcity. Yet, desalination requires large intakes of energy, which might come from fossil resources. Does the country/region own these resources? Does the country have this technology? Can the right investments be made? If so, desalination does increase the local quantity of water, but to the detriment of the global environment if renewable energies are not used. Both scenarios present multiple ecological, social, technological, and economical risks, which must be assessed in a holistic way to decide how to solve the problem of water scarcity in this hypothetical region. The work of the World Economic Forum (WEF), for example, which highlights the interconnected risks related to water, is one step further toward grasping and assessing the full picture (WEF, 2021).

Therefore, integrated modeling is required to provide a multidisciplinary framework for water risk assessment and corporate water management, as long as it takes into account the interrelated factors. As Laspidou et al. (2020) explain so well:

> The global community can promote a multisector integrated approach to complex trade-offs and challenges in order to strengthen resource security by developing governance processes that engage stakeholders from the nexus arena and empower them in analyses and management decisions, using appropriate tools. Such tools need to tackle the complexity of the system, introducing enough detail to capture the important trends, but also providing the data needed to enable informed decisions and quantified nexus analysis.

This approach is what scholars and public organizations are currently working on, with the aim to abandon what Laspidou et al. (2020, p. 14) call "silo-thinking" and "institutional fragmentation" in decision-making. In that sense, there is a belief that collective intelligence can address water risk (Diggle, 2013; Michel, 2020; United Nations Global Compact et al., 2020).

Vivid Economics (2020) reports that companies might be withdrawing more water each year despite increasing awareness on water risk.

Companies continue to exhibit the unsustainable behavior of resources competition that capitalism promotes by exploiting resources for their own benefits. Such a behavior leads to inaction toward reducing water use and water dependency in their supply chain, which is "due to a wholesale failure by investors, corporations and governments to price water risk into decisions, leading collectively to unsustainable behaviour" (Vivid Economics, 2020, p. 5). What this statement highlights is a collective failure. Without each party doing their part toward water efficiency, collective failure is bound to happen. In many cases, companies are failing to quantify their water use (Vivid Economics, 2020; WWF, 2013). In spite of this lack, some companies have improved assessing their water consumption, discussing water risk, or integrating water risk in their risk management assessment as a result of participating in research studies for organizations (Barton, 2010; Ceres, 2019). Albeit welcome news, what these studies imply is that companies need a nudge toward adopting sustainable behaviors, and if investors, corporations, and governments do not provide the necessary framework or ask questions, inaction is easier than action.

4 Contributions to Water Risk Modeling

As for all responses to major challenges, approaches to assess and manage water risk are ambiguous. On the one hand, they require specific knowledge on multiple physical, economic, and societal aspects, which can be obtained from experts in their field. On the other, the connectivity of these aspects must be accounted for to achieve an overall balanced solution. The modeling of water risk as well as the actions to manage it must comply with the contextual reference, and this is the attempt of this book.

More specifically, this book attempts to create what we would like to call a "micro-collective intelligence." The concept of collective intelligence argues that human beings have the resources—or can work toward obtaining these resources—to address global issues, thus arguing in favor of multidisciplinary and interdisciplinary work (Diggle, 2013). By sharing data, information, and knowledge, society moves one step forward in responding and solving global threats. In this book, we present a sample of collective intelligence regarding water risk and the financial sector. Even though the book focuses on a specific sector, the contributors come from diverse academic and non-academic backgrounds. Some of them are affiliated with research centers at universities, while others are renowned practitioners working at the forefront of water-related investments. Furthermore, the experts behind the chapters are working in different disciplines. From finance to engineering while passing through the physics and chemistry departments, the contributions in this book are wide-ranging, covering a multitude of topics and expertise. Accordingly, the types of contributions differ. Some chapters focus on a localized region or problem, while other chapters attempt to open a broader dialogue on a general issue. In essence, the book attempts to capture a micro-collective intelligence by featuring a variety of experts, disciplines, and angles.

Nonetheless, these chapters converse with one another. To accentuate these common topics, we have combined the chapters into two parts, with the first one titled *Assessing Water Risk: Frameworks, Models, and Tools*. This part is constituted of chapters that review some of the current approaches and models to assess the exposure from water risk, highlighting their benefits and their flaws. It is dedicated to discussing methods of assessment for water challenges and the barriers related to them. In addition to the chapters already mentioned, Zhou, McClamrock, Christianson, Krishnan, and Luo provide evidence of the difficulties in assessing the affectedness of power companies from water risk. Rayer, Haustein, and Walton provide a framework to assess the risk of financial investors from climate-driven events.

Since the first part addresses assessment, the second part focuses on management because we consider assessing and managing to be sides of the same coin. We called it *Managing Water Risk: Investing in the Future*. The chapters in this section bring the readers into the fields and hones in on material problems, with the goal of shedding light on practical issues related to their management. This part discusses the avenues for future management of threats and investment in opportunities. Sandhu, Weber, and Wood address the complex interaction of water-related economic and social impacts and ways to achieve interdisciplinary solutions for water security, thereby providing a case study for Ontario. In a similar direction, Mahajan and Porwal address the economic and social effects of virtual water, i.e., water that is embedded in the production of goods, and suggest how to account for hidden water in the case of India. The awareness and the management of drinking water systems is a challenge from both the perspectives of engineers and society. Karney and Gibson highlight the challenges of failures of these systems and initiate the conversation about ways to manage them. Finally, the chapter from Siegel,

Concha Larrauri, Bonnafous, and Lall provides a more comprehensive and forward-looking perspective as the authors discuss the correlated effects from climate-driven spatial water risk when it simultaneously hits a company's portfolio of assets. As extreme climate events tend to increase, bottom-up approaches are suggested to handle expected damages from tail water risks.

Figure 2 situates each chapter in relation to the dimensions of the water risk framework, thereby attempting to address both the contextual reference of the different approaches as regards their specific location in the overall framework (obtained from Fig. 1) and the relationship between the different approaches. It thus refers to the philosophy of a collective and contextual thinking.

Finally, we hope that these chapters will inspire every reader to become more acquainted with the various dimensions and drivers of water risk and to produce work that aims to raise awareness on this sustainability issue. More fundamentally, we hope that this book will be a part of a movement toward action rather than inaction, toward change rather than the business-as-usual mentality.

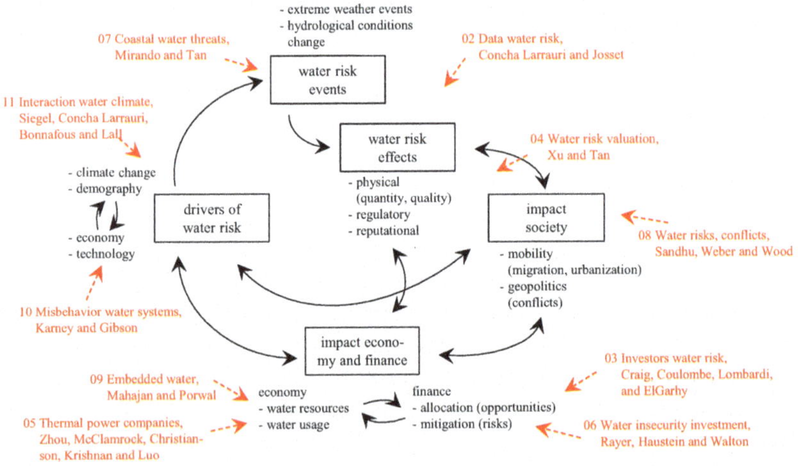

Fig. 2 Integrating the book contributions into the water risk framework

References

Ackerman, F., DeCanio, S. J., Howarth, R. B., & Sheeran, K. (2009). Limitations of integrated assessment models of climate change. *Climatic Change, 95*, 297–315. https://doi-org.lib-ezproxy.concordia.ca/10.1007/s10584-009-9570-x.

Alizadeh, M. R., Adamowski, J., Nikoo, M. R., AghaKouchak, A., Dennison, P., & Sadegh, M. (2020). A century of observations reveals increasing likelihood of continental-scale compound dry-hot extremes. *Science Advances, 6*. https://doi.org/10.1126/sciadv.aaz4571.

Barton, B. (2010). *Murky waters: Corporate reporting on water risk—A benchmarking study of 100 companies*. Ceres. www.waterfootprint.org/Reports/Barton/2010.pdf.

Boretti, A., & Rosa, L. (2019). Reassessing the projections of the World Water Development Report. *Npj Clean Water, 2*(1). https://doi.org/10.1038/s41545-019-0039-9.

CDP. (2015). *Accelerating action: CDP Global Water Report 2015*. https://www.cdp.net/en/research/global-reports/global-water-report-2015.

Ceres. (2015). *An investor handbook for water risk integration*. Ceres. www.ceres.org/sites/default/files/reports/2017-03_Ceres_ESG-WaterRisk_041515_Print.pdf.

Ceres. (2019). *Feeding ourselves thirsty: Tracking food company progress toward a water-smart future*. Executive summary. https://www.ceres.org/resources/reports/feeding-ourselves-thirsty-2019.

Christ, K. L., & Burritt, R. L. (2017). Water management accounting: A framework for corporate practice. *Journal of Cleaner Production, 152*, 379–386. https://doi.org/10.1016/j.jclepro.2017.03.147.

Darwish, S. Z., & Zubari, W. K. (2020). Strategic risk management (SRM): The future of desalination in the gulf cooperation council (GCC). *Academy of Strategic Management Journal, 19*(2), 1–14.

Diggle, T. (2013). Water: How collective intelligence initiatives can address this challenge. *The Journal of Futures Studies, Strategic Thinking and Policy, 15*(5), 342–353. http://dx.doi.org.lib-ezproxy.concordia.ca/10.1108/FS-05-2012-0032.

European Commission. (2020). *Sustainable Europe Investment Plan. European Green Deal Investment Plan*. https://eur-lex.europa.eu/legal-content/EN/TXT/?uri=CELEX%3A52020DC0021.

Garnier, M., Harper, D. M., Blaskovicova, L., Hancz, G., Janauer, G. A., Jolánkai, Z., Lanz, E., Porto, A. L., Mándoki, M., Pataki, B., & Rahuel, J. L. (2015). Climate change and European water bodies, a review of existing gaps and future research needs: Findings of the ClimateWater project. *Environmental Management, 56*, 271–285. https://doi.org/10.1007/s00267-015-0544-7.

Grigg, N. S. (2016). Framework and scenarios of IWRM. In *Integrated water resource management*. Palgrave Macmillan. https://doi-org.lib-ezproxy.concordia.ca/10.1057/978-1-137-57615-6_2.

Hester, R. E., & Harrison, R. M. (1998). *Risk assessment and risk management*. Royal Society of Chemistry.

Hogeboom, R. L., Kamphuis, I., & Hoekstra, A. Y. (2018). Water sustainability of investors: Development and application of an assessment framework. *Journal of Cleaner Production, 202*, 642–648. https://doi.org/10.1016/j.jclepro.2018.08.142.

IPCC. (2008). *Climate change and water* (IPCC technical paper VI). https://www.ipcc.ch/publication/climate-change-and-water-2/.

IPCC. (2018). Global Warming of 1.5°C. An IPCC Special Report on the impacts of global warming of 1.5°C above pre-industrial levels and related global greenhouse gas emission pathways, in the context of strengthening the global response to the threat of climate change, sustainable development, and efforts to eradicate poverty, Geneva.

Laspidou, C. S., Mellios, N. K., Spyropoulou, A. E., Kofinas, D. T., & Papadopoulou, M. P. (2020). Systems thinking on the resource nexus: Modeling and visualisation tools to identify critical interlinkages for resilient and sustainable societies and institutions. *The Science of the Total Environment, 717*, https://doi.org/10.1016/j.scitotenv.2020.137264.

Michel, D. (2020). Water conflict pathways and peacebuilding strategies. *US Institute of Peace*, 22–25. https://doi.org/10.2307/resrep26059.9.

National Oceanic and Atmospheric Administration. (2020). *Where is all of the Earth's water*? http://oceanservice.noaa.gov/facts/wherewater.html. Accessed 8 February 2020.

Nicholson, J., & Vespa, M. (2010). *Full cycle*. Water Canada. https://www.watercanada.net/feature/full-cycle.

O'Connell, E. (2017). Towards adaptation of water resource systems to climatic and socio-economic change. *Water Resources Management, 31*(10), 2965–2984. http://dx.doi.org.lib-ezproxy.concordia.ca/10.1007/s11269-017-1734-2.

PRI & WWF. (2018). *Growing water risk resilience: An investor guide on agricultural supply chain*. www.unpri.org/environmental-issues/growing-water-risk-resilience-an-investor-guide-on-agricultural-supply-chains-/2793.article.

Schneider, S. H. (1997). Integrated assessment modeling of global climate change: Transparent rational tool for policy making or opaque screen hiding value-laden assumptions? *Environmental Modeling and Assessment, 2*, 229–249. https://doi-org.lib-ezproxy.concordia.ca/10.1023/A:1019090117643.

UNECE. (2009). *Guidance on water and adaptation to climate change*. www.unece.org/env/water/water_climate_activ.html.

UNESCO. (2020). *The United Nations world water development report 2020: Water and climate change*. https://unesdoc.unesco.org/ark:/48223/pf0000372985.locale=en.

United Nations (UN). (2020). *The sustainable development goals report*. Department of Economic and Social Affairs Sustainable Development. https://sdgs.un.org/goals/goal6.

United Nations Global Compact, CDP, Suez, & WBCSD. (2020). *Corporate water resilience in an uncertain future*. www.ceowatermandate.org/resilience-report.

Vivid Economics. (2020). *The inevitable water finance response: Investor risks and opportunities*. https://www.unpri.org/the-inevitable-water-finance-response-investor-risks-and-opportunities/5908.article.

WaterAid, CEO Water Mandate, & World Business Council for Sustainable Development. (2017). *Corporate engagement on water supply, sanitation and hygiene: Driving progress on Sustainable Development Goal 6 (SDG6) through supply-chains and voluntary standards*. https://pacinst.org/publication/corporate-engagement-on-water-supply-sanitation-and-hygiene-driving-progress-on-sustainable-development-goal-6-sdg6-through-supply-chains-and-voluntary-standards/.

WEF. (2021). *Strategic intelligence: Water*. https://intelligence.weforum.org/topics/a1Gb00000015MLgEAM?tab=publications. Accessed 16 February 2021.

WWF. (2011). *Assessing water risk: A practical approach for financial institution*. http://awsassets.panda.org/downloads/deg_wwf_water_risk_final.pdf.

WWF. (2013). *Water stewardship: Perspectives on business risks and responses to water challenges*. https://wwf.panda.org/?210092/Water-Stewardship–Perspectives-on-business-risk-and-responses–to-water-challenges.

WWF. (2019a). *Drought risk: The global thirst for water in the era of climate crisis*. www.wwf.de/fileadmin/fm-wwf/Publikationen-PDF/WWF/DroughtRisk/EN/WEB.pdf.

WWF. (2019b). *Freshwater risks and opportunities: An overview and call to action for the financial sector*. https://wwfeu.awsassets.panda.org/downloads/wwf_waterrisk_financialvalue_part4_keypiece_web.pdf.

WWF & ABInBev. (2019). *Climate change and water: Why valuing rivers is critical to adaptation*. https://d2ouvy59p0dg6k.cloudfront.net/downloads/wwf_abi_water_climatechange__final_pdf.

WWF Germany. (2019). *Linking water risk and financial value—Part I: Considerations for the financial sector*. www.wwf.de/fileadmin/fm-wwf/Publikationen-PDF/WWF_WaterRisk_FinancialValue_Part1_BM1.pdf.

Assessing Water Risk: Frameworks, Models, and Tools

Data for Water Risks: Current Trends in Reporting Frameworks, Shortcomings, and the Way Forward

Laureline Josset and Paulina Concha Larrauri

1 Introduction

The attention to water risks in the private sector has been on the rise in the past decade. Climate extremes, such as floods and droughts, can cause immense losses to companies, investors, and society. Additionally, the shared nature of water and the complex interactions between its users can give rise to tension and conflicts. Society is more aware of the potential water scarcity and pollution that may emerge due to industrial activity. Consequently, to earn and maintain the social license to operate is vital for companies (Fraser, 2018; Prno & Scott Slocombe, 2012), and transparency is key (ICMM, 2014).

Investors have taken notice. Year after year the number of companies disclosing water risks is increasing, often initiated by investor requests. Reporting frameworks such as the Climate Disclosure Standards Board

L. Josset (✉) · P. Concha Larrauri
Columbia Water Center, Columbia University, New York, NY, USA

P. Concha Larrauri
e-mail: pc2521@columbia.edu

T. Walker et al. (eds.), *Water Risk and Its Impact on the Financial Markets and Society*, Palgrave Studies in Sustainable Business In Association with Future Earth,
https://doi.org/10.1007/978-3-030-77650-3_2

(CDSB), the Sustainability Accounting Standards Board (SASB), the CDP Water Questionnaire, or the Global Reporting Initiative (GRI) are creating standards and metrics related to water risks that are now disclosed or referenced by thousands of companies (CDP, 2020; SASB, 2020; TCFD, 2019).

1.1 The Many Dimensions of Water and Financial Risks

Despite an increased visibility and disclosure of water risks, their robust assessment remains tremendously challenging: Water risks assessments in the context of corporate activities are intrinsically trans-disciplinary. Water scarcity, floods, or pollution may impair or interrupt corporate activities and reciprocally, corporations may originate or amplify water risks (Table 1). Societal actions and impacts further compound these effects. Water risks thus manifest as physical, regulatory, or reputational impacts for the private sector. Their assessments and mitigation require a holistic approach at the intersection of physical (e.g., water availability, quality, and infrastructure), social (e.g., demographics, economics), and governmental (e.g., regulations, built infrastructure, financing) aspects.

In addition to the many dimensions underlying water risks and their links to financial impacts, their assessments span multiple scales (Josset et al., 2019; Lall et al., 2020). Temporally, water availability varies from day to day, while trends in groundwater levels or oscillation in climate and streamflow patterns occur over decades. Spatially, water withdrawals or contamination happen at very specific locations while their impacts might be felt over watersheds or aquifers spanning tens of kilometers and embedded in the water footprint of products that are then transported across the globe. Space and time are further compounded along the supply chain and by climate signals, which might lead to resiliency or amplify scarcity. In addition to being location specific, these attributes vary greatly from one industrial sector to another and across water risks.

Equally complex are the context and purpose of corporate water risk assessments, ranging from evaluating operational sustainability, informing capital investments, guiding risk mitigation strategies, and meeting regulatory obligations. Further, water risk reports are targeted to a variety of stakeholders: internal users, local communities, clients, authorities, and investors, and they have to be tailored accordingly.

The multiplicity of goals and actors may not necessarily translate into relevant information for the financial community. Indeed, the datasets,

Table 1 Non-exhaustive list of the water-related risks faced by corporations because of water risks

			WATER RISKS		
			SCARCITY	*FLOOD*	*POLLUTION*
CORPORATE RISKS	*PHYSICAL*	Asset-level	No or limited supply of water, investment for alternative sourcing.	Flooded assets, loss of production, deterioration or destruction of infrastructure.	Contamination of the supply of water, additional treatment cost (operational or infrastructural), contamination/loss of production.
		Supply chain-level	Interruption in production, limited supply, or interruption in distribution. Investment to secure alternative sources and loss of clients.	Interruption in production, supply, and distribution. Investment to secure alternative sources and loss of clients.	Interruption in supply, or distribution. Investment to secure alternative sources and loss of clients.
		Workforce	No water for workers' households. Incapacitated or stressed workforce.	Stranded workforce, loss of habitats. Flood may lead to pollution of drinking water resources.	Polluted drinking water supply. Workforce suffers from sickness and/or financial stress.
	REGULATORY	Governmental	Drought declaration, permit suspension or reevaluation, interruption in permitting process, limited diversification of water sources portfolio. (ESG) Certification: reevaluation, fine, loss of.	Permit suspension or reevaluation, interruption in permitting process, higher treatment regulation at intake, additional requirement for containment of risky material.	Wastewater permit reevaluation or suspension, higher treatment regulation at intake, additional requirement for containment of risky material, tighter control, suspension of production.
		Insurance	Higher premiums.	Higher premiums.	No coverage
		Certification	Reevaluation, fine, loss of permit for withdrawals.	If pollution ensues, fine, suspension, loss of permit.	Fine, suspension, loss of permit.
	REPUTATIONAL	Supply-distribution chain	Unreliable asset.		
		Local population and workers	Suspect corporation at the origin of scarcity/flood/pollution.		
		Governments	Blame company for scarcity/flood.		
		Investors	Fear of reputation being tainted by association.		
		Civil society/advocacy groups	Tighter scrutiny and echo suspicion.		
		Clients	Suspicion, loss of trust.		

tools, and metrics permitting the quantification of the risks are collected, designed, and developed by many entities, whose objectives may not align; this can lead to a loss or misuse of information. Effectively, the environmental aspect in environmental, social, and corporate governance (ESG) reporting is still largely dominated by carbon emissions disclosure and reduction targets. While there are water indicators in all the reporting frameworks, water reporting is not a top priority, with exceptions in water-intensive sectors such as mining (ICMM, 2017).

This can be explained in part because water stress is a complex notion that has taken many forms as acknowledged by numerous review articles on scarcity metrics in the scientific literature (Lloyd-Hughes, 2014; Mukherjee et al., 2018; Pedro-Monzonis et al., 2015; Vollmer et al., 2016). Investors have limited channels through which to understand and project how water can ultimately affect a company's performance or its portfolio. At the same time, companies have limitations to assess the water risks required in the reporting frameworks, beyond the risks directly related to operations. For example, a report by CDP (2020) finds that water pollution, which can have serious local impacts to communities and the environment, is not considered as a top risk for most companies. This is perhaps because of the difficulty to assess the sources and effects of water pollution or overuse, both internal and external to the company. In fact, there is indication of a negative trend in the quality and quantity of water metrics being reported even when more companies are disclosing ESG information (CDP, 2020; Sustainalytics, 2019).

In this chapter, we perform a diagnosis of the complex ecosystem of water risks to highlight data gaps and blind spots stemming from missing or misguiding data. We first map the data providers and users to understand fluxes of information, the roles of different actors, and the presence or absence of feedback loops (Sect. 2). Honing into the recommended water metrics and their relevance for the financial sector, we review the current challenges of ESG reporting (Sect. 3) and follow with a description of the tools available to quantify them (Sect. 4). We conclude with specific recommendations (Sect. 5), with the goal of creating an informative data ecosystem for water risk assessments.

2 At the Intersection of Water and Finance, a Complex and Fragmented Landscape of Data Stakeholders

2.1 *The Water and Finance Risks Information Ecosystem*

2.1.1 *Overview*

The many actors and overlap of scales in water risks assessment form a complex ecosystem of data stakeholders. The corporate sector itself is a data provider through their sustainability reports, which follow guidelines provided by ESG reporting standards (e.g., GRI, SASB, CDSB), which themselves are structured based on broader frameworks (e.g., Sustainable Development Goals). Furthermore, disclosure of risks often relies on tools developed by the civil society (e.g., the Water Risk Filter, Aqueduct), which embed global models developed by academics and consultants devised by aggregating datasets collected by governments or international agencies. Data aggregators and analysts blend this information to output ratings and recommendations to inform decision-making (e.g., via water stress indicators to guide investments).

Figure 1 provides an overview of the water and finance risk ecosystem, from the overarching frameworks developed by the international community to the investors and portfolio managers. We make the distinction between fluxes of data, metrics, and frameworks, and the nature of the actors (e.g., governmental, corporation). At the center of this ecosystem are the disclosures made by report issuers connected to three sub-ecosystems: the frameworks that guide these disclosures, the tools that facilitate the disclosures, and, more importantly for investors, the ways in which disclosures are used within the financial sector. Subsections below provide a more detailed description of these three sub-ecosystems.

2.1.2 *From Sustainability Reports/Corporate Disclosures to Actionable Financial Information*

Reports issued by corporations are used by multiple stakeholders in a myriad of ways. Specific to the financial community, they can be classified in five main categories:

- Directly by individual **investors or investment groups**, who rely on them to understand ESG risks and inform their investment or divestment decisions.

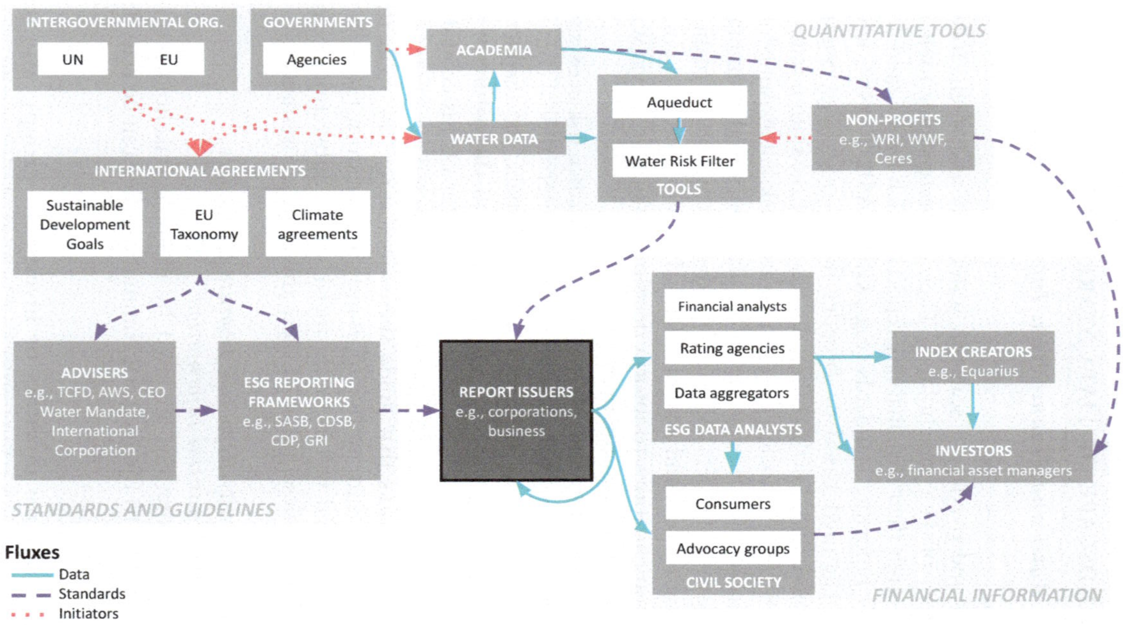

Fig. 1 The multi-agent ecosystem of water and finance risks. How structures relate to one another are described by arrows representing data, metrics, or frameworks

- **Financial analysts** inside investment banks or consulting groups, who review the reports and provide recommendations on the ESG-attributed financial risks for their clients.
- **Rating agencies and certifiers**, who, based on reports and frameworks, assess the level of compliance with a standard and certifies the company. This information is, in turn, used by portfolio managers, investment banks, and investors to meet clients' demands.
- **Data aggregators**, who, through text analysis and machine learning tools, recover and organize the information into datasets for further analysis.
- **Index creators**, relying on data aggregators and datasets from rating agencies, create ESG-rated financial products that are publicly traded.

The investors interest in ESG is twofold (Freyman et al., 2015):

(1) Understand their portfolio exposure to ESG risk and the potential impact on their financial performance.
(2) Meet their clients' demands on responsible and sustainable investing.

While the first objective is motivated purely by financial consideration, the second objective is motivated by the same societal desire that brought the Millennium Development Goals (MDGs) and Sustainable Development Goals (SDGs) to life. Therefore, corporations and developers of ESG frameworks alike reinforce these ties through additional feedback loops without direct involvement of the financial community:

- **Corporations** use their own sustainability reports for marketing and advertising purposes. This helps them position themselves as social and environmental stewards, which is seen positively by clients, shareholders, and investors. A favorable corporate sustainability image may also help companies create business opportunities, for example by having a competitive advantage as suppliers to other companies that value responsible ESG practices. It can also facilitate obtaining permits from regulators for expansions.
- **Developers of ESG frameworks** assess the adoption of their standards to improve their practices and engagement efforts. They also

conduct benchmarks of the adoption of other standards to identify opportunity areas, and to position and promote their approach and emphasis (Corporate Reporting Dialogue, 2019).

Additionally, two other actors may rely on sustainability reports through which reputational risks may be incurred:

- **Consumers** mindful of the ESG impacts of a corporation to inform their product choices.
- **Activists** and **advocacy groups** to assess and put in perspective the actions of a corporation when an environmental issue emerges in a community.

2.1.3 ESG Standards—The Guidelines for Sustainability Reports

Corporations disclose sustainability information following ESG standards. ESG frameworks are the natural vehicle as they translate societal objectives in quantitative, uniform, and structured information. The ESG frameworks themselves are constructed through many channels:

- ESG criteria originated from a United Nation (UN) initiative (Kell, 2018; The Global Compact, 2004). **International agencies** and **agreements** remain the main influencers to anchor ESG metrics onto societal goals. Three agreements in particular play a key role:
  - **Climate agreements** are central to ESG reporting with GHG (greenhouse gas) disclosures being one of the first and most followed quantifications. The Paris agreements in particular have advanced and informed current frameworks.
  - The **Sustainable Development Goals** (SDGs), introduced in 2015 by the United Nations in the continuity of the Millennium Development Goals, provide a global framework for development. Ratified by 193 countries, they form a worldwide agenda far beyond climate change, with, in particular, specific goals for water (SDG 6).
  - **EU Taxonomy Regulation** signed in June 2020 is at its core a classification system that identifies whether economic activities are sustainable from an environmental perspective (Ingman, 2020). It intersects with two other ESG regulations in the EU: The Non-Financial Reporting Directive (NFRD), which applies

to corporate disclosures; and the Sustainable Finance Disclosure Regulation (SFDR), which targets investment firms. The EU Taxonomy Regulation uses data generated from NFRD and enables investment firms to disclose in SFRD and also introduces new disclosures to companies reporting under NFRD (Ingman, 2020). As opposed to other frameworks used for voluntary disclosures, this is a unified system that is backed by law and will go into effect in 2021.

- **Consortia** of corporations, such as the CEO water mandate, which is a UN Global compact initiative that mobilizes business leaders to address global water issues aligned with the SDGs.
- **Non-profit organizations** that encourage the private sector to recognize environmental risks (e.g., Ceres, WRI, WWF), climate risks (e.g., Task Force on Climate-related Financial Disclosures) or water risks specifically (e.g., AWS), through direct partnerships and many white papers and reports on general matters or specifically for water (CDP, 2020; Freyman et al., 2015; Hofste et al., 2019a; ICMM, 2017; Morgan et al., 2020; Orr et al., 2009). While these organizations do not propose ESG frameworks themselves, they are key as intermediary between the international agencies and the corporations.
- **ESG frameworks boards**, constituted of multiple entities and individuals of varied backgrounds from the public, private, and civil sectors, provide recommendations on the content and format of the frameworks.

2.1.4 *The Quantitative Tools*

Unlike carbon emissions, disclosures of company-aggregated summative water volumes (e.g., withdrawals and discharges) are not informative for risk or sustainability assessments. This is because water risks are in their very nature, context specific. Though aggregated numbers are usually requested (e.g., company-wide water use), ESG frameworks recognize the necessity for more context-specific data and thus require distinct disclosures for areas that are identified as water stressed. This identification, however, is not straightforward and requires expertise, specialized methods, models, and datasets, often beyond a corporation's capability and means. A standardized process to determine and disclose water risks is necessary for the information to be meaningful, comparable, and trustworthy for the investors. Frameworks thus recommend tools to the report issuers to perform water assessments. It is important to note that these tools are not offered by the frameworks themselves, nor are they framework-specific, and constitute an ecosystem on their own:

- **Data collectors and providers** make available the datasets that are then harvested by the tools developers to populate their models.
- **Tool developers** build models and define metrics to compute water stress indicators.
- **Tool providers and aggregators**, such as WWF through the Water Risk Filter, or WRI with Aqueduct, host information on interactive platforms.

2.2 *Contextualization and Analysis of the Stakeholders Map*

2.2.1 *Climate Agreements, SDGs, and Financial Risks: A Narrow View of Water Risks*

Water stresses, whether scarcity-, flood- or pollution-related, can cause financial impacts. Climate change leads to and/or exacerbates water stress (UN Water, 2019), but not all water stresses stem from climate change.

However, historically, climate change has been at the forefront of corporate actions for sustainable development. First, reports were focused on reporting on greenhouse gas emissions, hence disclosing the corporation's contribution to climate change. The actions, behaviors, and disclosures had to evolve to take into consideration water risks (CDP, 2013).

In parallel, the incorporation of the SDGs anchored ESG disclosures in a broader international framework. Like climate change, the SDGs are tied to water stresses, though not necessarily systematically and some more directly than others (e.g., Table 1 in Morgan et al., 2020). Many frameworks refer to the UN's Human Right to Water and Sanitation and Sustainable Development Goal 6 (e.g., Ceres' Investor Water Toolkit; Ceres, 2021) and while other SDGs are considered for the ESG frameworks, the water implications behind them are rarely explicit.

Though the SDG framework and latest iterations of climate agreements are broad in scope and acknowledge the role of corporate actions, their objectives, targeted audience, and metrics were not intended to inform financial investments. Reports analyze and attempt to bridge SDGs and climate agreements (Dodwell et al., 2016), and some specifically look at the water dimension (UN Water, 2019). However, the divide with the financial community explains the many consultation groups, non-profit entities, and consortia that have fostered over the years to translate the frameworks and agreements into actionable investment information. However, the very indirect feedback mechanisms, as illustrated by the lack of stakeholders or flux of information (Fig. 1), prevent the consideration of the role of corporate actions on societal risks.

The challenges are not only limited to the conceptualization of these risks. Indeed, while SDGs are international, the metrics and datasets used are national, presenting a challenge in terms of uniformity to identify relevant data and define indexes within and across corporations.

This may result in blind spots with tremendous financial and societal repercussions. The full integration of societal, climate, and corporate risks and their linkages to water is yet to happen in a meaningful way to evaluate financial risks in terms of physical, regulatory, and reputational risks.

2.2.2 *The Absence of National-Level Action on Water Finance*

The presence of national institutions is limited to the peripheries of Fig. 1 and does not play a central nor explicit role in the ecosystem of sustainable finance information. Though fragmented, national governmental action is key in four aspects:

- On the regulatory side, national institutions, or regional agencies under national policies, have the authority to approve water withdrawal and wastewater permits, regulate land use, and authorize construction plans. These actions directly impact the regulatory risks faced by a corporation, having financial consequences through fines and determining the ability to operate and expand.
- Regarding physical risks, policies, regulations, and financial schemes enabling the construction or maintenance of key water infrastructure and/or water risks mitigation infrastructures are devised and managed by governments. In the case of a transboundary risk, the concerned governments are the ones involved with the resolution/mitigation of the risks. The built water-related infrastructure impacts the physical risks faced by a corporation, whether at the asset (e.g., flood damage, water available for production) or supply chain

level. Water infrastructure also relates to workforce safety, and to the health and economic growth of the community where a company operates, which translates to other social and operational risks.
- National pension funds and institutional investors were among the initial drivers for a growing demand for sustainable investments and have been key for a push toward ESG ratings (e.g., the Government Pension Fund Norway).
- Lastly, agencies at the national level are often the only entities given a mandate to collect, analyze, and disclose comprehensive water datasets. With the exception of remote sensing datasets (though they often stem from governmental programs), national agencies and their consortia are the main sources of water data behind the tools for water risks analysis.

These four aspects place governments at the national level in a strategic position to understand the water risks faced by a corporation within a country. Regional governments are key at the implementation and operational level for the mitigation of water risks, but planning tends to happen more often than not at the national level. The integration of these four aspects remains limited at the government level. Consequently, an assessment of their combined effects on financial risks remains non-existent. This can be in part explained by the mismatch between the geographical limits of national authorities and the international dimensions of corporations. It might be attributed also to a lack of awareness from governments, given that there is a strong tendency to understand and mitigate water risks regionally rather than having a national strategy.

However, within the specific context of sustainable finance, the role of national regulations is taking a sharp turn with the EU Taxonomy. Under the umbrella of the European Union, national governments are implementing financial regulations on the sustainability of investments. Whether the EU will implement a new set of policies and practices for corporations worldwide, whether other nations will come up with their own regulations, or whether individual governments will revise the role of their national agencies specifically to address the regulations of the Taxonomy remains to be seen.

2.2.3 *The Prominent Role of the Civil Society*

Society experiences water risks in different ways than corporations (Table 2). As opposed to national governments, civil society organizations

Table 2 Non-exhaustive lists of the risks faced by society because of water risks (WEF = Water-Energy-Food)

			WATER RISKS		
			SCARCITY	*FLOOD*	*POLLUTION*
SOCIETAL RISKS	*HEALTH*	Quantity	Reduced drinking water supply. Less water available for hygiene, cooking, sanitation.	*Flood volume:* Loss of lives, destruction of homes and essential infrastructure (hospitals, roads).	-
		Water supply quality	Less dilution of pollutant, higher contamination risk, having to change to another supply source of poorer quality.	*Contamination of water supply:* Intoxication through drinking and cooking, health issues through personal hygiene.	Intoxication through drinking and cooking, health issues through personal hygiene.
		WEF nexus	Less water for irrigation, food security implications, less water for energy, reduced to no AC and refrigeration.	Destruction of food production and supply, provision channels, destruction of power generation plant or grid, temporary interruption.	Pollution of soils (through irrigation), loss of yield and/or contamination of crops, contamination/loss of fish and seafood supplies.
	ECONOMIC	Quantity	*Water supply:* Interruption in or loss of supply for household and commercial activities, reduced or loss of income, additional cost to secure alternative supply at the household, business, communal, regional level.	*Flood intensity:* Loss of lives, destruction of homes and essential infrastructure (hospitals, roads).	-
		Water supply quality	Reduced quality impacts on production, cost of additional treatment might translate to loss of business.	*Contamination of water supply:* Impacts on production, cost of additional treatment might translate to loss of business.	Impacts on production, cost of additional treatment might translate to loss of business.
		WEF nexus	Impact on irrigation capacity and crop yields. Impact on energy availability and reliability.	Temporary interruption or destruction of food production and supply, provision channels. Temporary interruption or destruction of power generation plants or grid.	Loss of agriculture production from soil pollution or polluted irrigation water, inadequate water quality for F&B activity.

(continued)

Table 2 (continued)

	ENVIRONMENTAL	Surface water	Limited to no flow, reduced dilution of contaminants.	*Flood induced pollution*: Contamination of the supply, reduced concentration of dissolved oxygen.	Contamination of the supply, reduced concentration of dissolved oxygen.
		Groundwater	Depleting water table, subsidence, higher concentration of pollutant.	*Flood induced pollution*: Contamination of aquifer, mobilization of other pollutants.	Contamination of aquifer, mobilization of other pollutants.
		Surface water and groundwater dependent ecosystem	Drying out of flora and topsoils limiting recharge and augmenting run-off, resulting in potentially augmented flood. Loss of diversity in flora and fauna.	Drowning of the flora, losses and impacts on natural filtration, modification of run-off and recharge. Intoxication of fauna following contamination of food and water supply. Displacement of fauna, death by drowning or loss of food.	Contamination of flora, losses, impact on natural filtration, modification of run-off and recharge patterns. Intoxication of fauna.

are involved throughout the conceptual map (Fig. 1). From advocate, interpreter, intermediary, facilitator, aggregator, analyst, and definer of standards, the non-profit and non-governmental sectors are dominating the landscape of water data used in ESG reporting and investment analysis. It is noteworthy that the civil society does not generate data but defines and directs their fluxes; it does not define objectives but interprets overarching goals, and it does not create tools but reinterprets, adapts, and hosts them.

It is unclear whether the pivotal role of civil society has led for the national regulations to trail behind or whether it is the lack of government policies that gave no choice but for the civil society to step up. Regardless, the fact that the non-profit sector is not bound to a specific geography was critical to reach its prominent role. The lack of regulatory power from civil society also explains why growth in ESG disclosure stems from a voluntary adoption mostly driven by reputation. Though an approach devoid of regulatory constraints might have ensured a larger buy-in and encouraged a more consultative process, civil society's limited enforcing power also undermines the strength of the commitment to mitigate sustainability risks. Indeed, ESG framework developers are balancing the demands of the public and expert opinion without jeopardizing the continuing adoption of their standards by the corporations. Furthermore, civil society groups are not necessarily equipped with a sufficient level of expertise and insider knowledge of governmental datasets, as illustrated by the overwhelming reliance on academics for tools to estimate and interpret water stress.

Consequently, the breakdown of roles across many types of organizations prevents crucial feedback loops to incorporate corporate actions into water risks assessments. Not only is it a source of uncertainty in the quantification of corporate water risks, but it also hides the linkages between corporate action and the water risks induced to society. Government-led water risk mitigation actions, whether via regulatory, infrastructural, or financial actions, would benefit from a better integration of ESG-related corporation information.

3 ESG Disclosures and Investment Decisions

3.1 ESG Frameworks for Investors

The information voluntarily reported by companies using ESG frameworks is used by raters, index providers, ESG funds, and investment firms. Many frameworks exist, but, in general, investors tend to prefer frameworks that prioritize financial impacts and materiality, and that is why TCFD's recommendations seemingly dominate the ESG reporting requested by investment firms. The largest institutional investors have taken the recommendations from TCFD to create guidelines for the expectations of corporate performance on ESG (Freiberg et al., 2019). These recommendations are mostly aligned with existing reporting frameworks such as SASB and CDSB (Corporate Reporting Dialogue, 2019). A brief description of the three key players in finance oriented ESG reporting:

- **TCFD** focuses mostly on long-term climate risks associated with climate change and the transition to a low-carbon economy. The rationale is that present valuations do not adequately factor in climate risks, and this compromises long-term returns. TCFD's climate-related disclosures focus on four areas: governance, strategy, risk management, and metrics and targets (TCFD, 2017).

- **SASB** focuses on financially material environmental, social, and governance impacts. It has sector-specific guidelines and metrics covering seventy-seven industries.
- **CDSB** reporting principles and requirements guide companies on how to disclose climate-related financial metrics using data from the CDP questionnaire in compliance with TCFD's recommendations (CDSB, 2020).

3.2 Water Recommendations in ESG Frameworks

First, we look at the water-related information that ESG frameworks recommend companies to disclose. This is the base data that investors would be able to access.

Water disclosures in ESG reports mostly include estimations of water use (e.g., withdrawals, water used per unit of products, water returned, and similar volumetric measures), generalized measures of basin water stress, and qualitative water risk analyses. Information gathered through the CDP water questionnaire is routinely used to report in ESG frameworks (e.g., CDSB [2020] recommends specific questions from the CDP water questionnaire, categorizing them in the four focus areas of TCFD). Table 3 shows the recommended water disclosures in CDSB (taken from the food and beverage CDP water questionnaire) and in SASB (taken from the non-alcoholic beverages standard), aligned with TCFD's four focus areas (CDSB alignment to TCFD from CDSB [2020]; classified by the authors for SASB). In general, the recommendations from CDSB and SASB are very similar.

3.3 Challenges with ESG Reporting Compliance and Application in ESG Financial Performance

3.3.1 Context

Currently, the priority area for investors looking at ESG performance are risks associated with the transition to a net-zero carbon economy due to regulations and client pressure. Physical risks, where water is embedded, are not receiving the same attention. Besides the current focus on energy transition, Freyman et al. (2015) identify the main reasons for why the integration of water risks into investment decisions is lagging:

Table 3 Water disclosure recommendations in CDSB from CDP water questionnaire (CDSB, 2020) and SASB in compliance with TCFD's focus areas: Governance (G), Strategy (S), Risk Management (R.M.), and Metrics and Targets (M&T). Qualitative disclosures = "Q" and Quantitative disclosures = "#"

TCFD		Water Disclosures	Framework	
			CDSB	SASB
G	Q	Water policy scope and content.	■	
	Q	Leadership involvement in water issues.	■	
	Q	Water goals setting and monitoring.	■	
S	Q	Identification of water-related risks with potentially substantive financial or strategic impact, definition of those impacts and timelines.	■	■
	#	Number of facilities exposed to those water risks and their proportion.	■	
	#	Number of facilities exposed to water risks by basin and potential business impact.	■	
	Q	Details of water risks to direct operations by type (e.g., physical, regulatory, reputational), magnitude, likelihood, timeframe, and potential impact (e.g., fines, closure, brand).	■	■
	#	Potential direct financial impact and cost of mitigation.	■	■
	Q	Details of water risks to supply chains.	■	■
	#	Potential indirect financial impact and cost of mitigation.	■	■
	Q	Water related opportunities with substantial financial or strategic impact (e.g., efficiency, resiliency, markets, products and services) and timeframe.	■	■
	#	Financial impact of water opportunities.	■	
	Q	Water-related issues integrated into the short- and long-term strategic plan.	■	■
	#	Trend in water-related CAPEX and OPEX (as % change).	■	
	Q	Identification of water-related outcomes from climate-related scenario analysis/description of how climate change is expected to impact operations and revenues.	■	■
	Q	Tradeoffs in land use, energy production, and greenhouse gas (GHG) emissions from water management practices.		■
R.M.	#	Proportion of suppliers reporting water use, risks, and management and procurement spend they represent.	■	
	#	% by cost of supply chain materials sourced from water stressed areas.		■
	Q	Procedures for water risk assessments including tools used.	■	■
	Q	Risk mitigation efforts to achieve targets (e.g., water efficiency) and timeline.	■	■

(continued)

Table 3 (continued)

	Q	Contextual issues in water risk assessments (e.g., regulations, availability, quality, conflicts, ecosystems, stakeholders' perceptions of water withdrawals).	■	■
	Q	Stakeholders considered (e.g., investors, regulators, communities, NGOs, utilities).	■	
	Q	Process for identifying, assessing, and responding to water-related risks (all types) within the direct operations and the value chain.	■	■
	Q	Compliance with regulations related to discharges.		■
	Q	Activities and investments required to achieve strategic water plans, goals and/or targets, and risks that might affect their achievement.		■
M&T	#	Total volumes of water withdrawn, discharged, and consumed across all operations. % change versus previous year.	■	■
	#	Proportion of withdrawals/ consumption sourced from water stressed areas.	■	■
	#	Volumes withdrawn by source.	■	■
	#	Coordinates, water accounting and % change for facilities exposed to water risks that could have a substantive impact.	■	
	Q	Use of internal water price (Y/N).	■	
	Q	Water targets monitored.	■	■
	#	Number of incidents of non-compliance with water quality permits, standards, and regulations.		■

- The lack of consistent corporate disclosure of water metrics.
- The lack of platforms to easily access water data.
- The lack of a framework for investors to conceptualize water risks.

Although the financial sector and ESG are prone to many changes, little has changed over the past five years. Investors rely mostly on ESG reports, Securities Exchange Commission (SEC) filings, direct engagement with companies, and the CDP water disclosure program as sources of water data. Despite the data obtained from these sources, assessing water-related financial impacts remains challenging, due to the mostly qualitative information they contain (Table 3), and to other data issues discussed in Sect. 3.3.2. Water risks need to be reported as material impacts on revenue or returns to be used in financial risk assessments. The analysis of material impacts needs to consider the following aspects (Adriaens et al., 2014):

- The impact of water quality and quantity on the company's operations, reputation, and regulatory risks.
- The risk mitigation strategies a company has (e.g., infrastructure investment, insurance).
- The effect of risk management strategies in financial risk reduction (operational or valuation).

The guidelines, standards, and recommendations put forward by the creators of ESG frameworks touch on many relevant issues related to the points above, as seen in Table 3. However, companies are struggling to fully comply with them due to a variety of causes, leading to gaps in the data voluntarily reported. Additionally, there is lack of clarity on how investors can use this information when evaluating ESG performance. Here we comment on barriers that need to be removed to have clearer ESG corporate reporting and financial evaluations.

3.3.2 Data Availability, Complexity, and Costs

Companies are encouraged to describe when water-related financial impacts are expected (e.g., costs, revenues, assets and liabilities, capital and financing) considering climate variability, climate change, community relations, regulations, and the presence of areas with high biodiversity value (Table 3). These are complex tasks requiring data (internal and external), models, expertise, and monetary resources not available to all companies. Table 4 connects the water disclosure recommendations of SASB and CDSB described in Table 3 to the type of data needed to report them. As a result, compliance with the recommendations remains low. Some of the identified challenges are (TCFD, 2019):

- Companies are not sure how to use climate scenarios, seeing them as complex or too costly, and are struggling to use them for strategic planning and financial reporting.
- The lack of appropriately granular data and tools relevant to their businesses is a barrier.
- Companies cannot disclose confidential business information related to climate risk/water risk assumptions.
- The required metrics are not reflective of the risks or opportunities of some companies.
- Connecting water risks to finances is challenging.

Table 4 Examples of data needed to comply with disclosure requirements: Governance (G), Strategy (S), Risk Management (R.M.), and Metrics and Targets (M&T)

Category	*Data needed*	*TCFD area*
Physical risks—Drought, floods, water availability and ecosystems (asset level and key suppliers)	Historical precipitation, temperature, streamflow (for surface water), depth to water table (groundwater), water quality, regional water demand	S, R.M.
	Alternative water sources	S, R.M.
	Regional water demand projections, climate projections, basin water availability	S
	Monitoring of flora and fauna that could be impacted and their state of conservation	S, M&T
Reputational (asset level)	Demographics and dependence of the water source (direct use, economic, recreational, spiritual, etc.)	S
	Past and present water conflicts in the region (causes and consequences)	S
	Media coverage and activism in relation to water use and a specific sector	S
Communities (asset level)	Past, current, and planned community programs and agreements focused on water use and access	R.M., G
Regulatory (asset level and corporate)	Regulations related to water sources (e.g. permits, quality standards, thresholds, sanctions, reporting, etc.)	S
	Historical context of those regulations in the area and projections of how they may evolve	S, R.M.
	Water permit acceptance/rejections and timing of similar sectors	S
	Notices of non-compliance with water regulations	R.M.
Financial (asset and corporate)	Production losses due to lack of access ($loss/day) and due to the lack of raw materials (key suppliers)	S, R.M.

(continued)

Table 4 (continued)

Category	*Data needed*	*TCFD area*
	Historical information of water-related community compensation costs, remediation costs, legal costs, public relationships costs, and fees applicable to the types of impacts foreseen	S, R.M.
	Costs of community water-related programs and agreements including personnel	S, R.M., G
	Costs of monitoring and studies (water sources, water quality, quantity, flora and fauna)	S, R.M.
	CAPEX and OPEX for water sourcing and treatment, including current and projections	S
	Fluctuations in share value in response to water risks	S
Operational (asset level and corporate)	Water withdrawals by source and projected water withdrawals	M&T, S
	Water discharges volume and quality	M&T, S
	Historical water efficiency (water used/unit product)	M&T, S
	People involved in water management/water issues	G

Some companies and sectors have better compliance with water disclosures. For large companies in sectors such as food and beverage or mining, shortages in water supply and water contamination can be financially material, and understanding their water risks internally is vital. These companies are subject to scrutiny from regulators, investors, and community activists looking to their water management practices. As a result, ESG reporting for water-intensive companies tends to have a stronger water focus (examples in The Coca Cola Company, 2019) and collecting water quality and availability data at the location of their water sources is a routine activity. Some companies rely on groundwater commission hydrogeology studies to detect changes in the water table, to ensure groundwater resources remain available in the future for the continuation of their businesses, and to comply with regulations (examples in the mining sector in BHP, 2019; SQM, n.d.). Other sectors experience water risks indirectly, mostly through their supply chains. For them, the focus of disclosure is predominantly on water footprint accounting, tracking where and when water is used throughout the supply chain. These metrics are used to set reduction targets and to benchmark water use among peers.

For small companies with constrained resources, identifying, monitoring, and quantifying water risks may be more challenging. These companies rely mostly on publicly available tools and coarse resolution datasets to report water indicators, and they lack the capacity to perform more complex water risk analyses or to hire consultants for that purpose.

3.3.3 Lack of Clarity, Standardization, and Relevance

The purpose and function of the different frameworks and standards, and how they complement each other, is not clear to report issuers (Corporate Reporting Dialogue, 2019). The terminologies and methodologies are not well aligned and therefore reports are not compatible or comparable. Efforts on standardization and alignment between the frameworks, standards, and reporting recommendations are ongoing. A common thread is that sector specific standards within a framework are key for achieving comparability within a sector and across sectors.

Regardless of the sector and company size, current water disclosures cannot readily be translated to financial risk terms (e.g., probability, exposure, and the resulting impact on revenue, value at risk). For example, water footprint analysis is irrespective of the local context (e.g., water access, water infrastructure), and it does not include water quality measures, which can be more relevant, especially in the textile and agriculture-dependent industries (Chapagain & Tickner, 2012). Water footprints therefore do not measure risk beyond highlighting the reliance on water at the company or sector levels and are not directly linked to potential losses. Further, companies that carry out internal water-related financial risk analyses may not voluntarily disclose them because of issues with confidentiality and competitive advantage (TCFD, 2019). Consequently, the information included in ESG reports, even in major companies, does not necessarily reflect the true water risks and how they may evolve overtime.

Both report issuers and users are asking for more clarity on the connection between ESG disclosure and financial information, and a closer relationship with regulators to improve ESG reporting (Corporate Reporting Dialogue, 2019).

3.3.4 *Confusion in the Expectations of the Investment Community*

Companies are not clear on how investors are integrating ESG in their decision-making process. The term sustainable investment does not have a clear definition and although some classifications have been proposed (Ang, 2019; Chasan, 2020), they are not widely adopted yet. In addition, the construction of ESG rating indices is opaque and subjective which leads to indices that do not correlate across providers (Berg et al., 2020) as opposed to the highly correlated credit ratings. This adds a layer of mistrust (Hawley, 2017) that further weighs on capital allocation decisions.

Recognizing these issues, the EU has taken steps to formalize the definitions and classifications of sustainable investments with the EU Taxonomy Regulation. Investors are now facing regulatory pressure to increase transparency regarding how they classify an investment as sustainable to avoid greenwashing. The new regulations in the EU are shifting the ESG reporting world from voluntary to mandatory disclosures within a legal framework. Disclosures will apply not only to companies, but also to investors. Under the EU Sustainable Finance Disclosure Regulation (SFDR), investment firms will have to disclose the environmental sustainability of an investment according to the definitions of the EU Taxonomy Regulation. Investment firms will have to disclose how the ESG claims used to market the investment are defined, as well as the risks of the investment to ESG factors and the risk ESG factors have on the investment (Ingman, 2020). This poses a new challenge to investment firms in terms of data and analysis, needing new tools and data platforms.

To address this, efforts are underway to facilitate water disclosure and risk analysis. Non-profit organizations have been active in this space, creating guidelines and data products to help investors. For example, Ceres' Investor Water Toolkit provides recommendations and resources available to guide the integration of water risks into investment decisions, including standards, data, and tools created by other organizations such as SASB, CDSB, WWF, and WRI. The TCFD knowledge hub is another online platform that provides tools, resources, case studies, and insights to help organizations implement the TCFD recommendations. However, many challenges remain to have a water-relevant framework for investors, and the tools and data needed to implement it.

Overall, this is a work in progress where fundamental definitions of ESG reporting, the meaning to different stakeholders, data, and the process of implementation are rapidly changing.

4 Water Assessment Tools

In this section, we review the tools available to corporations and to investors. We differentiate two types of tools: those that are identified as water *risk tools* to disclose water in ESG reports, and tools that are aimed for *water valuation*. The tools reviewed rely on third-party models and data, which vary in objective, conceptualization, definition, accuracy, resolution, coverage, bias, and overall applicability to companies and investment firms.

4.1 Water Risk Tools

Risk is defined as the product of the probability of an event (i.e., the hazard), its consequences (i.e., the exposure), and the vulnerability of a system. The main focus of the best-known publicly available water risk tools is in reality the identification of the exposure, and some of them include measures of vulnerability. In general, the tools consider elements of basin and operational exposure, different temporal and geographical scales, risk persistence, and mitigation measures. Given the multiple dimensions of water risks, the recommendation is to use various tools and models to assess them (Morgan et al., 2020).

WRI's Aqueduct and the WWF's Water Risk Filter (WRF) (Table 5) are two of the most cited tools for reporting water in ESG, but companies use a mix to compensate for well acknowledged deficiencies to assess water risks at the local scale (Morgan et al., 2020). Both tools are based on global hydrologic models and datasets with global coverage that, although convenient for comparisons, are largely biased toward places where information for calibration and/or validation is available. Therefore, they carry uncertainties in the model inputs, model parameters, model structure, and in the observations for model validation and calibration (Döll et al., 2016). These uncertainties are not considered in the outputs of the tools, which are categorical indicators of low to high risk.

Water stress indicators. Having accurate estimations of water withdrawals and available supply at the global scale is very challenging (particularly groundwater) due to the lack of data available. Given that Aqueduct water stress indicators are used in many water valuation tools (Jorisch et al., 2018), we give a more thorough inspection to their construction. Aqueduct uses a gridded global hydrological model (Sutanudjaja

Table 5 Summary of Aqueduct and the WRF. Elaborated with data from Morgan et al. (2020), Hofste et al. (2019b), and WWF (2020b)

Parameter	*Aqueduct*	*Water risk filter*
Owner	WRI	WWF-DEG
Focus	Exposure	Exposure and mitigation actions
Indicators	Physical (10), regulatory and reputational (3)	Physical (13), regulatory (12) and reputational (7)
Main indicators	Baseline water stress (BWS)—withdrawal based Flood risk (FR)—not linked to extremes Drought risk (DR)—not linked to extremes or hydrological drought	Water depletion (WD)—consumption based Flood risk (FR)—recurrence with 1986–2019 data Drought risk (DR)—ratio of months when SPEI <=−1 with 10 years of data (2009–2019)
Indicator weights	Industry-specific	Industry-specific for 25 industry categories
Base models for the indicators	BWS: PCR-GLOBWB 2 (Sutanudjaja et al., 2018) FR: GLOFRIS: Ward et al. (Forthcoming) DR: FLOPROS (Carrão et al., 2016)	WD: WaterGAP3 (Brauman et al., 2016) FR: Archive of floods (Brakenridge, n.d.) DR: SPEI index (Vicente-Serrano, 2010)
Data[a]	*Water quantity*: Gridded modeled data (de Graaf et al., 2015; Schewe et al., 2014) *Water demand:* Modeled data Irrigation: MIRCA2000 (Portmann et al., 2010) Global Crop Water Model (Siebert & Döll, 2010) *Water withdrawals*: Modeled data (Schewe et al., 2014) Urban water (McDonald et al., 2014) Groundwater: MODFLOW (de Graaf et al., 2017) data from 1990 to 2014 (gridded monthly groundwater heads)	*Water quantity indicators*: Aridity indexes (Trabucco & Zomer, 2009), Water Depletion (Brauman et al., 2016) simulated data from 1971 to 2000 with WaterGap3, Baseline Water Stress (Aqueduct), Blue water stress (Mekonnen & Hoekstra, 2016), Moderate drought frequency SPEI index (Vicente-Serrano, 2010) *Operational data*: Company level data through a questionnaire

(continued)

Table 5 (continued)

Parameter	*Aqueduct*	*Water risk filter*
	Infrastructure/storage: GRanD dataset (Lehner et al., 2011) *Reputational risks:* RepRisk *Drought risk:* Baseline water stress, gridded precipitation from GPCC and other global indicators (Carrão et al., 2016)	*Reputational risks*: RepRisk and others (WWF, 2020b), Fresh water law and policy status (SDG6.5.1), Implementation status of water management plans (SDG6.5.1), Corruption perception index (Transparency International's data, 2018), Freedom participation in the world index (Freedom in the World Countries 2019, Freedom House), Water data monitoring (UN IGRAC, 2019 and the Global Runoff Database)
Geographic scope	Basin and sub-basin HydroSHEDS HydroBASINS Level 6 (mean area ~8,200 km^2)	Basin and sub-basin HydroSHEDS HydroBASINS Level 7 (mean area ~2,300 km^2) and Level 12 (mean area 130 km^2) available in 12 countries
Geographic coverage	Global	Global + higher resolution data in 12 countries
Temporal scale	BWS: Monthly	WD: Monthly
Temporal scope	Average past/recent conditions (e.g., BWS indicator: 1960–2014)	Average past/recent conditions (e.g., Blue water scarcity: 1996–2005)
Climate projections	Scenario analysis 2030 and 2040	Projected change to 2050 for water discharge, droughts and floods (2 °C warming)

Parameter	*Aqueduct*	*Water risk filter*
Climate models	CMIP5 GCMs. Scenarios: RCP4.5 and RCP8.5 Socioeconomic pathways, SSP2 and SSP3 (from IIASA) (Luck et al., 2015)	ISIMIP, Schewe et al. (2014) and Frieler et al. (2017)
Operational data	No	Yes—questionnaire

[a]Not a comprehensive list: datasets used in some of the models are not included

et al., 2018) to estimate water withdrawals, available water, and groundwater levels. Then those outputs are spatially and temporally aggregated to obtain water stress indicators in the basin and sub-basin scales (Hofste et al., 2019b).

One of the most cited indicators from Aqueduct is the baseline water stress, which is the ratio of water withdrawals to available surface and groundwater supply. This indicator does not consider environmental flows, water quality, water access, water governance, or "real" groundwater availability (i.e., global groundwater models have large uncertainty and bias and other shortcomings explained in Lall et al., 2020). The baseline water stress indicator is convenient for broad comparisons between regions, but it has limited application at a local level, which is acknowledged by its creators (Hofste et al., 2019a). Overall, beyond complying to ESG framework requirements, the indicator has limited use for assessing water risks, both for companies and investors. As Coca Cola points out, the baseline water stress in Aqueduct "measures chronic water stress rather than drought stress, and does not include the effects of deep groundwater, water governance or infrastructure in meeting industrial, domestic and agricultural demands for water" (The Coca Cola Company, 2019, p. 19).

Flood and drought risks. The flood and drought indicators in Aqueduct and the WRF do not consider extreme events, which are the ones that may pose the greatest financial risk. The riverine flood indicator in Aqueduct is the expected annual population affected by floods with some level of protection, assuming flood events are independent. This indicator does not account for the indirect impacts of floods nor has a measure of flood magnitude (Hofste et al., 2019b). The WRF flood indicator uses the number of floods that a location has experienced based on thirty-four years of observed flood events irrespective of the magnitude and existing flood protection strategies (e.g., a high-risk score is when 31–400 floods have occurred). The drought risk indicator in Aqueduct does not consider hydrological drought and is based on a combination of indexes (Carrão et al., 2016).

Regardless of the aforementioned limitations, companies and institutional investors have used Aqueduct indicators to disclose and assess current and future water risks at the portfolio level, for example in real estate investment (Deese et al., 2020), but the results can have little physical meaning. An alternative approach to incorporate the risk of extreme weather events at the portfolio level was proposed by Bonnafous et al.

(2017). Their water risk index can be used to explore the possible space and time clustering of exposure to extreme weather events across the sites in a geographically diversified portfolio (Bonnafous et al., 2017). Given that long climate records are needed to identify the temporal and spatial structure and clustering of extreme weather events, the index uses publicly available reanalysis climate data covering at least 100 years globally. Reanalysis data has issues with uncertainty and biases the further it goes back in time as described in Bonnafous et al. (2017), but the gains in robustness make it much more suitable for portfolio analysis than relying on climate projections.

Water quality. In terms of water quality, both Aqueduct and WRF include metrics of water quality external to the company, such as nitrogen and phosphorus loading, salinization, and a surface water contamination index. Given the lack of water quality data at a global scale (and even at a local scale), the water quality indicators (mostly based on models that have large bias and uncertainty) are not representative of the water quality issues a company faces or creates and how they deal with them.

4.2 Water Valuation Tools

This set of tools aims to quantify the value of water including externalities, and how water risks translate into financial impacts. We review three of them in this section (Table 6), all of which are recommended in Ceres Investor Water Toolkit (Ceres, 2021). A comprehensive list of valuation tools can be found in WWF's Valuing Water Database (WWF, 2020a).

The concept of shadow price is behind most of the water valuation tools, which is estimated using the total economic value (TEV) framework. The premise is that the market price of water underestimates the value of water so a higher shadow price for water can be used to modify financial or risk models (Freyman et al., 2015 Focus). TEV considers market use values (e.g., manufacturing), non-market use values (e.g., ecosystem services), and non-market non-use values (e.g., continued existence for future generations) (Morgan & Orr, 2015). For example, in Bloomberg's Water Risk Valuation Tool (WRVT), the shadow price is described as the cost per cubic meter of water considering projected ratios of supply and demand for freshwater resources, taking into account the agricultural and domestic supply (from Aqueduct baseline water stress), as well as the human health and environmental impacts (it is not clear from the tool's documentation what data is used to assess these). The

Table 6 Examples of water valuation tools. Data from individual manuals and methodologies

	Water risk valuation tool	*Water risk Monetizer*	*WaterVAR*
Focus	Revenue, costs, earnings, and share price impacts	Revenue and cost impacts	Price volatility related to water risks
Owner	Bloomberg	Ecolab—Trucost	Equarius
Objective	Assess the effect of a stranding asset scenario due to future physical water scarcity (gold and copper mining only)	To monetize water-specific business risks to protect against water quantity and quality constraints for growth	Estimate the value at risk Reduce stock volatility by investing in risk mitigation
Users	Investors	Corporates and investors	Investors
Methodology	Mapping mine assets against projected water scarcity indicators to calculate the supply-demand gap. Value of potentially un-extractable ore due to water scarcity. Shadow price with TEV framework	Calculate the water risk premium to the existing water price using the TEV framework Assessments of incoming water quantity and quality risks and outgoing water quality risks for surface water use	Estimation of VaR, estimation of the volatility risk premium for water (waterBeta^{TM}) Estimation of waterVaR^{TM} (Adriaens et al., 2014; Mogosanu et al., 2018)
Scope	Corporate	Corporate and portfolio	Corporate and portfolio
Water risks	Water scarcity.	Future water stress, regulatory and reputational	Physical (scarcity), reputational and regulatory risks (as intangible risks)
Output	Company-level value at risk with and without asset stranding	Water risk premium, revenue at risk and water risk ranking	waterVaR^{TM} and waterBeta^{TM}

(continued)

Table 6 (continued)

	Water risk valuation tool	*Water risk Monetizer*	*WaterVAR*
Sources and inputs[a]	– Aqueduct Baseline Water Stress – Expected production – Product water intensity	– Aqueduct Baseline Water Stress – Water Risk Filter – Water use/unit revenue – Basin population, GDP, economic activity – RepRisk—reputational risks – Internal data questionnaire: water use and discharges, water prices, production, revenues. Potential reputational and legal impacts, water sources and quality requirements	– Aqueduct Baseline Water Stress – Water Risk Filter Corporate data (Bloomberg) – ESG disclosures and SEC filings for data scraping (water intensity, water recycling, etc.)

[a]Not a comprehensive list. Refer to the tools manuals for the complete information

Water Risk Monetizer (WRM) from Ecolab-Trucost estimates a water risk premium relative to the market price considering water scarcity and water quality. It also uses the baseline water stress indicator from Aqueduct. In WRM the revenue at risk is estimated as the monetary value arising from the inability to produce due to water deficit, where water quality is also considered (Ecolab & Trucost, 2019). The calculation assumes that water allocations within a basin are proportional to the gross domestic product (GDP) a water user generates, namely the type of industry, local water stress, and water competition.

The **waterBeta™** and **waterVar™** measures by Equarius follow a different approach for approximating the shadow price as a function of stock volatility. WaterBeta™ is based on the market risk measurement "beta," which is a dimensionless ratio that captures deviations of a firm's performance from the market performance in a given industry. WaterBeta™ is an indicator of the sensitivity of revenue to water availability of a company versus a benchmark. The ratio is based on a water exposure metric called capital efficiency of water (i.e., the ratio of water withdrawals over revenue [Adriaens et al., 2014]). WaterBeta^{TM} also considers the fraction of economic output of a company in basins with high water stress (from Aqueduct) and an intangible risk factor related to the water intensity of an industry and total withdrawals (Mogosanu et al., 2018). Other intangible factors can include brand risk, supply risk, and regulatory or local issues, which are obtained through data scraping, although it is not clear from the tool's documentation how they are integrated into the metric. WaterBeta™ is intended to be used for portfolio management as a volatility risk premium that can be managed (Adriaens et al., 2014; Mogosanu et al., 2018).

The **value at risk model** or **VaR** is a probabilistic measure capturing the likelihood that an event will impact value in a company or portfolio at different confidence levels. This is adapted in Equarius' waterVaR™, which is defined as the opportunity cost from water risk exposures that may impact equity and portfolio volatility. It is the product of the waterBeta™ and the VaR signal of a company stock relative to its relevant industry sector benchmark (Adriaens et al., 2014).

The problem of the outputs of the tools is the underlying data and models, and the fact that they do not really consider the risks of floods or droughts (the water risk index by Bonnafous et al., 2017, aims to address this), neither the risk mitigation strategies a company has in place to cope with potential water quantity or quality issues (waterVaR™ attempts to do so indirectly [Mogosanu et al., 2018]), all of which will determine the "real" value at risk. There is also lack of clarity on how reputational and regulatory risks may emerge and result in financial impacts. These are common issues across water valuation tools.

5 Recommendations

5.1 *Reducing the Loss of Information*

With this chapter, we gave an overview of the fluxes of information, their contents, and the actors involved in reporting water metrics as part of ESG disclosures. We reviewed some of the most widely used frameworks, standards, and tools through the specific lens of data adequacy for risk assessments for investors. The complexity and fragmentation of the stakeholders map (Fig. 1) together with the multi-objectives of the ESG reporting frameworks and the limitations of the tools used for the disclosure are all sources of blind spots for water risks with important financial consequences.

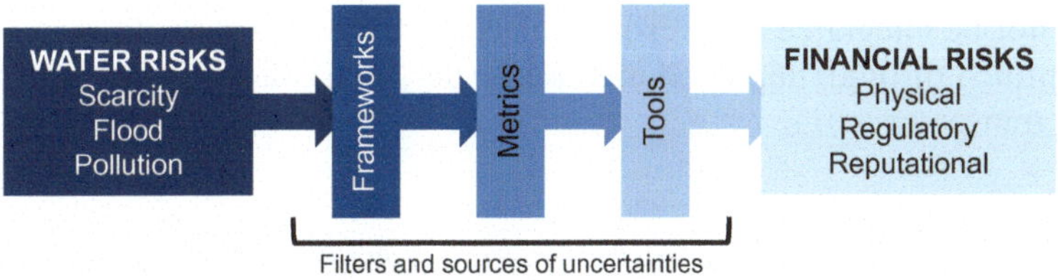

Fig. 2 Frameworks, metrics, and tools act as lenses through which water risks are perceived and translated into financial risks. Each stage of the process may hide, filter, and/or smooth the risk signals, reflecting only a subset of the water risks that may have financial impacts

Indeed, water risks (e.g., scarcity, flood, or pollution) are filtered out through the frameworks, metrics, and tools used to characterize them (Fig. 2). Moreover, the information that is propagated throughout these steps is only weakly correlated with the actual content of interest for the investment community, namely the physical, regulatory, and reputational risks that may lead to financial losses. We therefore organize our recommendations in two parts:

1. The blind spots in the current workflow and the associated sources of uncertainties that should be considered for a realistic quantification of water risks.
2. The developments in terms of data, tools, metrics, and frameworks that could address these shortcomings.

5.2 *Blind Spots in Water and Financial Risks*

Fund managers are not getting market signals to integrate water risks. Water is mostly perceived as an operating expense through the costs of infrastructure, distribution, and treatment, and in most situations is not expensive. The costs of water depletion, cumulative pollution, and low probability-high impact events such as prolonged droughts and intense floods are not visible from ESG reports. Furthermore, the data obtained from ESG reports are backward looking, and the investment community needs a forward-looking approach to assess risks. Therefore, water externalities slowly manifesting through environmental, social, reputational, and regulatory impacts are not reflected in valuations and ratings. Water will not be integrated in investment decisions until its externalities can be quantified through improved methodologies, data collection, regulations, and transparency.

5.2.1 *Scarcity*

- Drought metrics are numerous and have shown to be inconsistent. Moreover, the global hydrological models used to compute them do not represent, or only superficially does so, groundwater, the built infrastructure, and operations that determine the actual state of water resources (Lall et al., 2020). The estimation of the resulting uncertainties is a colossal endeavor beyond the resources of corporations, investors, and consultants alike.
- Indicators, if they do not take into account the specific infrastructure and use patterns at the asset and water body level, lead to assessments that may downplay or overestimate the water scarcity risks faced by the corporation.
- Similarly to the challenges corporations face when attempting to disclose carbon footprints, the water footprint at each step of the supply chain necessitates considerable information and an often-arbitrary estimate of what can be attributed to a product, a corporation, or a sector. The maybe pertinent information for investors is further diluted as water footprints aggregate their measure over the supply chain. Moreover, water footprint computation methods are poor proxies for water scarcity risks exposure as water use in itself is not informant of the risks if the local context is not taken into account.
- Furthermore, whether within drought metrics or water footprints, the impact of climate variability is rarely assessed, leading to the computation of an average exposure, when often only the events outside of the average are the sources of risks.

5.2.2 *Flood*

- Flood indicators in water risk tools do not represent the probability of floods nor the damage an asset or a supply chain could experience in the event of one based on flood intensity and protective measures.
- Compounded flood risks are not considered in any framework (i.e., flooding happening simultaneously at different assets and through supply chains).

- Extreme flooding events, which can have significant financial impacts, are not considered in risk quantification.

5.2.3 Pollution

- Water pollution is an afterthought. There is a lack of demand for quality data beyond compliance with effluent quality. This is true across ESG frameworks, which focus mostly on water quantity (RMF, 2020).
- The effects of cumulative water pollution on reputational, regulatory, and environmental risks are not visible. Cumulative pollution can be particularly risky for some sectors such as mining and can substantially increase closure costs or the risk of losing the social license to operate.
- Water pollution can also result from water scarcity and floods, and this link is not reflected in the reporting frameworks.

5.3 Developments for a Better Accounting of Water and Financial Risks

5.3.1 Data

Reassessing sustainability-focused data is pertinent to assess physical, regulatory, and reputational risks. An in-depth analysis of which data matters in the quantification of water-related financial risks is needed to inform investment decisions for all actors. Rather than solely focusing on harvesting all information and propagating it forward through metrics, models, and indicators, a backward design approach is more suited. *The identification of adequate datasets that relate the different aspects of water risks and their interdependencies to financial measures of interest needs to be undertaken.*

Transparency, uncertainty, and quantification of risks intersect. Water data analyses are sometimes misleading as illustrated by this case of water depletion in South India (Hora et al., 2019): A lack of understanding on data and its context has led to a complete misinterpretation of groundwater levels in the region. In this case, skilled researchers and analysts are not enough to warrant a good interpretation. A detailed understanding of why, how, and when data is collected, treated, and disclosed is necessary, which can be achieved through metadata, contextual information, and confidence intervals. Indeed, using data to quantify risks may lead to a false sense of confidence and to an underestimation of the risks. Consequently, *a comprehensive framework and standards to disclose, estimate, and propagate uncertainties for a robust quantification of the risks of interest is crucial. Important developments are needed for the construction and adoption of new standardized frameworks specific to the context of water financial risks.*

No miracle solution to the data challenge: the advantages and caveats of AI and remote sensing. Recent in their mainstream adoption, these two technological developments have led to a transformation in the evaluation of risks. In particular, text analysis and sentiment analysis are routinely used by consulting groups to process large volumes of pdf information or interpret reputational risks in social media. These methods present advantages and risks of misinterpretation (Mohammad, 2017), but maybe more importantly, they do not fundamentally collect information rather than extract underlying data of limited level of relevance or reliability in the context of water risks. A reconsideration on how to disclose the information so that it is relevant for the intended objectives would be more sensible. Similarly, remote sensing provides otherwise unavailable information for the prediction and mitigation of extreme events. It is also routinely used to identify water infrastructure and to evaluate water levels and use. These results may be prone to great uncertainty and be presented at an inadequate scale (e.g., GRACE groundwater data), and may overlap with more adequate information that is measured at a finer scale and on the ground. *The potential of these novel data collection and processing techniques will be fully exploited when they are paired with an adequate estimation of the robustness of their results in the specific context of water-related financial risks.*

Governments and governance are needed. Besides the data that is directly associated with the actions of a corporation, most data describing the local and global water context originates from governmental programs, putting public institutions in a privileged position to collect, provide access to, and interpret the data (Josset et al., 2019). Moreover, the development and maintenance of the cyberinfrastructure necessary to disclose and host the tools for the interpretation of the water risks is arguably a matter of national and/or international security.

5.3.2 Tools

No escape from going local. Each company and sector is exposed differently to water risks, so identifying these risks at the asset level and at the portfolio level is crucial to estimate material impacts. Existing water tools providing global indicators allow for an initial screening of regions that could have exposure to some water risks, albeit with many assumptions and uncertainty as discussed in Sect. 4. However, a deeper understanding of the exposure pathways and how they may translate into financial impacts remains missing. For a company, the analysis of water risks starts at the asset level, considering potential operational, regulatory, reputational, and environmental impacts in the region(s) where it operates. This analysis extends to the water risks faced by suppliers and to the compounded risks across a company's assets. Therefore, tools that focus on local water risks are more relevant for companies to estimate potential losses and opportunities.

Cause, effects, and interdependencies: uncovering water risk pathways and the value at risk. At the asset level, influence diagrams can be used to map the risk pathways leading to different operational, reputational, and regulatory water-related material impacts. These can manifest as higher capital, operating expenses, cost of capital, and even asset stranding (Concha Larrauri & Lall, 2018). The influence diagram approach considers water risks as internal and external to the company operations, which can be updated over time to reflect the dynamic nature of such risks. Given that influence diagrams are graphic representations of cause and effect, the interdependencies and complexity of water risks can be mapped and assessed. Influence diagrams are also useful to identify the data that is relevant to assess water risks for a particular asset, company, or sector. The outcome of this process is a probabilistic estimate of net present value (NPV), the value at risk, or other financial metric of interest. Risk mitigation actions and scenarios can be evaluated with influence diagrams to quantify the effect of different decisions. This methodology can be integrated with the approaches of Adriaens et al. (2014) and Bonnafous et al. (2017), to provide greater clarity of how water risks can manifest in time. It can also be a valuable tool to communicate material risks and justify risk mitigation actions within a corporation and to investors. Many software products with the required functionalities to perform risk analysis with influence diagrams are commercially available. Furthermore, the data and analysis resulting from the identification of risk pathways can be used to improve the estimations of water valuation tools such as waterVAR.

It will not be perfect. The data and modeling limitations of any tool need to be acknowledged and transparent so that the results obtained are framed within the scope and functionality of each tool.

5.3.3 *Metrics and Frameworks: Beyond Baseline Assessments and Climate Change Projections*

Governments and governance are needed here too. Beyond data and tools, legislation and regulation regarding water risks disclosures are needed to ensure transparency, accuracy, and compliance with the standards. These elements are needed to understand, facilitate, and regulate corporate action, prevention, and mitigation plans, and to embed them in national or international water security objectives. The governance structures to reach this objective remain to be invented and built, although the Energy Information Administration and the International Energy Agency are examples from the energy sector that can be followed to create equivalent structures for water.

Climate change is the tree that hides the forest of climate variability. Undoubtedly, changing patterns of climate will induce changes in water availability and demand. Furthermore, climate change projections and mitigation efforts are requested by clients and the public, and are thus necessary to limit reputational risks. However, human actions on the water cycle are such that climate extremes might not lead to water extremes, whereas specific climate patterns may lead to extreme water events, leading to high physical water risks, and potentially subsequent regulatory risks. Famously, California's water drought lasted longer than the climatic drought and led to new regulations. Comprehensive frameworks including human intervention on the water systems (e.g., land use, withdrawals, built infrastructure, return flows, management patterns and practices) are needed to understand what climate patterns lead to water scarcity risks, exacerbate floods, and indirectly impact pollution events. Corporations' water and land use and management must be included for any assessment to be meaningful due to the pivotal role that the industrial, energy, agricultural, mining, and food and beverage sectors play.

A multi-scale approach. The Water Toolkit from Ceres identifies the need for "an effective and broadly-recognized investor framework for conceptualizing water risks" and for "systematically capturing more location-specific corporate information" (Freyman et al., 2015, p. 10). We put forward that in order to do so, a multi-scale approach would (1) permit the integration of the relevant datasets while embedding their relative accuracy; (2) offer features to navigate from the watershed-level to the world-scale; (3) be equipped with tools that link water risks to financial risks; and (4) call for a transparent and robust uncertainty assessment. The features of the frameworks should be further accompanied by agile governance bodies to facilitate the water stewardship needed to reach water security.

Trans-disciplinary. We conclude this chapter with an invitation to any actors concerned by water and finance risks, to answer central, yet unanswered questions: What are the metrics, methods, and tools useful to achieve relevant, standardized, meaningful, robust, and investment-related quantification of water risks? And what are the feedback mechanisms that would help reach societal objectives?

References

Adriaens, P., Sun, K., & Gao, R. (2014). *Bridging physical and financial business water risk: waterVaR and waterBeta metrics for equity and portfolio risk assessment* (Working Paper No. 1237). Michigan Ross School of Business. https://ssrn.com/abstract=2445580.

Ang, G. (2019). Investment strategy and investment horizon: OECD work on ESG, the investment ecosystem and green finance. *ESMA Workshop on Short-Termism*. Paris, France. https://www.esma.europa.eu/sites/default/files/esma_workshop_on_short-termism_oecd_geraldine_ang_online.pdf.

Berg, F., Koelbel, J., & Rigobon, R. (2020). *Aggregate confusion: The divergence of ESG ratings* (Working Paper No. 5822–19). MIT Sloan School of Management. https://doi.org/10.2139/ssrn.3438533.

BHP. (2019). *Monitoring program—Groundwater Olympic dam version 16*. https://www.bhp.com/-/media/bhp/regulatory-information-media/copper/olympic-dam/0000/environment-protection-and-management-program-current-sa-govt/monitoring-program-groundwater.pdf.

Bonnafous, L., Lall, U., & Siegel, J. (2017). An index for drought induced financial risk in the mining industry. *Water Resources Research, 53*(2), 1509–1524. https://doi.org/10.1002/2016WR019866.

Brakenridge, G. R. (n.d.). *Global active archive of large flood events*. http://floodobservatory.colorado.edu/Archives/index.html.

Brauman, K. A., Richter, B. D., Postel, S., Malsy, M., & Flörke, M. (2016). Water depletion: An improved metric for incorporating seasonal and dry-year water scarcity into water risk assessments. *Elementa Science of the Anthropocene*, *4*, 83. http://doi.org/10.12952/journal.elementa.000083.

Carrão, H., Naumann, G., & Barbosa, P. (2016). Mapping global patterns of drought risk: An empirical framework based on sub-national estimates of hazard, exposure and vulnerability. *Global Environmental Change, 39*, 108–124. https://doi.org/10.1016/j.gloenvcha.2016.04.012.

CDP. (2013). *Moving beyond business as usual. A need for a step change in water risk management*. https://6fefcbb86e61af1b2fc4-c70d8ead6ced550b4d987d7c03fcdd1d.ssl.cf3.rackcdn.com/cms/reports/documents/000/000/637/original/CDP-Global-Water-Report-2013.pdf?1470314569.

CDP. (2020). *Cleaning up their act: Are companies responding to the risks and opportunities posed by water pollution?* https://6fefcbb86e61af1b2fc4-c70d8ead6ced550b4d987d7c03fcdd1d.ssl.cf3.rackcdn.com/cms/reports/documents/000/005/165/original/CDP_Global_Water_Report_2019.pdf?1591106445.

CDSB. (2020). *The building blocks. Connecting CDP data with the CDSB framework to successfully fulfill the TCFD Recommendations*. https://www.cdsb.net/sites/default/files/the_building_blocks_guidance_web_version.pdf.

Ceres. (2021). *Ceres investor water toolkit*. https://www.ceres.org/resources/toolkits/investor-water-toolkit.

Chapagain, A. K., & Tickner, D. (2012). Water footprint: Help or hindrance? *Water Alternatives, 5*(3), 563–581. https://doi.org/10.1080/0969160x.2013.820396.

Chasan, E. (2020, July 24). *Mutual fund industry tries to figure out how to define ESG funds*. Bloomberg. https://www.bloomberg.com/news/articles/2020-07-24/mutual-fund-industry-tries-to-figure-out-how-to-define-esg-funds?sref=fSOf3OlP.

Concha Larrauri, P., & Lall, U. (2018). *The value of water*. Columbia Water Center. https://drive.google.com/file/d/1aKtIgDdUoV6l9iZaOPit9HJUm_uxu-r5/view.

Corporate Reporting Dialogue. (2019). *Driving alignment in climate-related reporting*. https://corporatereportingdialogue.com/wp-content/uploads/2019/09/CRD_BAP_Report_2019.pdf.

Deese, B., Hildebrand, P., Bertolotti, A., & Suo, Y. (2020). *Troubled waters*. Blackrock. https://www.blackrock.com/us/individual/insights/blackrock-investment-institute/troubled-waters.

De Graaf, I. E. M., Sutanudjaja, E. H., Van Beek, L. P. H., & Bierkens, M. F. P. (2015). A high-resolution global-scale groundwater model. *Hydrology and Earth System Sciences, 19*(2), 823–837. https://doi.org/10.5194/hess-19-823-2015.

De Graaf, I. E., van Beek, R. L., Gleeson, T., Moosdorf, N., Schmitz, O., Sutanudjaja, E. H., & Bierkens, M. F. (2017). A global-scale two-layer transient groundwater model: Development and application to groundwater depletion. *Advances in Water Resources, 102*, 53–67. https://doi.org/10.1016/j.advwatres.2017.01.011.

Dodwell, C., Harries, J., Holdaway, E., Horrocks, L., Mittal, N., Saddington, D., Shabb, K., Smithers, R., Le-Cornu, E., Wartmann, S., & Monkhouse, C. (2016). Planning for NDC implementation. Quick start guide and reference manual. *Climate and Development Knowledge Network*. https://www.cdkn.org/ndc-guide/#.

Döll, P., Douville, H., Güntner, A., Müller Schmied, H., & Wada, Y. (2016). Modelling freshwater resources at the global scale: Challenges and prospects. *Surveys in Geophysics, 37*(2), 195–221. https://doi.org/10.1007/s10712-015-9343-1.

Ecolab and Trucost. (2019). *Ecolab water risk monetizer methodology*. www.waterriskmonetizer.com.

Fraser, J. (2018, February). Mining companies and communities: Collaborative approaches to reduce social risk and advance sustainable development. *Resources Policy*, 101144. https://doi.org/10.1016/j.resourpol.2018.02.003.

Freiberg, D., Rogers, J., & Serafeim, G. (2019). *How sustainability issues become financially material to corporations and their investors* (Working Paper No. 20–056). Harvard Business School. https://www.hbs.edu/faculty/Pages/item.aspx?num=57161.

Freyman, M., Collins, S., & Barton, B. (2015). An investor handbook for water risk integration: Practices and ideas shared by 35 global investors. *Ceres*. https://www.ceres.org/sites/default/files/reports/2017-03/Ceres_ESG-WaterRisk_041515_Print.pdf.

Frieler, K., Lange, S., Piontek, F., Reyer, C. P. O., Schewe, J., Warszawski, L., Zhao, F., Chini, L., Denvil, S., Emanuel, K., & Geiger, T. (2017). Assessing the impacts of 1.5°C global warming—Simulation protocol of the Inter-Sectoral Impact Model Intercomparison Project (ISIMIP2b). *Geoscientific Model Development, 10*(12), 4321–4345. https://doi.org/10.5194/gmd-10-4321-2017.

Hawley, J. (2017). ESG ratings and rankings. All over the map. What does it mean? *Truvalue Labs*. https://truvaluelabs.com/wp-content/uploads/2017/12/ESG-Ratings-and-Rankings-All-Over-the-Map.pdf.

Hofste, R., Kuzma, S., Walker, S., & Sutanudjaja, E. H. (2019a). Aqueduct 3.0: Updated decision-relevant global water risk indicators. *World Resources Institute*. https://www.wri.org/publication/aqueduct-30.

Hofste, R., Kuzma, S., Walker, S., & Sutanudjaja, E. (2019b). Aqueduct 3.0: Updated decision-relevant global water risk indicators. Technical note. *World Resources Institute*. https://www.wri.org/publication/aqueduct-30.

Hora, T., Srinivasan, V., & Basu, N. B. (2019). The groundwater recovery paradox in South India. *Geophysical Research Letters, 46*(16), 9602–9611.

ICMM. (2014). *Water stewardship framework*. https://www.icmm.com/en-gb/environment/water/water-management.

ICMM. (2017). *A practical guide for water reporting*. https://www.icmm.com/website/publications/pdfs/water/water-reporting_en.pdf.

Ingman, B. C. (2020, July 13). The EU taxonomy regulation: an overview. *Factset*. https://insight.factset.com/eu-taxonomy-regulation.

Jorisch, D., Frank, A., Schneider, M., Stadelmann, M., & Rosenthal, N. (2018). Water risks and financial market: Overview and analysis. *South Pole*. https://www.southpole.com/publications/water-risks-and-financial-market-overview-and-analysis.

Josset, L., Allaire, M., Hayek, C., Rising, J., Thomas, C., & Lall, U. (2019). The U.S. water data gap—A survey of state-level water data platforms to inform the development of a national water portal. *Earth's Future, 7*(4), 433–449. https://doi.org/10.1029/2018EF001063.

Kell, G. (2018, July 11). The remarkable rise of ESG. *Forbes*. https://www.forbes.com/sites/georgkell/2018/07/11/the-remarkable-rise-of-esg/#1712de011695.

Lall, U., Josset, L., & Russo, T. (2020). A snapshot of the world's groundwater challenges. *Annual Review of Environment and Resources, 45*(1), 1–24. https://doi.org/10.1146/annurev-environ-102017-025800.

Lehner, B., Reidy Liermann, C., Revenga, C., Vorosmarty, C., Fekete, B., Crouzet, P., Doll, P., Endejan, M., Frenken, K., Magome, J., & Nilsson, C. (2011). *Global reservoir and dam database, version 1 (GRanDv1)*. NASA Socioeconomic Data and Applications Center (SEDAC). https://doi.org/10.7927/H4N877QK.

Lloyd-Hughes, B. (2014). The impracticality of a universal drought definition. *Theoretical and Applied Climatology, 117*(3–4), 607–611.

Luck, M., Landis, M., & Gassert., F. (2015). Aqueduct water stress projections: Decadal projections of water supply and demand using CMIP5 GCMs. *World Resources Institute*. www.wri.org/publication/aqueduct-water-stress-projection.

McDonald, R. I., Weber, K., Padowski, J., Flörke, M., Schneider, C., Green, P. A., Gleeson, T., Eckman, S., Lehner, B., Balk, D., & Boucher, T. (2014). Water on an urban planet: Urbanization and the reach of urban water infrastructure. *Global Environmental Change, 27*, 96–105.

Mekonnen, M. M., & Hoekstra, A. Y. (2016). Four billion people facing severe water scarcity. *Science Advances, 2*(2). https://doi.org/10.1126/sciadv.1500323.

Mogosanu, I., Li, D., & Adriaens, P. (2018). Capital markets risk signaling: Linking financial asset risk pricing to water risk exposures. *University of Michigan*. https://uploads.strikinglycdn.com/files/5f6e3ea8-6daf-4e55-b165-c02595f94896/FinancialRiskAccountingwaterBeta.pdf.

Mohammad, S. M. (2017). Challenges in sentiment analysis. In E. Cambria, D. Das, S. Bandyopadhyay, & A. Feraco (Eds.), *A practical guide to sentiment analysis* (pp. 61–83). Springer International Publishing. https://doi.org/10.1007/978-3-319-55394-8.

Morgan, A., Laporte-Bisquit, A., Williams, T., & Maggo, D. (2020). Right tool for the job: Tools and approaches for investors to assess water risks and shared water challenges. *WWF & WBCSD*. https://www.wbcsd.org/Programs/Food-and-Nature/Water/Resources/Right-tool-for-the-job.

Morgan, A. J., & Orr, S. W. (2015). The value of water: A framework for understanding water valuation, risk and stewardship. *International Finance Corporation*. https://ceowatermandate.org/resources/value-water-2015/.

Mukherjee, S., Mishra, A., & Trenberth, K. E. (2018). Climate change and drought: A perspective on drought indices. *Current Climate Change Reports, 4*(2), 145–163.

Orr, S., Cartwright, A., & Tickner, D. (2009). Understanding water risks. A primer on the consequences of water scarcity for government and business. *WWF Water Security Series 4*. https://ceowatermandate.org/resources/understanding-water-risk-2009/.

Pedro-Monzonis, M., Solera, A., Ferrer, J., Estrela, T., & Paredes-Arquiola, J. (2015). A review of water scarcity and drought indexes in water resources planning and management. *Journal of Hydrology, 527*, 482–493.

Portmann, F. T., Siebert, S., & Döll, P. (2010). MIRCA2000-Global monthly irrigated and rainfed crop areas around the year 2000: A new high-resolution data set for agricultural and hydrological modeling. *Global Biogeochemical Cycles, 24*(1). https://doi.org/10.1029/2008GB003435.

Prno, J., & Scott Slocombe, D. (2012). Exploring the origins of 'social license to operate' in the mining sector: Perspectives from governance and sustainability theories. *Resources Policy, 37*(3), 346–357. https://doi.org/10.1016/j.resourpol.2012.04.002.

RMF. (2020). *Mining and water: Are operational concerns overriding public interest?* https://www.responsibleminingfoundation.org/research/water2020/.

SASB. (2020). *Global use of SASB standards*. https://www.sasb.org/about/global-use/.

Schewe, J., Heinke, J., Gerten, D., Haddeland, I., Arnell, N. W., Clark, D. B., Dankers, R., Eisner, S., Fekete, B. M., Colón-González, F. J., & Gosling, S. N. (2014). Multimodel assessment of water scarcity under climate change. *Proceedings of the National Academy of Sciences, 111*(9), 3245–3250. https://doi.org/10.1073/pnas.1222460110.

Siebert, S., & Döll, P. (2010). Quantifying blue and green virtual water contents in global crop production as well as potential production losses without irrigation. *Journal of Hydrology, 384*(3–4), 198–217. https://doi.org/10.1016/j.jhydrol.2009.07.031.

SQM. (n.d.). *Environmental monitoring plan*. https://www.sqm.com/en/planes-de-seguimiento-ambiental/.

Sustainalytics. (2019). *Water management and stewardship: Benchmarking corporate practices*. https://www.sustainalytics.com/esg-research/stewardship/water-management-stewardship-benchmark-2019/.

Sutanudjaja, E. H., Beek, R. V., Wanders, N., Wada, Y., Bosmans, J. H., Drost, N., Ent, R. J., De Graaf, I.E., Hoch, J. M., Jong, K. D., & Karssenberg, D. (2018). PCR-GLOBWB 2: A 5 arcmin global hydrological and water resources model. *Geoscientific Model Development, 11*(6), 2429–2453. https://doi.org/10.5194/gmd-11-2429-2018.

TCFD. (2017). *Technical bulletin*. https://www.fsb-tcfd.org/wp-content/uploads/2017/06/FINAL-TCFD-Technical-Supplement-062917.pdf.

TCFD. (2019). *Status report*. https://www.fsb-tcfd.org/publications/tcfd-2019-status-report/.

The Coca Cola Company. (2019). *The Coca-Cola Company CDP water security questionnaire*. https://www.coca-colacompany.com/content/dam/journey/us/en/policies/pdf/sustainability/2019-cdp-water-response.pdf.

The Global Compact. (2004). *Who cares wins. Connecting financial markets to a changing world*. https://d306pr3pise04h.cloudfront.net/docs/issues_doc%2FFinancial_markets%2Fwho_cares_who_wins.pdf.

Trabucco, A., & Zomer, R. (2009). *Global potential evapo-transpiration (Global-PET) and global aridity index (Global-Aridity) geo-database*. https://cgiarcsi.community/data/global-aridity-and-pet-database/.

UN IGRAC. (2019). *Global groundwater monitoring network GGMN*. https://www.un-igrac.org/special-project/ggmn-global-groundwater-monitoring-network.

UN Water. (2019). *UN-Water policy brief on climate change and water*. https://www.unwater.org/publications/un-water-policy-brief-on-climate-change-and-water/.

Vicente-Serrano, S. (2010). *The climate data guide: Standardized precipitation evapotranspiration index (SPEI)*. Retrieved February 18, 2020, from https://climatedataguide.ucar.edu/climate-data/standardized-precipitation-evapotranspiration-index-spei.

Vollmer, D., Regan, H. M., & Andelman, S. J. (2016). Assessing the sustainability of freshwater systems: A critical review of composite indicators. *Ambio, 45*(7), 765–780.

WWF. (2020a). *Valuing water database*. https://waterriskfilter.panda.org/en/Value/ValuationApproachFinder.
WWF. (2020b). *Water risk filter 5.0 methodology and documentation*. https://waterriskfilter.panda.org.

How Do Investors Assess Water Risks?

Milla Craig, Erica Coulombe, Charlotte Lombardi, and Nourhane ElGarhy

1 Water: The Investor Perspective

1.1 Water and the Responsible Investor

The inclusion of environmental, social, and governance (ESG) factors in investment decisions has gained momentum over the past several years. Increasingly, mainstream investors take ESG issues into consideration in both the evaluation and ongoing monitoring of potential and existing investments. This assessment can take the form of evaluating how a company manages their ESG issues through primary and secondary

M. Craig (✉) · E. Coulombe · C. Lombardi · N. ElGarhy
Millani, Montreal, QC, Canada
e-mail: mcraig@millani.ca

E. Coulombe
e-mail: ecoulombe@millani.ca

C. Lombardi
e-mail: clombardi@millani.ca

N. ElGarhy
e-mail: nelgarhy@millani.ca

T. Walker et al. (eds.), *Water Risk and Its Impact on the Financial Markets and Society*, Palgrave Studies in Sustainable Business In Association with Future Earth,
https://doi.org/10.1007/978-3-030-77650-3_3

research, or by engaging directly with a company (through proactive dialogue, meetings with management, etc.) to understand and influence their behavior with regards to environmental, social, and/or governance issues.

Recently, water, as an ESG risk, is gaining increased attention from a number of financial stakeholders. The main challenges associated with water are those relating to its quality, physical, chemical, or biological characteristics (OECD, 2005), and scarcity. Water scarcity is defined as "the point at which the aggregate impact of all users impinges upon the supply or quality of water under prevailing institutional arrangements to the extent that the demand by all sectors, including the environment, cannot be satisfied fully" (UN-Water Thematic Initiatives, 2006, p. 2).

Water risks are present in many ways, some not as apparent as others. For example, the mediatic portrayal of Canada, often as a vast landscape of pristine wilderness, may have contributed to the misconception that the country does not have water quality or scarcity issues. Canada is considered a freshwater-rich country, with almost 20% of the world's total freshwater resources (Environment and Climate Change Canada, 2018). However, as more than half of this freshwater drains northward into the Arctic Ocean and Hudson Bay, it is unavailable to 85% Canadians as they live along the country's southern border (Environment and Climate Change Canada, 2018).

Water issues in Canada also take on a social risk dimension, as more than 60 Indigenous communities are under long-term (lasting more than one year) drinking water advisories as of February 2020 (Indigenous Services Canada, 2020). Despite a commitment from Canada's Federal Government in 2016 to end all long-term drinking water advisories by 2021, Canada still faces systematic problems with respect to drinking water on First Nations reserves, meaning that the ability to deliver on this commitment is unlikely.

There are also opportunities associated with water. Ocean investing, or the "blue economy," is starting to attract more interest, with sovereign debt deals issued in the Seychelles and Mauritius in response to marine conservation needs (Gambetta, 2020).

This chapter examines the risks and opportunities associated to water quality and scarcity from both a global and a Canadian perspective. There are varying approaches to evaluating water risk, and several existing frameworks that aid both companies in their disclosure of water risk and investors in their risk assessments, some of which are set out below.

2 Assessing Water Risk

2.1 *Water Needs*

The extent of exposure to water risk can be assessed by evaluating the extent to which a company has water needs. Water dependency and needs vary greatly depending on a given company's sector, and may refer to the volume needed, the quality of water, the source type, the availability of access, etc. For instance, some sectors may require large volumes of water (e.g., power generation for cooling plant operations), while others may be dependent on the availability of high-quality water for operations (e.g., food and beverage production) (Ceres, 2018). Different sections of a business' value chain may vary in their exposure to water risk. Depending on the nature of the company, water risk exposure may only apply to a portion of its operations, or to the company as a whole.

2.2 *Water Risks Classification*

As an investor-driven initiative, the Task Force for Climate-related Disclosures (TCFD) developed a framework for climate change disclosures that provides a useful structure for climate-related risk assessment. The issue of water can be approached in the same way, by classifying risks into physical and transition risk categories (Task Force on Climate-Related Financial Disclosures, 2017).

2.2.1 Physical Risks

Physical water security addresses the ability of local water resources (surface or groundwater) to consistently meet demands, both in terms of volume and quality (Ceres, 2018). The following risks could be intrinsic (caused by poor governance, inefficiencies in operations, etc.) or extrinsic (dependant on the water basin context):

- Water scarcity: the availability of freshwater resources. Scarcity is a function of water consumption relative to water resources in a given area;
- Quality of water: the availability of adequate water quality for its intended use;

- Accessibility of water: whether individuals or companies are able to make use of the water resources that often depend on water infrastructure and affordability, among other factors;
- Climate change: the physical impacts of droughts and floods, rising sea levels, and extreme weather events (Ceres, 2018).

2.2.2 *Transition Risks*

Aside from physical risks, investors also face transition risks as the world shifts toward a low-carbon economy. Transition risks mainly stem from changing policies and regulations, as well as market demand and supply shifts. Typically, transition risks are categorized as operational, regulatory, financial, and reputational.

2.2.3 *Operational Risks*

Water stress could interfere with business operations and cause disruptions, leading to a loss in revenue and/or quality of product. Environmental changes can also bring about operational challenges, either due to not having access to sufficient water quantities, or to only having access to low-quality water that is unfit for use.

2.2.4 *Regulatory Risks*

Regulatory risks address the extent to which local laws and regulations can ensure water supply security in terms of volume and quality, including the treatment of wastewater and the maintenance of water health. This assessment also addresses the degree to which companies can predict water security. Ineffective and poorly implemented water governance could expose the company to significant risk. Likewise, changes in water laws and regulations can result in more stringent legislation, reduced allocations, fines, and penalties for water pollution (Ceres, 2018).

2.2.5 *Financial Risks*

Companies may risk disruptions to operations due to water stress, leading to a loss in revenue, increased costs due to competition for water, increased capital expenditure, and other costs related to the treatment of water, wastewater discharge, and the payment of fines. Financial losses can also be due to strategic risks related to water such as the inability to grow revenue, access new markets, or develop new facilities. The financial market implications of water risk may include the increased cost of capital due to higher perceived risk exposure and deteriorating credit ratings as investors become concerned about water risk.

2.2.6 *Reputational Risks*

In addition to impacting financial performance, ESG issues also impact intangible assets, which comprise the bulk of market valuations (Ocean Tomo, 2020). Water-related reputational risks include local opposition to industrial water use and withdrawals, water contamination events, real or perceived inequities in water use by the company or the industry, and other events that may lead to conflicts surrounding water. Reputational risks can be one of the largest and most volatile risks facing companies. Such risks could lead to the loss of a company's social license to operate (the acceptance of a company's operations by its employees, stakeholders, and the general public) (Kenton, 2019), which could lead to serious adverse financial repercussions.

3 Risks and Issues

Water risk is a complex and multi-faceted issue. Water issues may result from deep systemic risks such as climate change, and the mismanagement or misuse of water. Even local manifestations of water risk can have a global impact. The implications for investors become more complex with the understanding that water plays a major role in climate change. Such implications include changes in the valuation of physical assets, such as real estate and infrastructure; the risk profiles of water-dependent sectors such as seafood, tourism, and pharmaceuticals; and the opportunities that may emerge as a result of water-related conservation (Aviva Investors, 2020).

3.1 *Global Water Issues*

3.1.1 *Ocean Plastics*

Plastic pollution is one of the most pervasive threats to the marine environment, impacting ocean health, food safety and quality, human health, and even coastal tourism. Humans are responsible for the eight million tons of plastic that pollute oceans every year and make up 80% of all marine debris for both surface water and deep-sea sediments (International Union for Conservation of Nature, 2018). Due to ultraviolet radiation, wind, and water currents, plastic fragments break down into small particles called microplastics, which are smaller than 5 mm in size. Plastic particles of this nature have been detected in samples collected from all the world's oceans, as well as in freshwater, tap water, and even beer (Kosuth et al., 2018). The surface of plastic materials also gathers toxic contaminants from prolonged exposure to seawater. Plastic particles, along with toxic contaminants, can accumulate in the food network through ingestion by wildlife and humans. This ingestion can potentially cause adverse health and biodiversity effects (International Union for Conservation of Nature, 2018).

Ocean plastic pollution can be land-based or sea-based. Human activities on land cause the majority of marine pollution, which takes the form of urban and storm runoff from cities, solid waste leakage, and sewer overflows. Sea-based plastic pollution includes abandoned, lost, or discarded fishing equipment, nautical activity equipment, aquaculture, and other sea-based activities (International Union for Conservation of Nature, 2018; UN Environment COBSEA, n.d.).

3.1.2 Biodiversity

Biodiversity—the diversity within species, between species, and of ecosystems—is declining globally, faster than at any other time in human history. The world's 7.6 billion people represent just 0.01% of all living things by weight, but humanity has caused the loss of 83% of all wild mammals and half of all plants (Carrington, 2018). Of note, biodiversity loss is one of the top five risks in the World Economic Forum's 2021 Global Risks Report, both in terms of likelihood and impact (World Economic Forum, 2021).

Pollution from industry, increased water acidification, and agricultural runoff (containing contaminants, residues of fertilizers or pesticides, or other chemicals) threaten aquatic habitats that support a wealth of wildlife and natural ecosystems (Government of Canada, 2020).

3.1.3 Ocean Heating

Oceans are getting hotter, more acidic, and less oxygenated due to global warming. Part of the excess heat caused by the increase in greenhouse gases (GHGs) is absorbed by the atmosphere, but the majority ends up in the oceans. As ocean water heats up, its volume expands through a process known as thermal expansion. This process results in rising sea levels and the acceleration of the melting of sea ice, which in turn results in a further increase in sea levels (Aviva Investors, 2020).

Increased carbon absorption in oceans leads to a disruption of the water's pH balance, which threatens marine life. For example, in Canada, warmer than usual ocean temperatures off Newfoundland's southern coast in 2019 caused a drop in oxygen levels, which led to mass deaths in the Atlantic salmon population (Aviva Investors, 2020).

Warming seawater can also cause adverse health effects for humans through the consumption of seafood. Fish are the main source of human exposure to methylmercury, which can lead to long-term neurocognitive deficits. Rising seawater temperatures have led to higher levels of methylmercury in fish. Since the 1970s, there has been a 56% increase in tissue methylmercury in Atlantic bluefin tuna due to these higher seawater temperatures (The Ocean Foundation, n.d.).

3.2 *Climate Change and the Canadian Context*

Anthropogenic climate change causes an increase in likelihood of extreme weather events, which results in increased water-related physical risk. Although Canada is a water-rich country, it is still susceptible to water risks. With more frequent extreme weather events caused by shifts in both temperature and precipitation, climate change can affect the flow rates of waterways and is likely to severely impact Canada's watershed health (Semeniuk, 2017). Climate change effects could also lead to increased annual mean precipitation in Canada, and thus entail more events like the extreme precipitation that led to the 2013 southern Alberta flood (Bush & Lemmen, 2019).

3.2.1 *Changes in Snow, Ice, and Permafrost and Changes in Freshwater Availability*

Studies show that snow cover fraction (a parameter of snow depth and density), as well as seasonal snow accumulation and the portion of the year with snow cover, have decreased across most of Canada, and it is highly likely that snow cover duration will continue to drop (Bush & Lemmen, 2019; Swenson & Lawrence, 2012). This might present some economic opportunities for Canada, such as longer frost-free and crop-growing seasons, leading to new revenue streams for Canadian communities. However, there are some risks to consider. Among these risks are the thinning and decline of sea ice. Projections posit that the glaciers of the Western Cordillera (spanning the western coast of Canada and the US) will lose 74–96% of their volume by the late twenty-first century in a medium emissions scenario. This loss of mass will result in a decline of glacial meltwater supply to rivers and streams and may impact freshwater availability by mid-century (Bush & Lemmen, 2019).

Coastal flooding is expected to increase in many areas of Canada due to local sea-level rise, while loss in the mass of both glaciers and the ice cap in the Canadian Arctic is the largest contributor to sea-level rise on a global scale (Lemmen & Warren, 2016). Additionally, projected higher temperatures will result in shifts in the seasonal timing of peak stream flows. This change will set off a shift toward earlier floods associated with spring snowmelt, ice jams, and rain-on-snow events. How this shift affects the frequency or magnitude of those floods remains uncertain (Bush & Lemmen, 2019).

4 Sector-Specific Water Risks

Many sectors are affected by water, but not all in the same way. Physical, transition, and regulatory water risks are becoming financially material to companies, with variation in the duration, magnitude, and financial impact of exposure to those risks. Depending on the nature of the industry's operations, water risks can impact revenues, operating costs, capital expenditure, or physical and intangible assets. One way of categorizing those impacts is through understanding the upstream (effects of water on industry) and downstream (effects of industrial/commercial activity on water) challenges.

4.1 Agriculture

The agricultural sector is connected to water risks due to potential runoff (water moving over and through agricultural ground that picks up natural and human-made pollutants, and then deposits these in watersheds). This can promote algae growth and reduce available oxygen in water for other organisms. It is the leading source of impairments to rivers and lakes (U.S. Environmental Protection Agency, 2005).

On average, 70% of the freshwater withdrawn globally is associated with agriculture and its supply chains (World Bank, 2020). A comprehensive assessment of a company's agricultural supply chain water risk requires an assessment and understanding of the company's water dependence, an assessment of geographic water security, and a planned management response with mitigation actions (UN PRI, 2018a).

Several industrial sectors such as food and beverage production are reliant on agricultural commodities for raw materials from high water-stressed regions. As climate change leads to a rise in water stress levels globally, both irrigated and rainfed crops are increasingly exposed to water risk. Investors are thus rightfully concerned about water dependency and the risks that company supply chains are exposed to, and there will be more pressure on companies with high dependency on agricultural inputs (UN PRI, 2018a).

- **Physical water risks for agriculture supply chains**: case of insufficient water volume or poor quality for irrigation, affecting the ability to grow crops;
- **Regulatory water risks for agriculture**: the prioritization of water for one user over another in cases of water stress creating restrictions on water use and unpredictable changes to water allocations or discharge limits;
- **Reputational water risks for agriculture**: in instances where companies and their supply chains are associated with negative impacts on water due to poor water management (UN PRI, 2018a).

4.2 *Food and Beverage*

The $5 trillion global food and agribusiness sector is highly exposed to water risk (Ceres, 2019). Physical water risks for the food and beverage industry echo those of agriculture supply chains in terms of irrigation or water quality issues that could impact crop yields. This can lead to higher price volatility and/or an inconsistent supply of ingredients (Ceres, 2019). For example, in 2018, Olam International reported a 22% decline in profits as its peanut farming business in Argentina suffered from drought conditions (Olam International Limited, 2019). Regulatory risks for the food and beverage sector may include compliance issues from violations of water regulations, failure to predict regulatory action related to water, and legal actions or sanctions for failure to address negative environmental impacts (Ceres, 2019). Finally, reputational risks due to water-use and water management can damage brand equity (Ceres, 2019).

Food and beverage companies could improve the resilience of their supply chains by supporting sustainable farming practices in the watersheds where they source their most significant commodity inputs (Ceres, 2019).

4.3 *Consumer Goods*

A staggering 19% of water withdrawals come from industrial operations, such as textile, energy, and chemical production (UN-Water, 2018). Companies that depend on raw material inputs, such as cotton, wheat, and sugar, are subject to both downstream (impacts on others) and upstream (dependency on water source) water risks. According to a study by CDP Worldwide (formerly the Carbon Disclosure Project), physical and regulatory water pollution risks are the most commonly reported substantive risks by apparel and textile companies (CDP, 2020). However, fines and penalties associated with water pollution are still low in most countries and are likely to appear relatively unsubstantial on a company's financial statements. For example, a major apparel manufacturer estimated that fines associated with the discharge of untreated wastewater have not surpassed US$100,000. Nevertheless, as regulatory control becomes tighter on environmental issues, the materiality of these risks is likely to increase (CDP, 2020).

The CDP report also highlights the fact that addressing water pollution can also drive business opportunities. Respondents from the apparel and textile sector identified 34 opportunities related to water pollution reduction within their value chain. The combined possible financial gains were estimated for 15 of these opportunities to be US$184 million (CDP, 2020). The report suggests that if the negative environmental and societal externalities associated with the apparel and textile industry were addressed, the benefit to the global economy is estimated to be upwards of US$190 billion per year (CDP, 2020).

4.4 *Cannabis*

The Government of Canada legalized and regulated access to cannabis in 2018, which extends to both cannabis cultivation and production (Government of Canada, 2018). As such, the cannabis industry experienced considerable growth and received commensurate scrutiny. Outdoor cultivation is increasingly popular among producers, as it allows for savings on electricity and staffing costs (Deschamps, 2020). However, for these producers, water use can be a significant environmental risk. A single outdoor-grown cannabis plant requires up to six gallons of water daily during peak growing season (Madhusoodanan, 2019). Indoor cannabis growing also comes with challenges that may attract criticism. One kilogram of indoor-grown cannabis product requires an estimated 5,000 kWh, the same amount of energy an average Canadian household uses in four months (Lee-Anderson, 2018).

Cannabis legalization may also affect freshwater sources. Water is an interconnected and finite resource, meaning that wastewater is treated and released back into watersheds in order to be used safely again. Traces of chemicals (such as THC, the psychoactive compound in cannabis) not removed by sewage treatment plants, and thus ingested by humans, can be found in rivers and lakes. There are studies examining removal rates that indicate that as little as 20% of THC-COOH is removed through water processing plants (Park et al., 2016). While there are relatively few scientific studies on the potential effects on aquatic ecosystems and resident wildlife from cannabis-related chemicals, it is important to examine closely both the short- and long-term exposure effects in impacted environments (Palace, 2019). It is important to acknowledge that emerging industries (such as cannabis) may have water risks that have not previously been analyzed in-depth, so enhanced scrutiny from regulators and other stakeholders may be expected.

4.5 Energy and Utilities

Electricity systems are highly dependent on water. For example, hydroelectricity accounted for 60% of Canada's electricity generation in 2018 (Natural Resources Canada, 2020). Hydroelectric generation is directly tied to the availability of water. Without a steady and reliable source of water, energy cannot be harnessed and delivered to the grid.

Other types of power plants are also heavily water-dependent. Large steam-generation power plants, mainly fossil-fuel and nuclear-powered plants, need water for cooling purposes. Without adequate cooling water, power plants would have to cut back operations, or even shut down. Ultimately, as our demand for energy continues to rise, the water dependence of power plants exposes the energy sector, and consequently water assets, to a growing risk (Union of Concerned Scientists, 2011).

4.6 Mining and Extractives

Water is an essential component in mining and extractive operations as it is required for mineral processing, dust suppression (Managing Water Consumption in Mining, 2013), and fracturing operations. This leads to the sector's high dependency on water. As such, extractive companies face several ESG risks related to water stress, including water scarcity due to climate change, stricter regulations on water use, and competition over resources with local communities. The environmental impacts of mining are also a major stakeholder concern. Spills from mine tailings (leftover chemicals used in the mineral extraction process) can have long-term effects on the environment and can severely pollute water systems (Ochianu, 2019). In August 2014, a tailings dam containing mine waste from the Mount Polley copper and gold mine, located near Quesnel Lake in British Columbia, failed. The tailings spill dumped 18 million cubic meters of wastewater and sediments into the West Basin of Quesnel Lake (University of Alberta, 2020). Research from the University of Alberta shows that contaminants deposited during the mine spill six years ago may still be affecting the lake, as contaminated sediment is mixed into the water column each spring and fall (Hamilton, 2020).

Tailings deposition methods typically involve large volumes of water, and their failures in the past have often been attributed to water-related events, such as excessive rainfall (Owen et al., 2020). Waterways are also at risk due to potential acid drainage from mining activities. The acidity of drained fluids can damage water quality, human life, and marine life. Furthermore, while the impact of mining on water quality and related aquatic ecosystems is localized, it usually stretches over long periods of time, and therefore can cause disproportionately high levels of damage (Ochianu, 2019).

4.7 Oil Sands

While the oil and gas industry makes efforts to recycle and reuse water, the majority of the industry's water use is consumptive, meaning that water could be lost to steam, held in tailings ponds after use, or injected into oil reservoirs (Statistics Canada, 2016), thus being unavailable for further use. In Canada, oil sands mining operations have a zero-discharge policy, which details that all wastewater and runoffs must be contained on-site in order to avoid negative impacts to aquatic ecosystems (Natural Resources Canada, 2009).

Extracting bitumen from raw oil sands ore requires a considerable amount of water (Canadian Energy Research Institute, 2014). While on-site facilities require less than 0.3 barrels of freshwater per barrel of oil produced, a typical oil sands mine uses about three barrels of freshwater to produce one barrel of oil. However, about 80–95% of process-affected water is recycled back into the process and is always contained on-site, thus eliminating risks from its release back into waterways (Oil Sands Magazine, 2020). For example, Cenovus Energy (a Canadian integrated oil and natural gas company) is recycling wastewater from its oil sands operations. The company reports that 85% of the water used to produce steam in oil sands projects is recycled from operations, and that most of the additional water uptake is saline water that is unfit for consumption or agricultural use (Cenovus Energy, 2019). Therefore, it is essential to consider that not all water is equal (the initial quality of water used) and not all water use is equal either (the recycling and reuse of water). This allows for a more nuanced view when comparing different operations.

4.8 Real Estate

Flooding and more frequent and powerful storm surges are becoming an augmented risk to the real estate sector due to the increased risk of extreme weather events like hurricanes and cyclones. For example, Hurricane Katrina's storm surge caused catastrophic damage to New Orleans and its population in 2005. The National Hurricane Center (2018) estimates the damage at US$125 billion. The real estate sector in the US is increasingly assessing risk from storm surge for coastal residential and commercial assets. CoreLogic, a financial services company, estimates that there were over seven million single-family residences and 252,000 multi-family residences at risk of storm surge in 2020, worth over US$1.8 trillion in reconstruction cost (assuming complete destruction) (CoreLogic, 2020).

Due to the sheer weight of growing cities, the depletion of groundwater sources, and the weakening of underground support systems, cities around the world are starting to sink. Jakarta is one of the fastest sinking cities in the world, where the ground sank two and a half meters in less than a decade, and where 40% of real estate now exists below sea level (Kimmelman, 2017; Muggah, 2019).

London is experiencing similar risks, but due to different geological reasons. As the glaciers melt in the north of the UK, there is less weight bearing down on Scotland, and the country is thus slowly rising at a rate of 1 mm per year. The corresponding opposite effect impacts London in the south. London is protected through the Thames Barrier flood defense system, designed to defend the city from a once-in-100-year risk of flooding, by blocking North Sea tides and storm surges. Since its establishment in 1982, and for the first ten years of its operation, it was used once or twice a year. With higher sea levels, it is now used up to seven times a year (Aviva Investors, 2020; Kramer, 2018).

In Canada, almost two million households are at "very high risk" of flooding. One-in-100-year floods, meaning floods that have a one percent probability of occurring in any given year, are now happening much more frequently. In 2019, Canada's Minister of the Environment and Climate Change suggested that Canadian provinces are experiencing this type of flooding every two to five years, and that we can expect to see more flooding due to climate change (Dangerfield, 2019).

Flooding, coastal erosion, and other physical water-related risks can directly impact the market price of a property. From an investor's perspective, it is essential to consider the increased probability of the default that these risks trigger, and the subsequent effects on returns.

4.9 *Technology and Electronics*

Hardware and software for technology products require the mining of precious metals, which is highly water-intensive and generates toxic wastewater (S&P Global Ratings, 2019). Further exacerbating the exposure to water-risk for the electronics industry is the strong and growing manufacturing presence in Asian/Pacific-Rim countries, where water supplies are already under stress. As companies aim to keep up with the growing demand for electronic products, water scarcity or quality issues may prove to be a significant risk (The Pacific Institue and Ceres, 2009).

Water is also commonly used for cooling purposes in data centers and as technology use rises globally, demand for data centers grows. Data center water use has become a controversial subject, as news emerged that by 2021, Google will need as much as 1.46 billion gallons of water per year for a new data center in Red Oak, Texas (Sattiraju, 2020). This is in comparison to the average commercial building in the US, which uses 7.9 million gallons of water per year (EIA, 2017). Better water intensity management and improved technological innovation will be essential to support the sustainable growth of this sector. Such mitigations will also be crucial in managing reputational risks for technology companies.

4.10 *Tourism and Leisure*

Water pollution, scarcity, and flooding may present challenges in attracting tourists to affected sites, signifying severe risks for hotels and touristic attractions. Another facet of risk for the tourism industry is the above-average level of water use by hotel guests, especially in areas prone to water shortages. For example, in Zanzibar, Tanzania, tourists use on average 15 times more water per night than local residents, and in the Mediterranean, locals compete with tourists for freshwater (Becken, 2014).

Understanding and addressing water challenges, including both risks and notions of water equity, requires transparency and action from all sectors, as well as related regulatory bodies.

5 Assessing Water Information

In capital markets, issues are considered material if they are reasonably likely to have a significant impact on a company's operations, financial performance, prospects, and ultimately on its valuation.

An investor must be equipped with the appropriate process, tools, or framework to embark on a water risk and materiality assessment, or to discover investment opportunities. While most material issues are sector-dependent, the notions of water regionality and geography are factors that must be considered in materiality assessments across all sectors.

While water-related risks can be particularly material in the short term, some financial impacts tied to water may be less acute. Yet, over time such impacts negatively affect revenue, profit margins, and earnings. Three elements should therefore be weighed when considering financial materiality: duration of financial impact, the magnitude of the financial impact, and whether these impacts are related to revenue, operating costs, or capital expenditure.

An established information chain must be in place for investors to assess the water risk of an investment. The first link in the chain is company data that is both available and disclosed. The greatest challenge facing investors who are trying to assess water risks is the sourcing of detailed and reliable information and data from the companies in which they invest. Therefore, organizations need to employ specialists who can develop and apply the necessary frameworks to assess water risks, so that the organization can then properly disclose this information.

The next challenge for the investor is to determine which pieces of information are material and how to process these. There is a growing number of tools and frameworks available to guide both companies and investors in assessing water risk. The following section highlights some of these most utilized tools and frameworks.

5.1 *The CEO Water Mandate*

A United Nations Global Compact initiative, the CEO Water Mandate aims to mobilize business leaders in both water and sanitation activism. Endorsers of the CEO Water Mandate commit to continuous progress in six core elements of water stewardship. These six elements are (CEO Water Mandate, n.d.):

1. Direct operations
2. Supply chain & watershed management
3. Collective action
4. Public policy
5. Community engagement
6. Transparency.

5.2 *The Alliance for Water Stewardship*

The Alliance for Water Stewardship (AWS) is a global membership collaboration of businesses, NGOs, and individuals from the public sector. AWS members contribute to the sustainability of local water-resources through the adoption and promotion of a universal framework for the sustainable use of water—the International Water Stewardship Standard, or AWS Standard—that drives, recognizes, and rewards good water stewardship performance. The standard is based on five steps including gathering and understanding water-related data, committing to water stewardship, implementing a water stewardship plan, evaluating the performance, and communicating and disclosing progress with stakeholders. Following the steps of the AWS standard should result in improved performance in five areas (Alliance for Water Stewardship, n.d.):

1. Improved water governance
2. Sustainable water balance
3. Good water quality
4. Healthy status of important water-related areas
5. Access to water, sanitation, and hygiene (WASH) for all.

5.3 Ceres Investor Water Toolkit

The Investor Water Toolkit is a comprehensive resource that helps evaluate water risks in investment portfolios. Developed through a collaboration of more than 40 institutional investors, the Toolkit includes links to resources, databases, case studies, and other tools for all types of investors. Some of the topics covered are water risks, how investors should establish priorities, how to integrate water in buy/sell analysis, how to integrate water in portfolio and asset class analysis, and how to engage on water-related issues. The Toolkit is a valuable resource for any investor interested in learning more about the fundamentals of water risk evaluation (Ceres, 2018).

5.4 Climate Disclosure Project (CDP)—Water

The Climate Disclosure Project (CDP) is a not-for-profit organization that runs a global disclosure system for investors, companies, cities, states, and regions to manage their environmental impact. The CDP considers water security to be essential for tackling climate change and protecting the bottom line of organizations. Therefore, the CDP aims to motivate companies to disclose and reduce their environmental impacts (CDP, 2021).

The CDP helps organizations to improve measurement by providing a framework for understanding where their water comes from and where it goes, and eventually understand how secure their water supply is. According to the CDP, the first step toward building resiliency is to measure, monitor, and reduce water impacts. The next step is enhanced disclosure. The CDP helps organizations to benchmark progress against peers and to identify opportunities for change. One example of this assistance is by providing insight into how many companies are reporting risks linked to pollution, how many monitor the quality of their discharges, and how many have set pollution reduction targets. The CDP collects data that helps influence decision-makers to reduce risk, harness opportunities, and drive action toward better water sustainability (CDP, 2021).

5.5 *Metrics and Qualitative Judgement*

In order to thoroughly assess exposure to water risk, investors may find it useful to gather quantitative data points from issuers in order to include them in a company-specific or portfolio-wide risk exposure assessment. Investors can leverage information and data on their portfolio companies as they file reports on their environmental impact assessments, especially those in regulated industries, like utilities, mining, agriculture, etc. Investors could consider three types of indicators, as described by Ceres (2018, p. 29).

1. **Point-in-time**—these metrics provide snapshots or "red flags" of potential water risk where water dependency and risks are highest in the business value chain and potential materiality:
    - Water used per $ revenue (or percentage of revenue with high water-resource dependency);
    - Water required/wastewater discharged per unit of production for each major facility.
2. **Future Forecast**—these metrics provide information on possible future risk exposure to help determine the value at risk:
    - Percentage of future growth at risk due to water;
    - Value at risk in next five years due to water risk.
3. **Contextual**—these metrics provide geographic contextual information related to water resource security:
    - Percentage of operating regions in water-scarce areas;
    - News on water risks related to social license risks in key regions.

6 Considerations of Water in Investments

The concept of green and low carbon portfolio construction can be extrapolated to include "blue" components. There are several investment opportunities in water as a commodity, ranging from individual stocks (of firms in water-related businesses), indexes like the Dow Jones U.S. Water Index or the S&P Global Water Index, water exchange traded funds (ETFs), mutual funds, investment trusts, and others (Gray, 2019). Water Equity Funds are mutual funds with a focus on the stocks of companies operating in water and water-related industries like utilities, filtration, pipes, irrigation, desalination, purification, etc.

Water Indexes can be constructed by selecting companies from water supply sectors such as utilities, infrastructure, and technology and creating a shortlist of companies that derive 30% or more of revenues from water-related business activities. Similarly, companies that provide innovative solutions to the global water challenge are included based on an assessment of water efficiency and water impact. This can also apply to organizations with strong water efficiency that operate in any high water-intensity sector, such as agriculture or semiconductor production (Calvert Investments, 2016). Examples include the KBI Institutional Water Fund, Fidelity Water Sustainability Fund, AllianzGI Water Fund, and Calvert Global Water Fund. Water indices can be used to track the performance of water supply sectors and water-related business activities.

Water joined other commodities traded on Wall Street in December 2020. "Farmers, hedge funds and municipalities alike are now able to hedge against – or bet on – future water availability in California," highlighting fears of future water scarcity (Chipman, 2020). Water "futures" may provide a hedge against water price fluctuations and a water scarcity gauge for investors.

6.1 *Water in Equity Portfolios*

For investors, it is important to evaluate aggregate water risk at the portfolio level and across different asset classes to ensure their holdings are not overweighed to high water-risk industries or regions. For equity portfolios, Ceres developed a four-step approach for investors to work through portfolio water risks:

- Step 1: Conduct a high-level heat-mapping or foot printing to analyze water risk on an industry level and to prioritize portfolio companies based on their water risk. This entails analyzing data on sector and geographic exposure by security;
- Step 2: Conduct an investor influence assessment relating to the ability of the investor to influence, change, or mitigate holdings' risks, as measured by data such as percentage ownership and ownership structure;
- Step 3: Conduct a regional risk-mitigation opportunities assessment. This step involves mapping the geographic water risk exposure of an entire portfolio and assessing whether there are risk mitigation synergies available to the investor;
- Step 4: Prioritize security-level water risk analysis. Based on the identified subset of portfolio holdings that potentially have high water risk and materiality, the investor would conduct water-risk stress testing and value at risk analysis for individual securities and prioritize action accordingly (Ceres, 2018, pp. 44–49).

It is possible to mitigate water risk through strategic water management, stakeholder engagement, and climate adaptation. However, some macro-level water risks are not easily diversifiable and may have larger impacts on an investment portfolio. These may include:

- The lack of climate change-ready wastewater storage infrastructure in industries like mining, which may lead to increased frequency of catastrophic accidents;
- The accumulation of contaminants and the degradation of freshwater systems and associated dead zones in coastal areas, leading to changes in soil and water management, adding costs to businesses, and increasing risks associated with social licensing to operate;
- The lack of access to clean freshwater and sanitation, resulting in loss of productivity and economic activity (Ceres, 2018).

6.2 *Private Equity*

Private equity investing is more long-term focused than its public equity investing counterpart. Therefore, private equity managers often carry out a more comprehensive due diligence process before investing. Such due diligence should rigorously evaluate the level of exposure and type of water risks, to help inform and prepare mitigation strategies. The Private Equity Water Due Diligence Decision Tree developed by Ceres provides a useful preliminary resource for investors:

- Step 1: Assess the level of current and future water dependency of a project;
- Step 2: Conduct a regional analysis of physical and regulatory water risks as well as social (community and stakeholders) water risks and issues. Also, ensure those risks can be addressed to guarantee that needs are met, and the social license to operate can be easily obtained;
- Step 3: Evaluate project water risk including physical, governance, and stakeholder water resilience (Ceres, 2018).

Assessing water dependency can be achieved by first mapping the water needs and risks of the company (supply chain, direct operations, wastewater discharge, or product lifecycle); then measuring the proportion of sales with high water needs/risks and conducting a stress test to assess the value of earnings exposed to extreme water events; and finally, understanding the duration of water risks and how it can evolve over time considering the impacts of climate change among other factors (Ceres, 2018).

6.3 Fixed Income

There are several debt financing mechanisms for local capital investments related to water. Municipal bonds are considered the primary financing source for water utilities and distributed water infrastructure projects. Municipalities in Canada are increasingly assessing the bond market (green, blue, and transition bonds) to finance environmentally sustainable projects (Investment Industry Association of Canada, 2019). Municipal bonds offer a promising option for financing both sustainable water and climate change adaptation projects.

7 Assessing Water Risk

Thus far, we have highlighted the complexity of water risk and its potential impact on different sectors, as well as at the national and global levels. Having established the potential for water risk for an investee company, investors can then apply different methods in order to assess potential impacts: stress testing, scenario analysis, and impact measurement.

7.1 Stress Testing

Stress testing helps to understand regional water security by identifying the tipping points where water risks can have a material impact. Understanding this can help in the development of mitigation strategies, which may require a collective effort from all stakeholders involved—namely regional water managers, investors, and companies. For example, a company that uses water in an area where its operations may lead to the deterioration of the water quality region-wide should conduct a stress test to establish additional capital expenditures and operating costs that may be associated with having to source lower quality water instead. This exercise helps in developing viable alternatives so as to avoid reaching potential tipping points established during the stress test. The company can then work with its stakeholders to invest in implementing appropriate watershed land use protection measures and sustainable practices to help mitigate any potential water risk (Ceres, 2018).

7.2 Scenario Analysis

Scenario analysis is a planning tool that allows for long-term resilience and effective strategic development. This forward-looking approach helps avoid data availability bias as it informs decision-making without being constrained to historical information. It is commonly used in climate change modeling. Around the world, banks, companies, investors, and even governments are using scenario analysis to model water risks and integrate existing climate change and weather variability data into developing water scenarios (Ceres, 2018). This allows them to assess the risks' associated durations, and probabilities of occurrence, along with each scenario's potential financial impact and total value at risk. For example, the WWF estimates that 46% of global GDP could come from areas facing high water risk by 2050. Considering the magnitude of potential impact, the organization created a proprietary Taskforce on Climate-related Financial Disclosure-aligned tool—the *Water Risk Filter*—which incorporates potential physical, regulatory, and reputational risk to its assessment. The aim is to help with water-related scenario analysis. It follows a four-step approach (Explore, Assess, Value, and Respond) based on two types of risk: water basin risk, which is determined from the user's geographic location of operations, and operational risk, which is defined by the user's water usage, performance, and potential impacts on the basin in which it operates. This assessment provides users with tailored best practices and approaches to adequate water stewardship by evaluating various potential scenarios based on the data input (WWF, 2020).

7.3 Impact Measurement

Impact measurement can be defined as the measurement of "positive and negative, primary and secondary long-term effects produced by a development intervention, directly or indirectly, intended or unintended" (OECD, n.d.) and serves to better assess the outcomes of impact investments. To help companies and investors improve their impact measurement analyses, the United Nations Principles of Responsible Investment (UN PRI) developed a Market Map based on 10 thematic investment sectors which were created through an analysis of the UN Sustainable Development Goals (SDGs) and the PRI Reporting Framework. Water was identified as an overarching theme within this map and as a multi-impact investment since it is intrinsically linked to several other PRI impact themes like food supply, industrial production, health, productivity, and the environment (UN PRI, 2018b). Several other water-related measurement tools have also been developed in order to better quantify potential impact. The UN provided a definition for water and sanitation businesses; MSCI designed a global index covering sustainable water and sanitation companies; and impact investing funds developed new approaches to measure impact from companies in this field (MSCI, 2020; UN PRI, 2018b).

8 Investor Engagement and Corporate Water Stewardship

Investors need applicable frameworks, materiality identifiers, and accessible metrics to assess any risk. They depend on initiatives, suppliers, and corporations to be as transparent and insightful about water as they would about any financial subject. In order to encourage and advance dialogue around specific issues such as water, investors can directly or, with other investors, collaboratively engage with the companies in which they invest and request better disclosure on relevant and material information. For example, it is important to know whether the companies in which they invested have mapped out the impact of water issues on their business, as well as the value at risk, and how management is addressing those potential impacts and managing those risks.

Investors who take an active approach in engaging with their portfolio companies on water risks can raise important issues and instigate a change of corporate practices and policies. By doing so investors can improve investment returns, create long-term value, and mitigate water-related and other risks. Investors should consider developing clear guidelines and engagement policies concerning ESG issues (including water issues) and setting engagement strategies and proxy voting guidelines to align their activities to them (Ceres, 2018).

8.1 The Importance of Corporate Water Stewardship

It is important for a company to responsibly manage their water. Not only does it help them lower their water risk, and therefore potentially lower operational interruptions and loss of equipment, but it also allows for the possibility of identifying additional opportunities and efficiencies. For companies to have effective water stewardship process in place, they need to build their knowledge of global water challenges and establish which ones are relevant for their operations, supply chains, customers, and communities. Companies also need to better understand their dependence on water and where that dependence is highest within the value chain (UN PRI, 2018a).

Following that, companies must then understand the two types of potential water impact: impact on the company's internal level, meaning the effect of water risks on operations and the supply chain, and the impact on the basin level and the upstream and downstream water supply. This understanding is essential in order to grasp the materiality of water as an issue, and to effectively integrate water into the management strategy (UN PRI, 2018a).

Once the materiality of water is established, investors need to have access to this information in order to assess risks appropriately. Therefore, companies need to disclose how they are managing the issue, what their approach is, and who is accountable for implementing water-related programs. Almost half the constituents listed on the S&P/TSX Composite Index include a dedicated water section in their 2019 Sustainability Report, and this statistic is expected to increase (Millani, 2020). Pressure for water quality disclosure also comes from institutional investors. BlackRock, with US$7.2 trillion in assets under management, released a report in July 2020 on water stress risk, signaling that institutional investors will increasingly be engaging with companies on their water management and strategy (BlackRock Investment Institute, 2020).

Water stewardship then takes the form of ongoing action and improvement. Internally, the company should consider improvements of water governance systems, water efficiency, and reduction of pollution, as well as better monitoring, data tracking, and reporting. External or collective action entails that companies work within their communities, their peers, the public sector, and non-governmental organizations (NGOs) to address water issues (UN PRI, 2018a). Companies can also implement water considerations into their supplier code of conduct, sustainability standards, and/or certification standards.

As water resources are shared, mitigation strategies may require collective action by investors, firms, and other local stakeholders in the basin or watershed region. Collective action by an entire industry, when industry practices are modified, maybe another option to mitigate a particular water-related risk threatening that specific sector (Ceres, 2018).

Finally, companies can influence local authorities on their water governance strategies and invest in water basin sustainability. Influencing the governance of water, meaning the rules and institutions which determine how water is managed, can be an essential aspect of water risk management. Companies need to understand the policy landscape (local, basin, regional, and national) and build coalitions of support with other companies and/or stakeholders to develop shared policy advocacy positions in key water risk areas (UN PRI, 2018a).

From carrying out impact assessments, disclosing on water management, fulfilling responsible stewardship duties, and wielding stakeholder influence, companies have the potential to play a major role in water resource management and the achievement of United Nations Sustainable Development Goal (SDG) 6, which aims to ensure the availability and sustainable management of water and sanitation for all (UN, n.d.).

9 Resilience and Opportunity

A full assessment of water-related issues highlights the need for mitigation and reveals possible resilience strategies and can also reveal new opportunities for business model adjustments. The principles of a circular economy, advancements in smart technology, and innovative finance solutions are examples of opportunities that may lead to heightened efficiency, successful conservation efforts, and increased financial returns. These opportunities are discussed in more detail below.

9.1 Circular Economy

Circular economy principles can be applied to water systems management in order to change the currently short-sighted, linear approach to water withdrawal, use, and disposal. The goal of a circular economy is to keep resources in use for as long as possible, extract the maximum value from them while in use, then recover and regenerate products and materials at the end of each service line (WRAP, n.d.).

A circular approach to water management can address local and global water demand in a more sustainable manner and offer opportunities for economic advantage from reduced costs, or provide additional revenue streams from recycled water. Such an approach to water management could be realized by following the Ellen McArthur Foundation's steps to building a circular economy (Ellen McArthur Foundation, 2018):

- Avoid water use, eliminate ineffective withdrawal, and rethink products and services with high water intensity;
- Reduce use through continuous improvement of water use efficiency and resource allocation;
- Close the loop and pursue opportunities to reuse water;
- Recycle water for internal operations or external purposes;
- Replenish water efficiently and effectively to the basin.

9.2 *Smart Water Networks*

From a single household to a transnational scale, smart water management can allow for significant advances in water solutions. Integrating smart water networks into the broader utility network has the potential to reduce the need for new infrastructure, allow for real-time detection and response in case of water issues, and contribute to increasing knowledge and security of our water systems. Smart water management can help improve current use based on future planning by leveraging localized real-time data and climate scenarios.

Several of the risks and impacts caused by water-related issues could be better mitigated through smart water management, including increasing agricultural efficiencies as well as improving coastal water management and preparedness in case of extreme physical water risks, like flooding or droughts. For example, smart water management solutions can generate water stocks in the case of a drought, therefore lessening the severity of the impact of such an event and allowing essential activities, like the irrigation of agricultural crops, to continue undisrupted (K-Water & International Water Resources Association, 2018).

9.3 *The Concept of "Blue" Investments*

Following the success of green finance, financial institutions are looking at opportunities arising from the "blue economy" (Gambetta, 2020) which encourages better ocean stewardship, namely the preservation of the marine environment. The "blue economy" is defined by the World Bank as the sustainable use of ocean resources for economic growth, improved livelihoods and jobs, and ocean ecosystem health. It encompasses activities from various sectors including maritime transport, renewable energy, fisheries, tourism, climate change mitigation/adaptation, and waste management (World Bank and United Nations Department of Economic and Social Affairs, 2017).

In December 2020, the Government of Canada announced its endorsement of the High Level Panel for a Sustainable Blue Economy's *Transformations for a Sustainable Ocean Economy: A Vision for Protection, Production and Prosperity*:

> *Transformations* outlines a set of priority ocean actions that countries can take to build a sustainable blue economy across five areas: ocean wealth,

> ocean health, ocean equity, ocean knowledge, and ocean finance, as well as a commitment to sustainably manage 100 percent of the ocean area under national jurisdiction by 2025, guided by sustainable ocean plans. (Fisheries and Oceans Canada, 2020)

Canada also joined the United Kingdom's Global Ocean Alliance, advocating for the adoption of a global target of 30% marine conservation by 2030 (Fisheries and Oceans Canada, 2020).

As Gambetta (2020) discusses, an example of blue economy investing is NatureVest, which raised US$15.2 million in impact capital loans (and US$5 million in charitable grants) to buy back US$21.6 million of the Seychelles' sovereign debt from creditors in 2016. Under the renegotiated terms, the government was required to take marine conservation and climate adaptation measures. Those commitments have now been met, with almost a third of the Seychelles' ocean now designated as "protected." More investment products are being announced, with the twin benefits of financial returns and environmental conservation. The economic value of global ocean assets is already estimated to surpass $24 trillion, making the oceans the seventh-largest economy by GDP in the world (Gambetta, 2020).

10 Conclusion

The integration of environmental, social, and governance (ESG) issues is increasingly becoming mainstream for financial stakeholders in markets around the world. Water is recognized as an ESG issue and is gaining more attention, both in relation to climate change, but also as an independent issue.

Water risks are apparent in a multitude of sectors and in nations across the world, and can relate to water quality, scarcity, accessibility, etc. Even emerging sectors such as cannabis production and software development have their own complex water risks. Accordingly, companies have a corporate responsibility to manage their specific water needs and to disclose their approach to responsible water use. To assess their actions, investors have access to many tools, such as the Ceres Water Toolkit, to better understand exposure to water risks, whether they are company-specific, industry-related, or global systemic risks.

Assessing water risks may also lead to opportunities for investors, companies, and society in general, as it can lead to increased profits, lower costs, water usage improvement, conservation efforts, and/or the development of effective ways to recycle water. Following the recent increase in interest from the financial world in the mitigation of climate change, water and the "blue economy" may be the next areas to draw scrutiny, require regulation, and invite innovation.

Using the frameworks and tools described in this chapter and analyzing potential risks and opportunities associated with water, companies and investors can reframe the dialogue around water so as to instigate positive change and responsible stewardship of our world's shared, finite and, precious resource: water.

References

Alliance for Water Stewardship. (n.d.). *About the alliance for water stewardship.* Retrieved October 5, 2020, from https://a4ws.org/about/.

Aviva Investors. (2020). *Hot water—The threat of warming oceans.* https://www.avivainvestors.com/en-ca/views/aiq-investment-thinking/2020/02/sea-level-change/.

Becken, S. (2014). Water equity—Contrasting tourism water use with that of the local community. *Elsevier—Water Resources and Industry, 7*, 9–22.

BlackRock Investment Institute. (2020). *Troubled waters, water stress risk to portfolios.* https://www.blackrock.com/us/individual/insights/blackrock-investment-institute/troubled-waters.

Bush, E., & Lemmen, D. S., (Eds.). (2019). *Canada's changing climate report.* Government of Canada. https://www.nrcan.gc.ca/climate-change/impacts-adaptations/canadas-changing-climate-report/21177.

Calvert Investments. (2016, September). *Constructing a smart water index.* UN PRI. https://www.unpri.org/constructing-a-smart-water-index/169.article.

Canadian Energy Research Institute. (2014). *Oil sands environmental impacts.* https://ceri.ca/assets/files/Study_143_Full_Report.pdf.

Carrington, D. (2018, May 21). Humans just 0.01% of all life but have destroyed 83% of wild mammals—study. *The Guardian.* https://www.theguardian.com/environment/2018/may/21/human-race-just-001-of-all-life-but-has-destroyed-over-80-of-wild-mammals-study#:~:text=The%20world's%207.6%20billion%20people,livestock%20kept%20by%20humans%20abounds.

CDP. (2020). *Interwoven risks, untapped opportunities.* https://www.cdp.net/en/research/global-reports/interwoven-risks-untapped-opportunities.

CDP. (2021). *Water.* https://www.cdp.net/en/water.

Cenovus Energy. (2019). *Water stewardship.* https://www.cenovus.com/responsibility/environment/water.html.

Ceres. (2018). *Investor water toolkit*. https://www.ceres.org/resources/toolkits/investor-water-toolkit.

Ceres. (2019). *Feeding ourselves thirsty: Tracking food company progress toward a water-smart future*. https://feedingourselvesthirsty.ceres.org/.

CEO Water Mandate. (n.d.). *Six commitment areas*. https://ceowatermandate.org/about/six-commitment-areas/#1529600036526-75d9e107-daea.

Chipman, K. (2020, December 6). *California water futures begin trading amid fear of scarcity*. Bloomberg. https://www.bloomberg.com/news/articles/2020-12-06/water-futures-to-start-trading-amid-growing-fears-of-scarcity?sref=oUjKJw8m.

CoreLogic. (2020). *2020 storm surge report*. https://www.corelogic.com/downloadable-docs/storm-surge-report-20200528-screen-2.pdf.

Dangerfield, K. (2019, April 26). '100-year floods' are increasing in Canada due to climate change, officials say—Is this true? *Global News*. https://globalnews.ca/news/5206116/100-year-floods-canada-increasing/.

Deschamps, T. (2020, July 5). Interest in outdoor cannabis cultivation grows among Canadian producers. *Global News*. https://globalnews.ca/news/7140924/cannabis-outdoor-cultivation/.

EIA. (2017, February 19). *2012 commercial buildings energy consumption survey: Water consumption in large buildings summary*. https://www.eia.gov/consumption/commercial/reports/2012/water/.

Ellen McArthur Foundation. (2018). *Water and circular economy*. https://www.ellenmacarthurfoundation.org/assets/downloads/ce100/Water-and-Circular-Economy-White-paper-WIP-2018-04-13.pdf.

Environment and Climate Change Canada. (2018, August 13). *Water and the environment*. Retrieved October 14, 2020, from https://www.canada.ca/en/environment-climate-change/services/water-overview/frequently-asked-questions.html.

Fisheries and Oceans Canada. (2020, December 3). *Canada commits to growing the blue economy at home and around the globe*. https://www.canada.ca/en/fisheries-oceans/news/2020/12/canada-commits-to-growing-the-blue-economy-at-home-and-around-the-globe.html.

Gambetta, G. (2020, October). *A deep dive into investing in the blue economy*. Responsible Investor. https://www.responsible-investor.com/articles/a-deep-dive-into-investing-in-the-blue-economy.

Government of Canada. (2018). *Cannabis act, S.C. 2018, c. 16*. https://laws-lois.justice.gc.ca/eng/acts/c-24.5/.

Government of Canada. (2020, January 24). *Agriculture and water quality*. https://www.agr.gc.ca/eng/agriculture-and-the-environment/agricultural-practices/agriculture-and-water/watershed-protection/agriculture-and-water-quality/?id=1371491033072.

Gray, T. (2019, July 11). As fresh water grows scarcer, it could become a good investment. *The New York Times*. https://www.nytimes.com/2019/07/11/business/fresh-water-shortage-invest.html.

Hamilton, A. K. (2020, August 11). Seasonal turbidity linked to physical dynamics in a deep lake following the catastrophic 2014 Mount Polley mine tailings spill. *Water Resources Research, 56*. https://doi.org/10.1029/2019WR025790.

Indigenous Services Canada. (2020, February 17). *Ending long-term drinking water advisories*. https://www.sac-isc.gc.ca/eng/1506514143353/1533317130660.

International Union for Conservation of Nature. (2018, May). *Marine plastics*. https://www.iucn.org/resources/issues-briefs/marine-plastics.

Investment Industry Association of Canada. (2019). *Opportunities in the Canadian green bond market v3.0*. https://iiac.ca/wp-content/uploads/Opportunities-in-the-Canadian-Green-Bond-Market_June-2019.pdf.

Kenton, W. (2019, August 23). *Social license to operate*. Investopedia. https://www.investopedia.com/terms/s/social-license-slo.asp.

Kimmelman, M. (2017, December 21). Jakarta is sinking so fast, it could end up underwater. *The New York Times*. https://www.nytimes.com/interactive/2017/12/21/world/asia/jakarta-sinking-climate.html.

Kosuth, M., Mason, S. A., & Wattenberg, E. V. (2018). Anthropogenic contamination of tap water, beer, and sea salt. *Plos One, 13*(4). https://doi.org/10.1371/journal.pone.0194970.

Kramer, K. (2018, October 5). *Sinking cities, rising seas*. https://www.christianaid.org.uk/sites/default/files/2018-10/Christian-Aid-Sinking-cities-rising-seas-report.pdf.

K-Water & International Water Resources Association. (2018). *Smart water management—Case study report*. https://www.iwra.org/wp-content/uploads/2018/11/SWM-report-final-web.pdf.

Lee-Anderson, S. (2018, October 16). *Spotlight on cannabis—Part 2: Taking a closer look at the environmental costs of cannabis cultivation*. McCarthy Tetrault. https://www.mccarthy.ca/en/insights/blogs/canadian-era-perspectives/spotlight-cannabis-part-2-taking-closer-look-environmental-costs-cannabis-cultivation.

Lemmen, D. S., & Warren, F. J. (2016). *Synthesis; in Canada's marine coasts in a changing climate*. Government of Canada. https://www.ouranos.ca/publication-scientifique/Coastal_Assessment_Synthesis_en.pdf.

Madhusoodanan, J. (2019). Can cannabis go green? *Nature, 572*, S8.

Managing Water Consumption in Mining. (2013, August 13). Retrieved from Mining Technology. https://www.mining-technology.com/features/feature-managing-water-consumption-mining-global-shortage/.

Millani. (2020). *A Canadian perspective: A study of the S&P/TSX composite index and their ESG disclosures in 2019.* https://corostrandberg.com/wp-content/uploads/2020/09/millani-annual-esg-disclosure-study-2020.pdf.

MSCI. (2020, May). *MSCI global sustainable water index (USD).* https://www.msci.com/documents/10199/6cbe6af7-2788-46d9-8d53-5388ee09fbe3.

Muggah, R. (2019, January 16). *The world's coastal cities are going under. Here's how some are fighting back.* World Economic Forum. https://www.weforum.org/agenda/2019/01/the-world-s-coastal-cities-are-going-under-here-is-how-some-are-fighting-back/.

National Hurricane Center. (2018, January 26). *Costliest U.S. tropical cyclones tables updated.* https://www.nhc.noaa.gov/news/UpdatedCostliest.pdf.

Natural Resources Canada. (2009, March). *Water use by the natural resources sectors.* https://www.nrcan.gc.ca/sites/www.nrcan.gc.ca/files/www/pdf/com/resoress/publications/wateau/wateau-eng.pdf.

Natural Resources Canada. (2020, October 6). *Energy data and analysis—Electricity facts.* Retrieved October 9, 2020, from Science and Data: https://www.nrcan.gc.ca/science-data/data-analysis/energy-data-analysis/energy-facts/electricity-facts/20068#L3.

Ocean Tomo. (2020). *Intangible asset market value study.* https://www.oceantomo.com/intangible-asset-market-value-study/.

Ochianu, O. (2019, February 1). *Water risks in extractive industries.* Sustainalytics. https://www.sustainalytics.com/esg-blog/extractive-industries-water-risk/.

OECD. (2005, December 2). *Glossary of statistical terms.* https://stats.oecd.org/glossary/detail.asp?ID=2907.

OECD. (n.d.). *Development results.* https://www.oecd.org/dac/peer-reviews/Development-Results-Note.pdf.

Oil Sands Magazine. (2020). *Water usage.* https://www.oilsandsmagazine.com/technical/environment/water-usage?rq=water%20usage.

Olam International Limited. (2019, November). *Investor presentation.* https://www.olamgroup.com/content/dam/olamgroup/investor-relations/investor-relations-pdfs/Olam-InvestorPresentation-Nov2019.pdf.

Owen, J. R., Kemp, D., Lèbre, É., Svobodova, K., & Pérez Murillo, G. (2020). Catastrophic tailings dam failures and disaster risk disclosure. *International Journal of Disaster Risk Reduction, 42.* https://doi.org/10.1016/j.ijdrr.2019.101361.

Palace, V. (2019, February 4). *How will cannabis legalization affect our fresh water?* IISD. https://www.iisd.org/articles/cannabis-legalization-fresh-water.

Park, Y., Mackie, A., & Gagnon, G. (2016, November 22). A critical review of the occurrence, detection, and treatment of Δ9-tetrahydrocannabinol in aquatic environments. *Environmental Reviews, 25*(3), 255–268. https://doi.org/10.1139/er-2016-0061.

S&P Global Ratings. (2019, May 21). *ESG industry report card: Technology.* Ratings Direct. https://www.spglobal.com/_media/documents/esg-industry-report-card_technology.pdf.

Sattiraju, N. (2020, April 2). The secret cost of Google's data centers: Billions of gallons of water to cool servers. *Time.* https://time.com/5814276/google-data-centers-water/.

Semeniuk, I. (2017, June 11). Charting Canada's troubled waters: Where the danger lies for watersheds across the country. *The Globe and Mail.* https://www.theglobeandmail.com/news/national/canada-fresh-water-review-1/article35262579/.

Statistics Canada. (2016). *Human activity and the environment, freshwater in Canada.* https://www150.statcan.gc.ca/n1/pub/16-201-x/16-201-x2017000-eng.pdf.

Swenson, S. C., & Lawrence, D. M. (2012, November 7). A new fractional snow-covered area parameterization for the community land model and its effect on the surface energy balance. *Journal of Geophysical Research: Atmospheres, 117*(D21). https://doi.org/10.1029/2012JD018178.

Task Force on Climate-Related Financial Disclosures. (2017). *Final TCFD recommendations report.* https://www.fsb-tcfd.org/wp-content/uploads/2017/06/FINAL-2017-TCFD-Report-11052018.pdf.

The Ocean Foundation. (n.d.). *Ocean and climate change.* Retrieved October 5, 2020, from https://oceanfdn.org/ocean-and-climate-change/.

The Pacific Institue and Ceres. (2009, February). *Water scarcity & climate change: Growing risks for businesses & investors.* https://pacinst.org/wp-content/uploads/2009/02/growing-risk-for-business-investors-2.pdf.

U.S. Environmental Protection Agency. (2005). *Protecting water quality from agricultural runoff.* https://www.epa.gov/sites/production/files/2015-09/documents/ag_runoff_fact_sheet.pdf.

UN. (n.d.). *The 17 goals.* https://sdgs.un.org/goals.

UN Environment COBSEA. (n.d.). *Marine litter and plastic pollution.* Retrieved October 5, 2020, from https://www.unenvironment.org/cobsea/what-we-do/marine-litter-and-plastic-pollution.

UN PRI. (2018a). *Growing water risk resilience: An investor guide on agricultural supply chains.* https://www.unpri.org/environmental-issues/growing-water-risk-resilience-an-investor-guide-on-agricultural-supply-chains-/2793.article.

UN PRI. (2018b, August 19). *Impact investing market map—Water.* https://www.unpri.org/impact-investing-market-map-water/3544.article.

Union of Concerned Scientists. (2011). *The energy-water collision—Power and water at risk.* https://www.ucsusa.org/resources/energy-water-collision.

University of Alberta. (2020, August 13). *Contaminants from Mount Polley tailings spill continue to affect Quesnel lake*. ScienceDaily. Retrieved February 3, 2021, from www.sciencedaily.com/releases/2020/08/200813134555.htm.

UN-Water. (2018). *Sustainable Development Goal 6, synthesis report on water and sanitation*. http://www.unwater.org/app/uploads/2018/12/SDG6_SynthesisReport2018_WaterandSanitation_04122018.pdf.

UN-Water Thematic Initiatives. (2006). *Coping with water scarcity: A strategic issue and priority for system-wide action*. https://www.preventionweb.net/publications/view/1770.

World Bank. (2020, May 8). *Water in agriculture*. https://www.worldbank.org/en/topic/water-in-agriculture.

World Bank and United Nations Department of Economic and Social Affairs. (2017). *The potential*. World Bank. https://openknowledge.worldbank.org/bitstream/handle/10986/26843/115545.pdf?sequence=1&isAllowed=y.

World Economic Forum. (2021). *The global risks report 2021*. https://reports.weforum.org/global-risks-report-2021/.

WRAP. (n.d.). *WRAP and the circular economy*. https://wrap.org.uk/about-us/our-vision/wrap-and-circular-economy.

WWF. (2020, October 6). *New water risk filter scenarios will help companies and investors turn risk into resilience*. https://wwf.panda.org/wwf_news/?933366/New-Water-Risk-Filter-Scenarios-will-help-companies-and-investors-tur.

The Developing Field of Water Risk Valuation for the Financial Industry

Yuanchao Xu and Debra Tan

1 Water Availability Impacts on the Profitability of the Coal and Power Sectors

Research and recent events suggest that limited water availability can affect the profitability of power stations, especially when it is combined with the rise of temperatures. For instance, in 2015, heatwaves in Germany and Poland forced coal-fired power plants to reduce their power generation, and in 2003, EdF had to curtail the equivalent of four to five nuclear reactors for the same reason (Roehrkasten & Schaeuble, 2015). A combination of factors can help explain these temporary/partial shutdowns: warmer rivers, lower river flows, and strictly enforced regulations pertaining to water discharge temperature.

In India, the coal sector has been hit by low rainfall in 2015–2016 and droughts more recently. In the period between March and May 2016,

Y. Xu (✉) · D. Tan
China Water Risk (CWR), Hong Kong, China
e-mail: ycx@chinawaterrisk.org

D. Tan
e-mail: dt@chinawaterrisk.org

T. Walker et al. (eds.), *Water Risk and Its Impact on the Financial Markets and Society*, Palgrave Studies in Sustainable Business In Association with Future Earth,
https://doi.org/10.1007/978-3-030-77650-3_4

over 4GW of coal power capacity was shut down several times due to insufficient cooling water (Fernandes & Krishna, 2016).

The extent to which China's thermal power plants have so far suffered water-related shutdowns is unclear. However, a 2016 Nature Climate Change article anticipates warmer temperatures and decreased river flows along the Yellow River (van Vliet et al., 2016). The same article foresees reductions in usable capacity for 81–86% of the thermal power plants worldwide for 2040–2069.

It should be noted that renewable energy generation, such as solar PV and wind turbines, are significantly less exposed to water stress given their minimal water use to operate. A diversified power generation mix can therefore reduce exposure to water risk.

Water plays a vital role in the mining sector, such as in transporting and processing ore, cooling and lubricating heavy drilling equipment, managing waste trailings, and suppressing dust (UNEP FI, 2012). Therefore, a company's profitability can be affected by water risks due to production disruption and rising water costs (CDP, 2013). These impacts can then result in financial risks as was shown by Columbia's research on drought-induced financial risk in the mining industry (Columbia Water Center, 2015).

These physical as well as regulatory risks have already been recognized by the financial sector. For example, HSBC studies the growth limitations caused by water in a 2015 report (HSBC, 2015), and CLSA recommends that, "coal and coal-related industries as well as providers of capital to such sectors, whether in equity or credit, factor in water risk in China" in a report (CLSA, 2013, p. 5).

In order to assess the impact of water risks on both the coal and power sector in China and the progress of their valuation, CWR conducted water risk assessment case studies using two methodologies: (1) shadow price model and (2) balance sheet exposure, on ten listed energy majors (CWR, 2016).

These ten listed Chinese energy majors included the then five largest coal mining companies ("Coal-5") and the five largest power generation companies ("Power-5") in China. The Coal-5 accounted for 14% of China's coal output in 2014 whereas the Power-5 was responsible for 19% of the nation's thermal electricity generated.

Data used in the analysis of these companies were taken from publicly available sources such as annual reports, stock exchange filings, sustainability and/or CSR reports, company websites, and official government

Table 1 Ten listed companies analyzed by CWR in 2016

Coal	*Power*
China Shenhua Energy Company Limited (SSE: 601088)	Huaneng Power International, Inc. (SSE: 600011)
China Coal Energy Company Limited (SSE: 601898)	Datang International Power Generation Co., Ltd. (SSE: 601991)
Yanzhou Coal Mining Company Limited (SSE: 600188)	Huadian Power International Corporation Limited. (SSE: 600027)
Inner Mongolia Yitai Coal Co. Ltd. (SSE: 900948)	GD Power Development Co., Ltd. (SSE: 600795)
Datong Coal Industry Co. Ltd. (SSE: 601001)	China Power International Development Limited. (SEHK: 2380)

Note 1 Datong Coal Industry Co., Ltd. Is now named as Jinneng Holding Shanxi Coal Industry Co., Ltd.
Note 2 This figure has been reproduced for this chapter with the kind permission of CWR
Source CWR (2016)

data from 2014 to 2015 as this was when the CWR report was published. Note that these listed companies will hereafter be referred to by a shortened version of their names (Shenhua, China Coal, Yanzhou, Yitai, Datong, Huaneng, Datang, Huadian, Guodian, and CPI). These should not be mistaken with their respective parent groups which hold the same name.[1] The ten listed majors are listed in Table 1.

2 Shadow Price Model

Current water pricing does not reflect the "real value of water" should water scarcity and competition for water be taken into account. Therefore, a "shadow price" of water can be used as a proxy to estimate the potential economic impact of water risks on a company's profitability. Using a shadow price can help integrate water risks into decision-making as the "real risks" from water can be evaluated.

The shadow price of water is based on the so-called Total Economic Value (TEV) of water, as recommended by the World Business Council for Sustainable Development (WBCSD, 2013). This TEV is supposed to capture the full economic value of water to reflect supply constraints and

[1] Data used in this analysis are based on 2014 annual reports of the ten listed companies.

demand pressures. The TEV includes external benefits that water provides to society and ecology, in addition to the private benefit enjoyed by water consumers.

As the shadow price of water is location-specific, the average shadow price of water for a company is computed as a weighted average of their operations' water use across various locations. Therefore, for this case study, each company's assets were mapped, and the local water use for each of their operations was estimated by CWR to determine the company's average shadow price.

In CWR's water shadow pricing case study, two tools were adopted, respectively the Corporate Bonds Water Credit Risk Tool (CBWCRT) and the Water Risk Monetizer (WRM) as these are both accessible by the public and available to use for free. These two tools associate a shadow price of water for every location, based on local water stress. CBWCRT accounts for population density, whereas WRM does not. WRM does, however, adjust to account for the country's purchasing power parity (PPP). The CBWCRT tool does not adjust for this, partly explaining why the shadow price values obtained by the CBWCRT tool are often significantly higher than with the WRM tool.

This shadow price of water can then be used as a proxy to reflect exposure to water risk by introducing it into the listco's financial valuations via increased operating costs. The additional OPEX line, inserted in the P&L statement, is obtained by multiplying the amount of water used with the shadow price vs the then current price of water obtained from the government (WRI, 2016) (Fig. 1).

To assess the impact of water risk on the coal and power sector, CWR used the CBWCRT and WRM tools to provide shadow prices of water to approximate its "real value" and therefore the impact of the company's profit & loss account (P&L) on the ten listed Chinese energy majors discussed above (CWR, 2016).

The impact on margins, the range of water shadow prices, as well as the water use for each of the ten energy listed companies are discussed in the following part in the respective sector peer groups.

2.1 Impacts on Profitability

The charts (Fig. 2) from CWR's report show the potential impacts of water risk on profitability margins if the "true cost of water" was included through the use of the two shadow pricing tools in the P&L for the

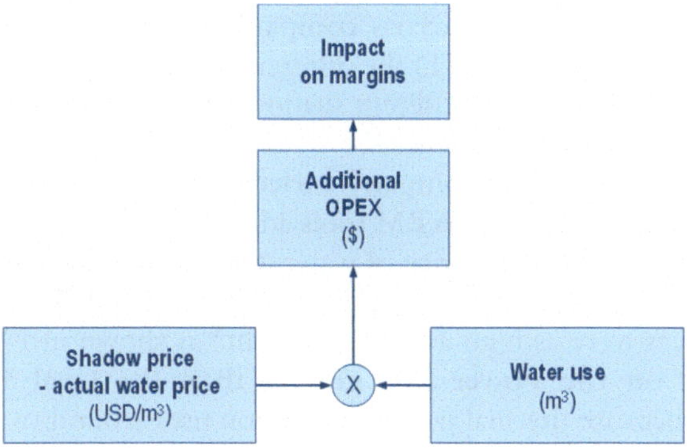

Fig. 1 Shadow price of water and water use and their impact on margins (*Note* This figure has been reproduced for this chapter with the kind permission of CWR. *Source* CWR [2016])

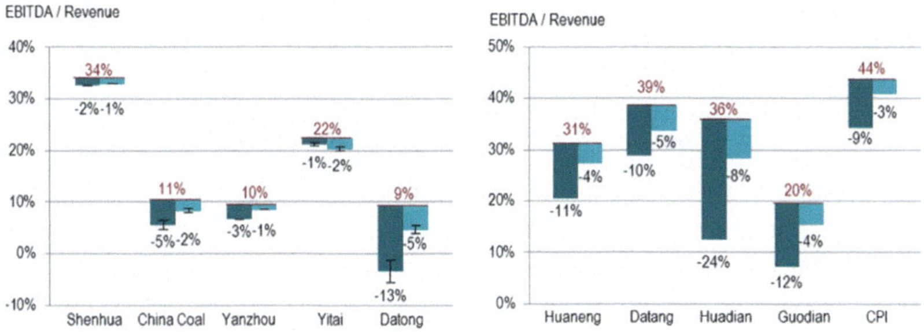

Fig. 2 Shadow price of water and its impact on the EBITDA of Chinese coal (left) and power (right) companies (*Note* This figure has been reproduced for this chapter with the kind permission of CWR. *Source* CWR [2016])

Coal-5 (left) and the Power-5 (right). Profitability margins are presented in terms of the EBITDA margin (a company's operating profit as a percentage of its revenue).

The charts show that the EBITDA margins are negatively impacted for all ten companies, and this ranges from −1 to −13% for the Coal-5 and from −3 to −24% for the Power-5.

Key result observations noted in the CWR report for the two sectors are:

- **Impact varied widely across companies in the same sector peer group**: This was due to (1) the different locations of each company's operations resulting in different shadow prices of water, as well as (2) different water use efficiencies;
- **Impact on the same company varied widely across tools**: This is because CBWCRT and WRM tools adopted different correlations to determine the shadow price of water. On top of this, WRM caps the shadow price of water[2] at USD 4.5/m^3whereas CBWCRT does not and prices were as high as USD 12.1/m^3 as shown in Fig. 3; and
- **Impact on the Power-5 is greater than the Coal-5**: This was mainly because thermal power generation uses around twice as much water per unit of revenue than coal mining (including mine water drainage). Therefore, despite lower shadow prices, the Power-5 exhibited higher exposure than the Coal-5.

We now dive into the two factors causing these wide-ranging results (1) the shadow price of water, and (2) the amount of water used in each company's operations.[3]

2.2 *Shadow Water Prices: Wide-Ranging Values*

The average shadow prices of water for the ten energy companies are shown in the charts below extracted from CWR's report. It is important to note that (1) the shadow price of water is solely dependent on the location regardless of the industrial sector, and (2) prices shown per company

[2] According to Water Risk Monetizer Methodology (Ecolab, 2015), the water risk premium is capped with the highest water market price existing in the world then, which is USD 4.5/m^3 in Aarhus, Denmark.

[3] Only the water use of each company's core business was accounted for: coal mining for Coal-5 and thermal power generation for Power-5. The potential loss of profitability is therefore only related to this core business segment. However, the current EBITDA margin (indicated in red in the above charts) refers to the company's overall profitability from all segments like coal mining, coal to chemicals, power generation, and transportation.

For example, Shenhua's higher margin is attributable to the fact that coal sales represent only 66% of Shenhua's revenue compared to 97% for Yanzhou and Datong. Since only the impact on water for coal is accounted for, Shenhua's downside impact is thus underestimated relative to its peer group.

The Power-5 companies on the other hand, are more directly comparable, with power generation representing from 90 to 100% of their entire revenue.

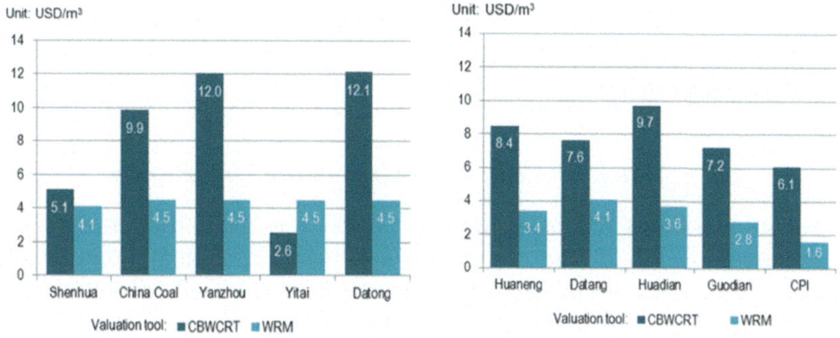

Fig. 3 Average shadow price for water (*Note* This figure has been reproduced for this chapter with the kind permission of CWR. *Source* CWR [2016])

is the average, weighted by the freshwater use at every facility rather than by the production of coal or electricity.

Key result observations as per the CWR report for the two sectors are:

- **Shadow prices of water were higher for the Coal-5 than Power-5**: This was due to the fact that a larger proportion of coal mines are located in highly or extremely high water-stressed regions, compared to thermal power plants (see Sect. 3 for which regions are most at risk);
- **Shadow prices can be more than 100 times the actual price of water**: The shadow price can be more than 100 times the actual water price paid by companies which was estimated to be at times less than USD 0.10/m^3, which prompted much discussion around whether shadow pricing is realistic (see Sect. 2.5 Challenges); and
- **Shadow prices calculated using CBWCRT are higher than WRM**: This held true for both sectors. Exposure variations were reflected by the wide-ranging results: CBWCRT values ranged from USD 2.6–12.1/m^3 while the range for WRM was only USD 1.6–4.5/m^3. The main reason was that WRM caps the shadow price of water at USD 4.50/m^3, the highest water tariff observed globally at the time.[4] (Note: Yitai's shadow prices illustrate a different outcome but they

[4] WRM is designed to stay within actual market conditions, so the water price is capped at the highest water tariff observed globally.

are calculated correctly. Yitai had a higher shadow price with WRM than CBWCRT due to location approximation of its assets to the closest city in WRM.)

2.3 *Water for Coal: Data Gaps Bring Uncertainty*

A key input of this risk assessment by CWR was the actual amount of water used in each of the company's facilities. Unfortunately, none of the ten companies disclosed such information at the facility level at the time. Such uncertainty makes it difficult to conduct proper benchmarking. From the companies in the Coal-5, only Yanzhou disclosed water use at a facility level. Meanwhile, Shenhua is the only company that disclosed both its freshwater use and its mine water drainage amounts. The other three in the coal peer group disclosed neither water use nor drainage. Since water use for coal mining remains a mystery[5] and companies in the peer group are only comparable when using the benchmark of water use per ton of coal mined, at that time Shenhua and Yanzhou disclosed to derive m^3/ton benchmarks for coal mining and processing.

In the analysis, due to the missing data, assumptions were made by CWR; as Shenhua disclosed its average water use for thermal power generation, it was assumed that all the water not used for thermal power generation was used for its coal segment. This resulted in an estimated 0.04 m^3 per ton of coal mined and processed for 2014.

As for China Coal, Yitai and Datong, their values were assumed in the CWR report to be the average of those for Shenhua and Yanzhou. The resulting ranges are represented by the black bars in Fig. 4. However, since there is a lack of disclosure on this front, the charts from the CWR

[5] Another uncertainty pertains to the type of water that should be accounted for. The most straightforward answer would be to consider only the amount of freshwater used for coal mining and processing (e.g., for cleaning equipment and dust control). However, one may also argue that the drained mine water (the water seeping into the mine that is pumped out) represents a water loss for local competing users and hence should be accounted for as well.

Since companies are already paying a mine water drainage fee, mine water drainage was included in the computations to derive the impact on EBITDA margins. Since only Shenhua disclosed this drainage amount, this level was used across all companies and drained mine water is represented by the gray bars in the above chart. These can be compared to national benchmarks set out in the NEA-NDRC action plan.

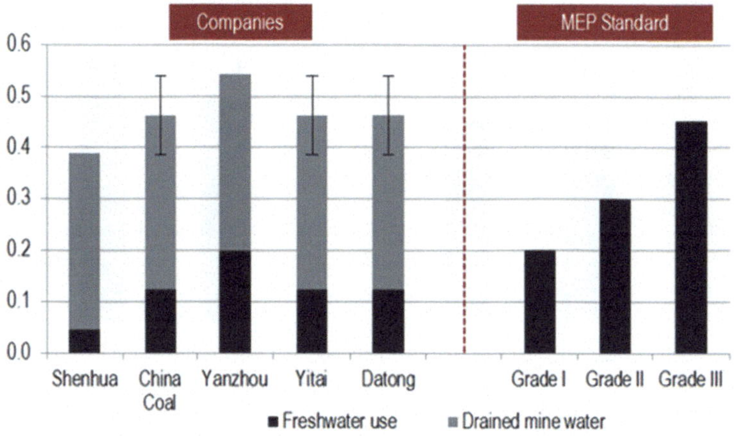

Fig. 4 Estimated water requirement for coal production (unit: m^3/tonne) (*Note* This figure has been reproduced for this chapter with the kind permission of CWR. *Source* CWR [2016])

report show the results alongside national standards or targets. This standard was a guideline for cleaner coal production devised by the Ministry of Environmental Protection (MEP); while not strictly mandatory, this indicator can help benchmark the Coal-5's water performance.

2.4 *Water for Power: Bottom-up Approach*

On the power front, it was a different story. All five companies, except Huadian, disclosed the average water consumption of their thermal power generation units. These levels are shown in Fig. 5 taken from the CWR report.

Two main factors explain the difference between companies: (1) the overall efficiency of their fleet and (2) more importantly, the use of different cooling types.

Water use/consumption levels vary depending on location, different cooling types, and water stress (and hence shadow water prices). Therefore, ideally to analyze water risk exposure, knowledge of each power plant's water use, specific location, and cooling type is required.

Using the average values across all power plants as per Fig. 5 fails to account for such variability, thereby undermining the reliability of

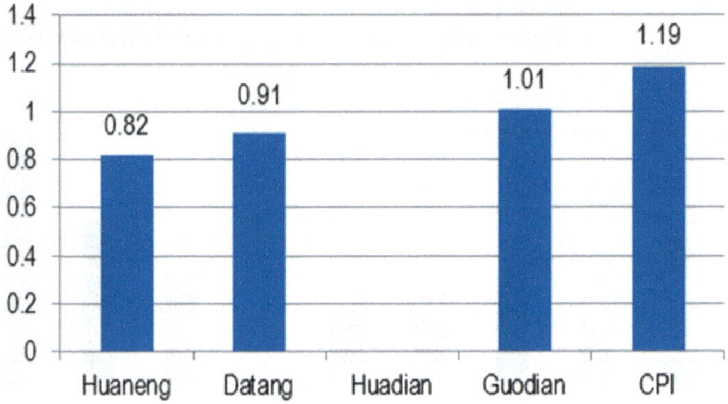

Fig. 5 Average water consumption for thermal power generation (unit: m^3/tonne) (*Note* This figure has been reproduced for this chapter with the kind permission of CWR. *Source* CWR [2016])

the results. As such, to establish a baseline, a bottom-up approach was adopted by CWR in deriving the amount of water usage for each power plant.

As water consumption varies significantly depending on cooling technology type to estimate water usage at the power plant level, CWR ascertained the cooling types across the entire fleet for each of the Power-5 companies.

Three types of cooling systems are typically used:

1. Once-through cooling: water is used only once and discharged afterward, which presents a low risk in coastal locations where seawater is used but can present a high risk if freshwater is the only source in a water-stressed location;
2. Closed-loop cooling: water is recycled and thus the water risk is medium; and
3. Air cooling systems do not consume water, so the water risk is low. However, this is highly energy intensive and so there could be carbon or energy risks attached.

Given the different technologies, water consumption can be as low as 0.3 m^3/MWh for air cooling to 2.5 m^3/MWh for closed-loop (Qin et al., 2015). Different cooling technologies imply different types and degrees of water risks, both physical and regulatory. Power plants with once-through technologies in inland areas are the most exposed, while air cooling equipped power plants bear minimal risks on the waterfront.

The cooling types of the 135 power plants of the Power-5 were derived from online databases, satellite images as well as industry data, and assumptions were made by CWR when only partial information was available. The estimated results are shown in Fig. 6.

Using this bottom-up approach and the average peer group values per cooling type, CWR estimated Huadian's average water consumption to be 1.4 m^3/MWh.

From Fig. 6, it seems that Guodian is most at-risk with a high proportion (32%) of electricity generated using once-through freshwater for cooling. However, this cannot be viewed on its own as location matters. In mapping the location of Guodian's thermal fleet, CWR's analysis revealed that 34% of it lies in low-water stress areas (see Sect. 3, Balance Sheet Exposure). It would thus be unfair to over-penalize Guodian for its large once-through fleet. Cooling type and location therefore need to be viewed together when analyzing a company's water risk exposure.

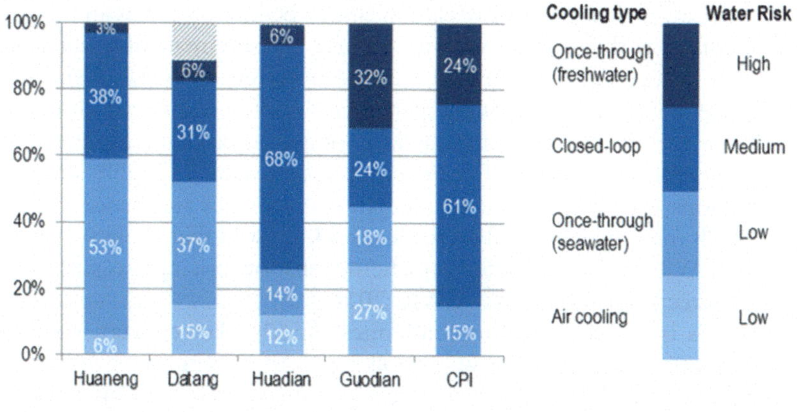

Fig. 6 Power generation by cooling type (unit: TWh/TWh) (*Note 1* Grayed areas refer to power plants whose location is unclear and cooling type cannot be known from available sources, *Note 2* This figure has been reproduced for this chapter with the kind permission of CWR. *Source* CWR [2016])

2.5 *Challenges*

Using shadow pricing for water is a rather new approach with the ability to approximate potential impacts on profitability and to benchmark companies on water risks. However, there are still serious concerns over the underlying assumptions. These concerns mean that 82% of the investors surveyed by CWR thought that the use of shadow pricing methods to value water risk was "a good first step, but needs complementary analysis" (CWR, 2016). It is worth noting here that 77% of those that provided feedback were from mainstream finance, with 23% identifying themselves as practicing Socially Responsible Investment (SRI).

Key limitations, usefulness, and challenges of this method highlighted through investor feedback in the CWR report were:

- **Still much debate on theoretical shadow prices vs reality**: There was much doubt on whether shadow pricing is sufficient to screen out companies' exposure to water risk. E.g., will China ever raise industrial water tariffs to USD 12.10/m^3, the highest shadow price? However, to be fair to the shadow pricing tools, their developers are not claiming that water tariffs will ultimately reach these values. Instead, such shadow prices are meant to be interpreted as an "upper bound" as they can be seen as a proxy measurement of water risks that go beyond the potential increase of water tariffs. High shadow prices could indicate that companies could suffer from a real shortage of water due to the physical lack of water or from government intervention in terms of reducing access to water in highly stressed regions, which would curtail operational activities and hence impact profit margins. Financial sector feedback showed that those surveyed by CWR remained divided on whether shadow pricing was sufficient to screen out water risks (CWR, 2016).
- **Tools ignored pollution challenges**: Another key concern was the lack of consideration for water pollution. Addressing water pollution in water risks assessment is crucial for many sectors given China's ongoing "war on pollution" and the issuance of the Water Pollution Prevention & Control Action Plan. This held particularly true for coal mining rather than power generation.
- **Use with a grain of salt—tool limitations and pervasive disclosure challenges**: The use of shadow pricing tools is still limited to a very small group of professionals due to their complexity. In

addition, whether to trust the disclosure of water use and drainage numbers remains a question.

- **Binary risks were more concerning than cost increases**: In some cases, if water is depleting, companies may face binary risks rather than gradual cost increases, i.e., no water to use. What happens if the access to water itself cannot be guaranteed anymore, either for ongoing operations or planned expansion? For example, in some water-scarce regions in China, no new water use permit will be issued anymore. Investors surveyed by CWR found binary risk of water scarcity from droughts and competition for water or regulations, which restrict water use, more concerning than water price increases (CWR, 2016). We discuss it in more detail in the next section.

3 Balance Sheet Exposure

When looking at water risk exposure from an asset perspective, local water availability matters, so it is important to understand local water stress. There are different indicators for water stress (Brown & Matlock, 2011). The commonly used indicator is the *Baseline Water Stress* index, defined as follows by WRI: "Baseline Water Stress measures total annual water withdrawals (municipal, industrial, and agricultural) expressed as a percent of the total annual available flow" (Gassert et al., 2014, p. 8).

The Baseline Water Stress index assumes that high water stress begins when 40% or more of the total annual available flow is withdrawn. Extremely high-water stress is reached when this figure exceeds 80%. Arid areas are those with both low water resources and low water use.

WRI mapped out these Baseline Water Stress (BWS) values globally in its Aqueduct BWS Map.[6] Although there are other water stress maps available, CWR used WRI's Aqueduct BWS map to assess balance sheet exposure of assets to water risks as the Aqueduct BWS map is also the underlying water stress map for CBWCRT, WRM, and Bloomberg shadow pricing tools. It therefore provided a consistent platform to view both P&L and balance sheet impacts.

The same ten listed energy companies as in the last section were selected to conduct the analysis on balance sheet exposure of water risks.

[6] The CWR case studies in this chapter were based on WRI Aqueduct 2.0 map. In August 2019, WRI released its new Aqueduct 3.0 map in which China's baseline water stress profile changed.

3.1 *Coal Mining Companies: Up to 100% Exposure to Water Stress*

In China, most of the coal is mined in the Northern regions, and particularly in the Yellow River basin, where water resources are limited and under pressure. The Yellow River Commission, established by the Ministry of Water Resources (MWR), estimates that the water deficit in this watershed reaches 9.5 billion m^3 a year (Yellow River Conservancy Commission, 2013). The commission also estimates that this deficit will increase to 13.8 billion m^3 in 2030, mainly due to the development of coal and coal power bases.

CWR found that of the 79 mines of the Coal-5 mapped using WRI's Aqueduct BWS map, over half of the mines are located in high to extremely high water-stressed regions. However, the true level of exposure depended on the volume of coal produced by each mine, which resulted in a much higher exposure. Factoring in coal production volumes, CWR's analysis revealed that the exposure to water stress is significant: three of the five companies source more than four-fifths of their coal from extremely high water-stressed regions: Yanzhou (87%), China Coal (90%), and Datong (100%). In turn, Shenhua and Yitai appear relatively less exposed, with only 2 and 0%, respectively, located in high water-stressed regions.

As can be seen from Fig. 7, 80% of the Coal-5's total coal output is from the Ordos-Datong-Baotou region. This region, at the border between Inner Mongolia and Shanxi, is clearly water-stressed.

CWR's analysis also illustrated a high degree of provincial exposure. Two regions, namely Inner Mongolia and Shanxi represented 83% of all Coal-5 companies' coal output—57% of their coal is extracted from Inner Mongolia and 26% from Shanxi. It should be noted that these two provinces also account respectively for 26 and 24% of China's national coal output. Regional production exposure is illustrated in Fig. 8.

This pronounced geographical concentration means that provincial focus matters. Different provinces have different benchmarks for water use/drainage as well as local regulations pertaining to water use permits, resource fee, and so on. Since Inner Mongolia and Shanxi accounted for half of China's output and are water-stressed, they face stricter regulations and therefore regulatory risk.

The CWR report highlighted three key areas of regulatory risk: (1) allocation of water; (2) tighter requirements leading to increased compliance costs in both CAPEX and OPEX as well as (3) violations and fines.

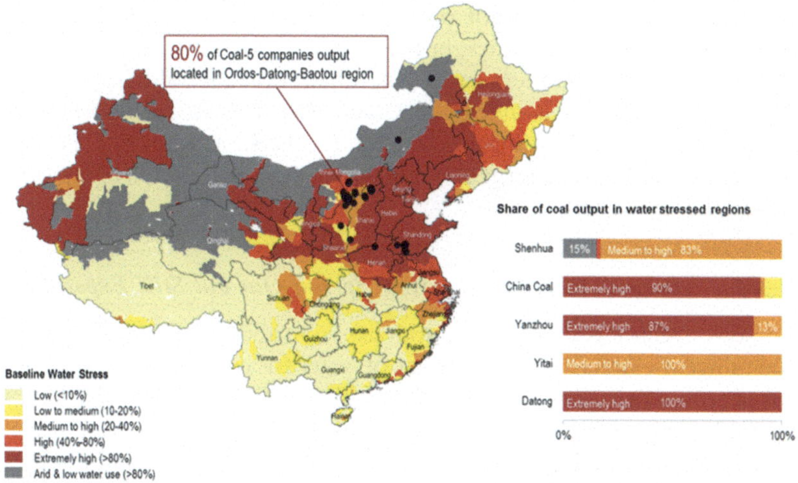

Fig. 7 Coal mining operations of Coal-5 companies (79 mines) (*Note* This figure has been reproduced for this chapter with the kind permission of CWR. *Source* CWR [2016])

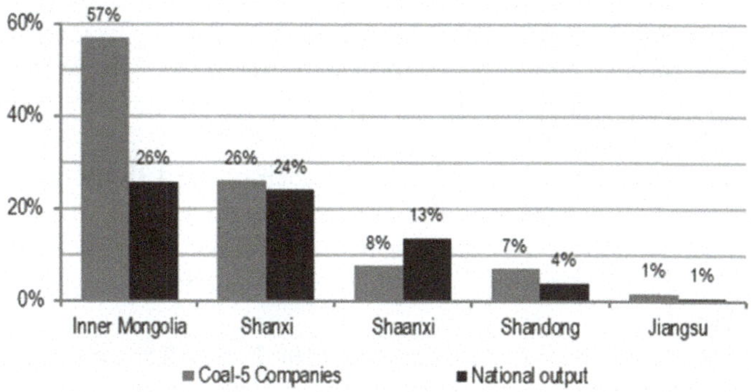

Fig. 8 Top-5 coal mining provinces (2014) (*Note 1* Percentage numbers presented are rounded, *Note 2* This figure has been reproduced for this chapter with the kind permission of CWR. *Source* CWR [2016])

Therefore, companies face not only more uncertainties in dealing with inherent locational water issues, but also local regulations.

We have not included regulatory risk implications on valuations in this chapter as the analysis is unique to China. However, it is important to consider them as they do impact valuations in the short term but alleviate water stress in the long term as regulations are usually implemented to manage the watershed.

3.2 Thermal Power Generation: Spread Out but Still Exposed

The power sector is exposed to, and also contributes to water stress in China since it extracts a huge amount of water for cooling. This is particularly true in the Northern provinces which hold only 25% of China's renewable water resources but account for 51% of its national thermal power generation, 82% of coal production, and 86% of its coal reserves (CWR & IRENA, 2016). The water-stressed provinces of the North are also home to nearly half of China's farmland (Hao et al., 2003). As climate change worsens and water becomes scarcer, conflicts between food and energy security could originate in these provinces.

China's ambition to develop a dozen large coal power bases—mines and power plants—in these Northern provinces further exacerbates the exposure of thermal power generation to water stress. The water stress around these bases appears to be even more severe than the provincial average (Zhang et al., 2016).

CWR analyzed the power generated by each of the 135 thermal power plants of the Power-5 against the WRI's Aqueduct BWS map. Results showed various degrees of exposure to water stress: from 34% of power generation located in extremely high water-stressed regions for CPI to 81% for Huadian.

Figure 9 clearly illustrates that assets of the Power-5 were more spread out across low water stress regions than those of the Coal-5. This geographical spread of assets was also reflected in the lower shadow prices across the companies for this peer group.

As flagged in the previous section, the water risk profile varies depending on cooling type and location of the plant. The water stress exposure shown below for each company should therefore be viewed together with the cooling type of their respective fleets. To recapitulate, if we take the case of Guodian as an example, although 32% of electricity generated used water-intensive once-through freshwater cooling, it may

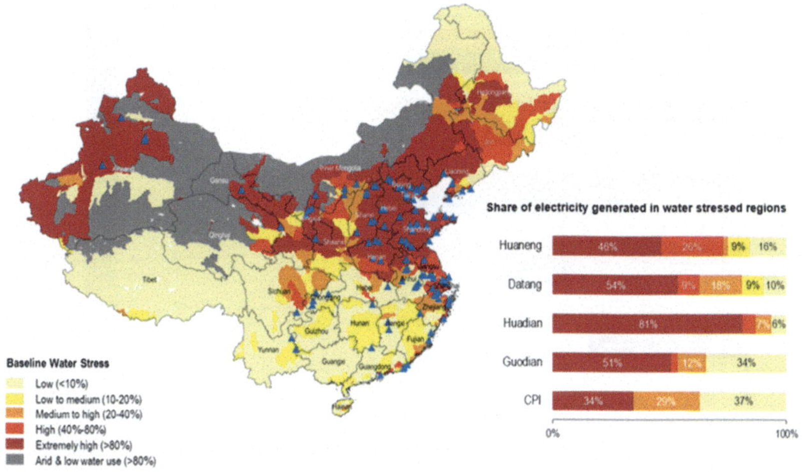

Fig. 9 Thermal power plants of Power-5 companies (135 power plants) (*Note* This figure has been reproduced for this chapter with the kind permission of CWR. *Source* CWR [2016])

not prove to be an issue, as these plants may be part of the 34% of the fleet located in low water stress areas.

Matching cooling types to water availability is therefore important. This cross-analysis of at-risk plants vis-à-vis their cooling types was conducted by CWR so that the Power-5's real exposure can be ascertained. CWR found that 6% of Datang and 2% of Huadian's electricity generated used once-through freshwater cooling in high and extremely high water-stressed areas (Fig. 10).

Similar to the Coal-5, provincial exposure mattered as there were also regulatory risks to consider such as the banning of once-through cooling in some provinces and the requirement of all new power plants in water-scarce region to adopt air-cooling.

3.3 *Challenges*

Baseline water stress maps are indeed a convenient way to assess and compare exposure to water stress across companies, sectors, or commodities. However, there are challenges in using them, and one should bear

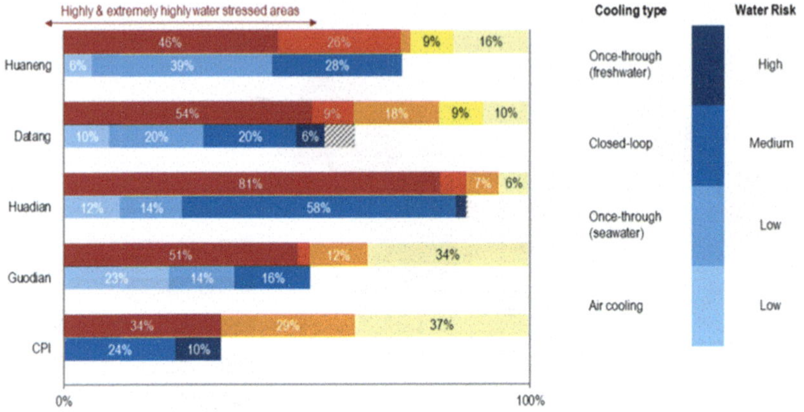

Fig. 10 Power generation by cooling type in highly water stressed areas (unit: TWh/TWh) (*Note* This figure has been reproduced for this chapter with the kind permission of CWR. *Source* CWR [2016])

these in mind. Those highlighted by CWR regarding its analysis above are:

- **Limited accuracy of underlying data may lead to under/over-estimate the degree of water availability challenges**. In this instance, CWR highlighted the case of the Kuye River catchment. The Kuye River is the largest tributary in the middle reaches of the Yellow River and its catchment is under threat with rising water demand and fluctuating resources. Vegetation change, coal mining, and hydraulic projects have significantly affected the river flow: the average flowrate over the last decade (2000–2010) was 73% lower than during 1956–2000, according to the Yellow River Institute of Hydraulic Research (Yellow River Conservancy Commission, 2013). In 2010, the actual water demand exceeded the water supply by 33% according to the same institute.

  Yet, the Kuye River catchment includes Ordos, a key area for China's coal industry, and despite its small size (only 20% larger than Shanghai), the Kuye River catchment holds 49% of Shenhua's coal output and 70% of Yitai's coal output.
- **The granularity of the maps affects the results**. Apart from the global Aqueduct BWS map, WRI also developed a new map

dedicated to China with more accurate data (Wang et al., 2016). The global Aqueduct BWS map uses FAO AQUASTAT country-aggregated water withdrawal data and spatially disaggregates it to sub-national level. In contrast, the new China version, known as the China BWS map, uses water withdrawal and consumption data of 345 administrative subdivisions nationally; thus, it is more granular.

Although, no obvious differences between the two maps are visible at a glance, the China BWS map does capture the levels of water stress in the Kuye River basin. Accordingly, the water stress levels in this region increase from "medium to high" to "high." As a consequence, for all ten companies, the exposure to water stress shifts. The changes resulting from CWR's re-mapping of the Coal-5 and Power-5 assets onto the new map are set out in Fig. 11.

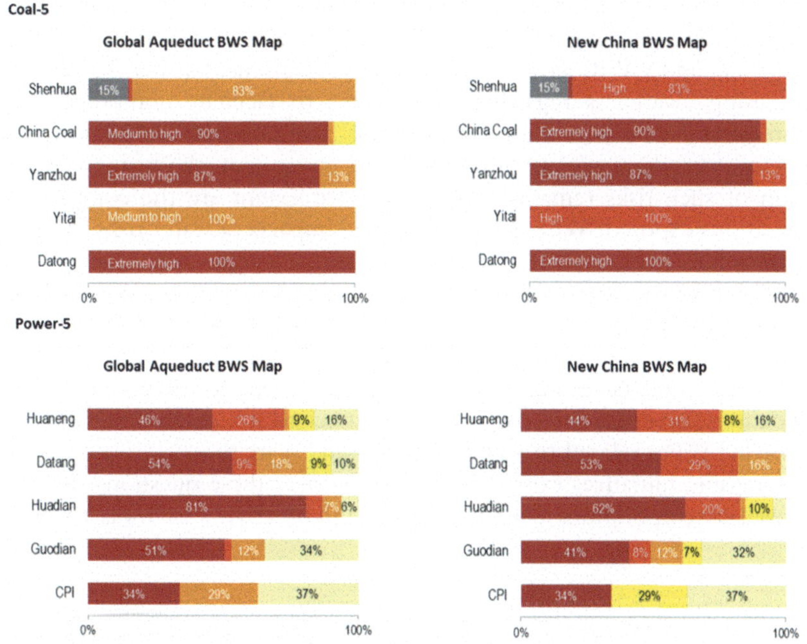

Fig. 11 Water stress levels in the Kuye River basin—Increase from New China BWS map (*Note* This figure has been reproduced for this chapter with the kind permission of CWR. *Source* CWR [2016])

It becomes clear that Coal-5 output was more exposed when using the China BWS map whereas the opposite could be argued for the Power-5.[7]

- **Water stress indicators should be analyzed along with government action.** High water stress can indicate possible regulatory risks which could impact operations in the short term. However, this action could also protect water resources in the region in the long term so water stress indicators must be analyzed along with this government action. For example, Singapore is extremely water-stressed; however, it is not at risk due to the government's recognition of the risks and management of its resources.

4 Conclusion

Over the last few years, much research has highlighted the growing water-related risks for industrial sectors. Power and mining sectors are on the front line and CWR, along with other organizations, has led the charge in raising awareness regarding water risks to facilitate comprehensive management of water issues. Various tools have also been developed.

However, despite years of research and investment, the financial valuation of such risks remains elusive. To what extent are the companies' operations exposed? What are the potential impacts of water risks to their bottom line?

With investors increasingly concerned about water risks (CWR, 2016), financial institutions say it is time to take a step forward—from raising awareness to quantifying geo-spatial exposure to water-related risks (UBS, 2016a). By understanding these risks and accounting for them in financial valuations, investors and corporations can act to mitigate them, thereby resolving issues related to water pollution and scarcity.

Ongoing initiatives exist that can help answer these questions. NGOs as well as corporations have developed different types of tools to help

[7] The China BWS map should only be used to compare assets within China. When comparing assets located outside China on Aqueduct BWS to assets in China mapped on China BWS, the China assets may be unfairly penalized due to increased granularity in water use data. It is also important to note here that the shadow pricing tools discussed in the previous section are based on the global Aqueduct BWS map and not the China BWS. Given the shift of the assets toward more water stress in the new map, shadow prices should be even higher for the Coal-5.

financial institutions, investors, and businesses better gauge their portfolio or company's water risks. These tools range from disclosure and water risk management guidelines, the mapping of water risk and water stress, to those that drop out dollar values for the risk exposure using the concept of shadow pricing (CWR, 2016). While still in their infancy, these tools can collectively provide a convenient framework to screen water risks across portfolio companies.

As water risks vary across industries, the easiest starting point is to assess the most water-intensive industries first and benchmark assets/companies against each other after. The power and mining sectors were analyzed in this chapter as they are large industrial users of water, and their exposure to water risks is relatively better understood. Indeed, since 2016, various banks and asset managers have assessed and factored in water risk in their power asset valuations, e.g., UBS (2016b) and BlackRock (2020).

However, all companies operating in water-stressed locations face risks as water issues are locational. They are exposed to local hydrological characteristics, water use competition, and regulations.

The approaches mentioned in this chapter can help kick start water risk assessments from different angles:

- The shadow pricing method focuses more on the profit or loss of a company. Different factors such as population, water use competition, and economy are quantified to obtain a shadow price, which is then reflected in the income statement in form of increasing operating expenses. As the calculated shadow price is currently high compared to reality, the reliability of this method is questionable (CWR, 2016). However, shadow pricing still can be used as a warning signal of possible impending water scarcity or regulations affecting P&L instead of a real water price.
- The balance sheet exposure assessment focuses more on the water risk of a company's assets, which are geographically mapped on water risk maps. Different types of maps can be used, such as baseline water risk maps, water pollution maps, flood maps, etc. As most of the analysis is based on the map used, the quality of the map matters, especially the granularity.
- In addition, local regulatory risks faced by the sector must also be evaluated as they can be disruptive in the short term but could provide long-term water stress alleviation as discussed previously.

Indeed, investors surveyed regarding the coal and power sectors in China were more concerned over regulatory risk with 59% wanting to commission/conduct more research on such risks over physical risks (41%) (CWR, 2016).

Finally, the quality of corporate disclosure underpins both approaches: water risk is location-specific, yet companies seldom disclose facility level water data. There is a long way ahead before this level of information is disclosed. Until then, the best available data must be used for estimations.

Nevertheless, these methodologies can be a good first step to help identify key physical risks and determine regulatory water risks and benchmark exposure of companies within a sector. It should be noted that despite the varying views on materiality and "usefulness" of both methodologies, feedback from the financial sector, as a peer group, revealed that they were keen to see the impact on the companies. Such water risk analysis performed on the peer group allowed them to compare the companies' exposure vis-à-vis each other and gauge their relative risk.

The ability to compare and benchmark companies' exposure to water risk within a sector, as well as the opportunity to cross-reference exposure across sectors, are just as, if not more important than the absolute impact on profitability. Such analysis thus provides a sound start for better-informed investment decision making and for engaging corporations in improving risk management strategies. Further research and development on the various tools will only help improve and streamline this in the future.

References

BlackRock. (2020). *Troubled waters*. https://www.blackrock.com/us/individual/insights/blackrock-investment-institute/troubled-waters.

Brown, A., & Matlock, M. (2011). *A review of water scarcity indices and methodologies*. http://www.sustainabilityconsortium.org/wp-content/themes/sustainability/assets/pdf/whitepapers/2011_Brown_Matlock_Water-Availability-Assessment-Indices-and-Methodologies-Lit-Review.pdf.

CDP. (2013). *Metals & mining: A sector under water pressure*. https://www.chinawaterrisk.org/research-reports/metals-mining-a-sector-under-water-pressure-2/.

CLSA. (2013). *Water for coal: Thirsty miners will share the pain*. https://www.chinawaterrisk.org/research-reports/water-for-coal-thirsty-miners-will-share-the-pain/.

Columbia Water Center. (2015). *Mining and water risk: Diagnosis, benchmarking and quantitative analysis of financial impacts*. http://water.columbia.edu/2015/11/04/mining-and-water-risk-diagnosis-benchmarking-and-quantitative-analysis-of-financial-impacts/.

CWR. (2016). *Toward water risk valuation—Investor feedback on various methodologies applied to 10 energy ListCo's*. https://chinawaterrisk.org/wp-content/uploads/2016/08/China-Water-Risk-Report-2016-Toward-Water-Risk-Valuation-WebM.pdf.

CWR & IRENA. (2016). *Water use in China's power sector: Impact of renewables and cooling technologies to 2030*. http://chinawaterrisk.org/notices/cwr-irena-water-use-in-china-power-sector/.

Ecolab. (2015). *Water risk monetizer methodology*. https://waterriskmonetizer.com/wp-content/uploads/2015/08/Water-Risk-Monetizer-Methodology-August-2015.pdf.

Fernandes, A., & Krishna, J. (2016). *Investor briefing—Water shortages threaten coal company revenues*. http://www.indiaenvironmentportal.org.in/files/file/Investor%20Briefing%20Water%20and%20Coal.pdf.

Gassert, F., Landis, M., Luck, M., Reig, P., & Shiao, T. (2014). *Aqueduct metadata document—Aqueduct global maps 2.1*. http://www.wri.org/publication/aqueduct-global-maps-21.

Hao, F., Chang, Y., & Ning, D. (2003). 中国耕地资源面临的挑战与可持续利用对策. *环境保护*, *4*, 30–33.

HSBC. (2015). *No water, more trade-offs—Managing China's growth with limited water*. HSBC commissioned CWR to provide water risk analysis.

Qin, Y., Curmi, E., Kopec, G., Allwood, J., & Richards, K. (2015). China's energy-water nexus—assessment of the energy sector's compliance with the "3 Red Lines" industrial water policy. *Energy Policy, 82*, 131–143.

Roehrkasten, S., & Schaeuble, D. (2015). *Secure and sustainable power generation in a water-constrained world*. https://www.iass-potsdam.de/en/output/publications/2015/secure-and-sustainable-power-generation-water-constrained-world-policy.

UBS. (2016a). *Are investors pricing in water risk? A geospatial perspective*.https://neo.ubs.com/static/login.html?origin=%2f&utm_source=China+Water+Risk&utm_campaign=15a33d5547-China_Water_Risk_November_2016_Newslette10_19_2016&utm_medium=email&utm_term=0_caee821f95-15a33d5547-363521569.

UBS. (2016b). *Is China consuming too much water to make electricity?* https://neo.ubs.com/static/login.html?origin=%2f&utm_source=China+Water+Risk&utm_campaign=15a33d5547-China_Water_Risk_November_2016_

Newslette10_19_2016&utm_medium=email&utm_term=0_caee821f95-15a33d5547-363521569.

UNEP FI. (2012). *Chief liquidity Series III: Extractives sector*. https://www.chinawaterrisk.org/research-reports/chief-liquidity-series-extractives-sector/.

van Vliet, M., Wiberg, D., Leduc, S., & Riahi, K. (2016). Power-generation system vulnerability and adaptation to changes in climate and water resources. *Nature Climate Change*, 375–381. https://doi.org/10.1038/nclimate2903.

Wang, J., Zhong, L., & Long, Y. (2016). *Baseline water stress: China*. World Resources Institute. https://files.wri.org/s3fs-public/Baseline_Water_Stress_China_0.pdf.

WBCSD. (2013). *Business guide to water valuation*. http://www.wbcsd.org/Pages/EDocument/EDocumentDetails.aspx?ID=15801.

WRI. (2016). *Water resource management in China coal industry: Policy summary*. https://www.wri.org.cn/sites/default/files/coal%20water%20policy_CN_S.pdf.

Yellow River Conservancy Commission. (2013). *Yellow river basin integrated planning (Chinese)*. http://www.yellowriver.gov.cn/special/gwypf/.

Zhang, C., Zhong, L., Fu, X., Wang, J., & Wu, Z. (2016). Revealing water stress by the thermal power industry in China based on a high spatial resolution water withdrawal and consumption inventory. *Environmental Science and Technology, 50*(4), 1642–1652. https://doi.org/10.1021/acs.est.5b05374.

Financial Implications of Parched Power: Insights from an Analysis of Indian Thermal Power Companies

Lihuan Zhou, Jack McClamrock, Giulia Christianson, Deepak Krishnan, and Tianyi Luo

1 Introduction

Climate change poses risks to investments across asset classes, which can result in financial losses and economic disruption; as climate change advances, these risks are becoming more urgent and apparent. Climate change risks can be divided into two major categories: transition risks related to the transition to a lower-carbon economy and physical risks related to the physical impacts of climate change (TCFD, 2017). As climate change unfolds, the exact timing, location, and severity of future physical impacts are difficult to estimate, which makes measuring

---

This chapter was adapted from the working paper, *Financial Implications of Parched Power: Insights from an Analysis of Indian Thermal Power Companies*, published by the World Resources Institute.

L. Zhou (✉) · J. McClamrock · G. Christianson · D. Krishnan · T. Luo
World Resources Institute, Washington, DC, USA
e-mail: Lihuan.zhou@wri.org

T. Walker et al. (eds.), *Water Risk and Its Impact on the Financial Markets and Society*, Palgrave Studies in Sustainable Business In Association with Future Earth,
https://doi.org/10.1007/978-3-030-77650-3_5

and pricing physical risks uniquely challenging for both companies and investors.

An essential element of investment decision-making is pricing risk to support informed capital-allocation decisions. Therefore, institutional asset owners—pension plans, insurance companies, sovereign wealth funds, large family offices, endowments, foundations—and the asset managers who frequently manage their money have started to demand and use available data to put a price on climate-related risks. However, despite recent progress with climate-related financial disclosures, it is not yet common for companies to disclose information about physical risks, much less quantitatively measure them.[1]

Water-related risks, including constraints on water availability, accessibility, and quality, are prominent physical risks that affect a wide range of industries. However, not all sectors are affected by water-related risks in the same way. In this paper, we focus on the electricity generation industry, which is particularly exposed to acute water shortage risks because thermal and hydro electricity production requires a reliable, abundant, and predictable source of water (Feeley & Ramezan, 2003).[2] Water shortages can cause costly outages for thermal and hydro power plants, which can lead to reductions in revenues, increased capital expenditures, and constraints on future growth.

The primary objective of this paper is to enable investors to better measure water shortage risks and to support their engagement efforts to influence corporate decision-making around water issues. To accomplish this objective, we independently developed a methodology to assess the financial impacts of water shortage-induced outages, with outages defined as instances when a normally operating generation unit is unable to produce electricity.[3] By producing historical and forward-looking analyses, the paper also provides insights to companies on how to analyze and disclose decision-useful information in line with recommendations of the Task Force on Climate-related Financial Disclosures (TCFD).

[1] An analysis of 124 ASX200 companies in the most exposed sectors to climate-related risks found that none of the 124 companies provided quantitative disclosure around physical risks (EY, 2018).

[2] A thermal power plant generates electricity from heat (thermal) energy, including coal, petroleum, natural gas, solar thermal energy, geothermal energy, and nuclear energy (Occupational Safety & Health Administration, 2002).

[3] We did not engage with companies in developing and testing this methodology.

The remainder of this paper is structured as follows: Sect. 2 outlines our approach. Section 3 provides the historical and forward-looking analyses. Section 4 discusses implications of the analyses for investors and thermal power companies. Section 5 concludes our findings.

2 Approach

2.1 *Process and Scope*

Our methodology relies on publicly available data on power plants' daily outages and their causes. This type of high frequency data is seldom made public by companies or regulators. The Indian electric regulator, Central Electricity Authority, makes these data publicly available, and we leverage the data already collected by Luo et al. (2018) to apply our methodology to Indian thermal power companies. Our methodology deploys traditional financial and data analysis techniques to advance the findings of Luo et al. (2018).

First, in our historical analysis, we analyze the effects of water shortages on Indian thermal power companies' profits in the context of each company's historical financial performance, providing insights into the size of the impacts relative to overall financial performance. We refer to this section of the analysis as the historical financial analysis. Our historical financial analysis covers India's FY 2014–2017 (the financial year being April 1 through March 31).[4] We cover five companies that are publicly listed, with sufficiently detailed financial disclosures to apply our methodology. The Securities and Exchange Board of India (SEBI) regulates publicly listed companies in India and requires them to disclose quarterly and annual reports with audited financial statements. These documents served as the chief source of financial information used in this paper.[5]

We also analyze power plant cooling technologies and baseline water stress for all thermal power plants in India. This analysis provides further context for our financial analysis because these two variables are

[4] For example, FY 2014 is April 1, 2013, through March 31, 2014. Where possible, we have updated the outage and generation data from Luo et al. (2018), resulting in a historical analysis through the end of India's 2017 financial year.

[5] The five companies are Adani Power Limited, NTPC Limited, Reliance Power Limited, The Tata Power Company Limited, and Torrent Power Limited.

Table 1 General overview of companies in the analysis

Company	*Brief description*	*Headquarters*
Adani Power Limited	Power subsidiary of Indian conglomerate Adani Group	Gujarat, India
NTPC Limited	Indian's largest power generator	New Delhi, India
Reliance Power Limited	Power subsidiary of Indian conglomerate Reliance Anil Dhirubhai Ambani Group	Maharashtra, India
The Tata Power Company Limited	Power subsidiary of Indian conglomerate Tata Group	Maharashtra, India
Torrent Power Limited	Power subsidiary of Indian conglomerate Torrent Group	Gujarat, India

Note While all five companies are publicly listed on both the Bombay Stock Exchange and the National Stock Exchange of India, NTPC Limited is the only company for which India's central government is the majority owner.
Source Company websites, adapted by WRI

commonly associated with water shortage-induced outages (Luo et al., 2018).

Then, in our forward-looking analysis, we estimate the impact of potential future water shortages. We use regression analysis to quantify the relationship between drought and water shortage-induced outages using historical data and then estimate the impact of future drought conditions using climate projection data included in IPCC AR5. The forward-looking analysis provides projections for two future periods, 2019–2035 and 2036–2050, relative to the reference period 2006–2018.[6] These two periods were chosen based on the relative shorter-term time horizons of investment decision-making and relative longer-term periods required to show impacts from climate change.[7]

The historical and forward-looking analyses primarily cover a subset of the companies analyzed in Luo et al. (2018): Those five that are publicly listed, with sufficiently detailed financial disclosures to apply our methodology (see Table 1).

[6] We use relative changes to a reference period to mitigate the uncertainties between projections generated using different models or model versions, which is consistent with IPCC AR5 (Kirtman et al., 2013).

[7] Compared to natural internal variability.

We analyze plants across the broader Indian thermal power sector in several instances where more data were available. Our analysis of power plant cooling technologies and baseline water stress levels cover all thermal power plants in India. In our regression analysis, we analyze water shortage-induced outages for every Indian thermal power plant that had water shortage-induced outages during FY 2014–2017. Covering a wider sample for these research components provides more robust results than if we had only analyzed the five companies.

2.2 Data Sources

Our methodology uses a combination of thermal power company data at the sub-asset level (power plant generator unit-level), asset level (power plant-level), and corporate level, and both historical and forward-looking data on precipitation, evaporation, and drought indexes.[8] Table 2 outlines our data sources.

2.3 Methodology

2.3.1 Historical Analysis

In our historical financial analysis, we consider a counterfactual scenario in which no water shortage-induced outages had occurred at the five companies over the four-year period of the analysis. We use that counterfactual scenario to estimate both the potential losses in revenue and the potential losses in EBITDA due to water shortage-induced outages.[9]

To estimate potential losses in EBITDA, we calculate the gross margin on the sale of power and multiply the forgone increases in revenue (potential losses in revenue) by this margin to account for forgone increases in variable costs (fuel and power purchased) from the counterfactual scenario. We calculate the forgone increases in revenue by multiplying water-shortage-induced outages by the rate of sale. Fuel and power purchased are the two largest components of variable costs, which account for 50–75% of overall costs.

[8] A power plant can be formed by more than one unit (generator). Each unit can generate independently.

[9] EBITDA is a widely used financial earnings indicator that backs out the effects of corporate financing and accounting decisions seen in the interest, taxes, depreciation, and amortization line items in the income statements of companies (Verriest et al., 2018).

Table 2 Data used in the analysis

Data	*Time frame*	*Frequency*	*Availability*	*Sources*
Unit-level and plant-level rate of sale of power data	Financial Year 2014–2017	Annual	Public	CEA data compiled by WRI authors
Unit-level outage data (water shortages and other)	FY 2014–2017	Daily	Public	CEA data compiled by WRI authors
Unit-level capacity, built year, operating status, fuel, business type, and ownership data		Fixed	Proprietary	The Platts World Electric Power Plants Database
Unit-level cooling technology		Fixed	Public	Luo et al. (2018)
Corporate-level financial statements	FY 2014–2017	Quarterly	Public	Corporate annual reports and quarterly financial statements compiled by WRI authors
Drought indexes	FY 2014–2017	Monthly	Public	Aadhar and Mishra (2017)
Precipitation and evaporation projections under RCP scenarios	Calendar year 2006–2050	Monthly	Public	Knutti (2013a, 2013b, 2013c, 2013d)

Source WRI

We then compare potential losses in revenue and EBITDA to actual historical figures. This comparison provides an estimate of the financial impacts of water shortage-induced outages. Figure 1 outlines the key steps in our historical financial analysis methodology.

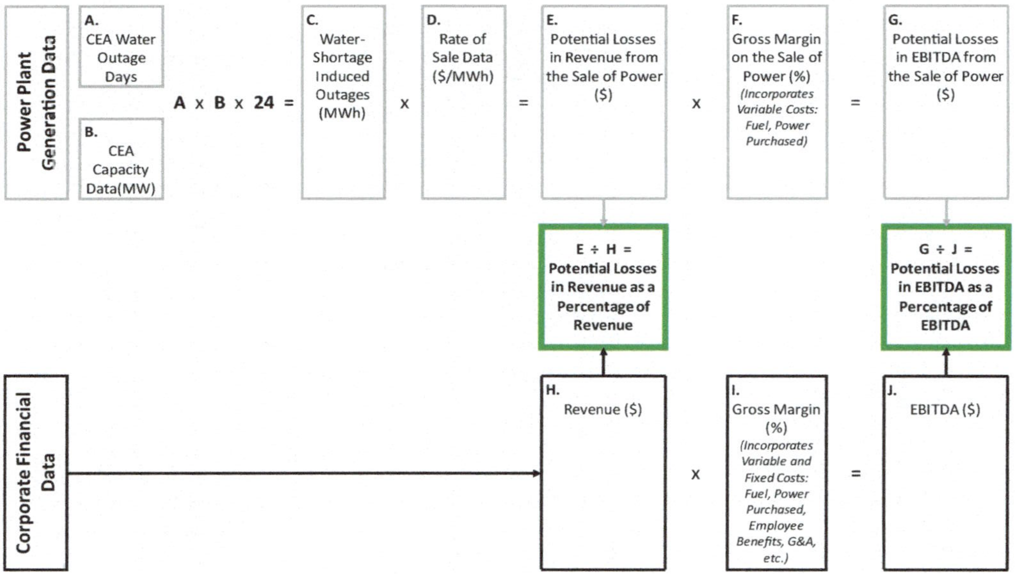

Fig. 1 Overview of historical financial analysis methodology to estimate the financial impacts of water shortage-induced outages at thermal power plants (*Note* CEA refers to India's Central Electricity Authority, which was our source of information for outage days and plant capacity data. EBITDA refers to earnings before interest, taxes, depreciation, and amortization. We assume that during days of forced water shortage-induced outages, power plants could have operated at full capacity for 24 hours. This assumption aligns with the methodology used in Luo et al. [2018]. *Source* WRI)

Table 3 Descriptive statistics for the forward-looking analysis dataset

Variable	*Mean*	*St. dev.*	*Min*	*Pctl(25)*	*Pctl(75)*	*Max*
Days of outages	1.9	6.7	0	0	0	31
Installed capacity	365.2	194	110	210	600	660
Age	13.8	13.3	0	2	28	44
$\log_e$(population)	14.6	0.7	13.3	14.3	15	15.9
12-month SPEI	0.5	0.8	−2.2	−0.1	1	2.6
Number of power plant units	53					
Total observations	2,287					
Number of companies	17					
Once-through cooling units	189%					
Recirculating cooling units	811%					

Source WRI

2.3.2 *Forward-Looking Analysis*

We first conduct regression analysis to quantify the relationship between a drought index and water shortage-induced outages.[10] We include all power plant units in India that experienced water shortage-induced outages in FY 2014–2017. Our dependent variable is the number of days of water shortage-induced outages in each month for each power plant unit. Our key explanatory variable is a drought index (the 12-month SPEI), which is based on water balance (precipitation minus evapotranspiration) and is commonly used to analyze drought events (Pepler, 2015; Vicente-Serrano et al., 2010; World Meteorological Organization, 2012).

We use monthly observations of the drought index in the analysis. Table 3 provides descriptive statistics for our analytic dataset. Thirty-seven of the fifty-three power plant units were built before April 1, 2013, and we have no missing observations for them from April 2013 to March 2017 (Platts, 2018). Sixteen power plants were commissioned between 2014 and 2017 (Platts, 2018); therefore, these units were not operational in some time periods and we have missing values (an unbalanced panel), which create potential sample selection problems.

The main model of our regression analysis is an unbalanced panel Poisson model with fixed effects using all available observations from the

[10] At 0.05° spatial resolution.

fifty-three power plant units.[11] We chose a Poisson specification because of its attractive properties for estimating integer and non-negative data (count data) compared to Ordinary Least Squares regressions.[12] We use power plant unit-level fixed effects to control for unobserved, time-constant power plant unit characteristics, such as power plant management quality and water management quality, to solve potential omitted variable problems.[13] To further test the robustness of the main model, we also estimate random effects and pooled Poisson models and obtain very similar results.

The fixed-effects model includes month dummy variables, τ, the 12-month SPEI, *SPEI*, the age of power plant unit, *age*, and the error term, ε. Using the exponential as the mean function, we have:

$$\ln Days\,of\ Outage_{it} = \beta_0 + \beta_1 SPEI_{it} + \beta_2 age_{it} + \tau_m + \varepsilon_{\text{it}}$$

where i is for power plant unit, i and t measure year-month.

Using the exponential as the mean function, we have Poisson models with random effects and pool Poisson models in the following form:

$$\begin{aligned}\ln\,(Days\,of\ Outage)_{it} \\ &= \beta_0 + \beta_1 SPEI_{it} + \beta_2 age_{it} + \beta_3 mw_i + \beta_4 coolingtype_i \\ &\quad + \beta_4 \log_e(population)_i + \beta_5 company_i + \tau_m + \varepsilon_{it}\end{aligned}$$

where i is for power plant unit, i and t measure year-month.

The model includes month dummy variables, τ, the 12-month SPEI, *SPEI*, the age of power plant units, *age*, the installed capacity in MW, *mw*, the cooling type, *coolingtype*, the natural log of local district population in 2011, $ln(population)_i$, the owner, *company*, and the error term, ε. We include month dummy variables in all models to control for water-demand factors, such as seasonal agriculture water use.

[11] A Poisson regression is a generalized linear model form of regression analysis used to model outcomes that are count data (e.g., the number of days of water shortage-induced outages). Fixed-effect Poisson models are used to control for certain fixed effects (e.g., power plant unit characteristics).

[12] It is normally called count data in econometric literature.

[13] If there is an omitted variable problem, we cannot consistently estimate coefficients, nor will we be able to tell if there is a problem. However, we are not able to include time-constant power plant characteristics, such as cooling technology, in fixed-effects models.

We estimate these fixed-effects models using the balanced data to supplement our main model (an unbalanced panel Poisson model with fixed effects) as the unbalanced dataset creates potential sample selection problems. In the unbalanced dataset, the sample selection is correlated with the dependent variable, the age of the power plant unit. The newer the power plant unit, the more likely that it was not operational at the beginning of our observation period and has missing values. As a result, the missing observations are not random, which might cause more weight to be given to the plants where there is a complete dataset. To mitigate potential sample selection problems, we estimate models using the balanced data only, which are a subset of the unbalanced data and include 37 power plant units or 1,776 observations. We obtain very similar results to the unbalanced models.

We use a robust variance matrix estimator to calculate standard errors for the coefficients estimated in all models. Specifically for the fixed-effects model, the robust variance matrix estimator allows for any deviations from the Poisson distribution and arbitrary time dependence, except for the conditional mean assumption of strict exogeneity (Wooldridge, 2010).

We considered adding additional water demand variables in the models, such as seasonal agriculture water use and residential water use, but we could not find data with high-enough spatial and temporal resolutions. To control for these water demand factors, we include monthly dummy variables in all models.

After quantifying the relationship, we estimate the impact of future water shortages for the five publicly listed companies using climate projection data included in IPCC AR5. We take the following steps to estimate the future impact, corresponding to the right side of Fig. 2.

1. We calculate historical simulations and projections of water balance (precipitation minus evaporation) from 2006 to 2050 for the location (longitude and latitude) associated with each power plant from 40 climate models for all four RCP scenarios: High Mitigation Scenario (RCP2.6), Medium Mitigation Scenario (RCP4.5), Low Mitigation Scenario (RCP6.0), and Business-as-Usual (BAU) Scenario (RCP8.5) (Knutti, 2013a, 2013b, 2013c, 2013d; Vicente-Serrano et al., 2010).

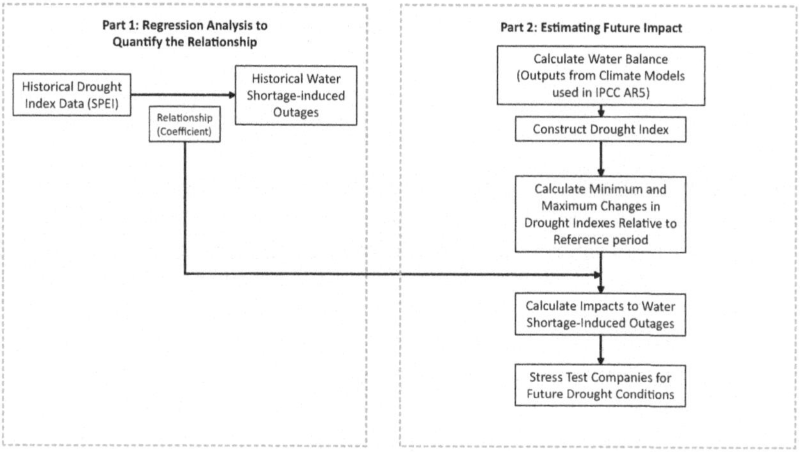

Fig. 2 Overview of forward-looking analysis methodology to estimate the impacts of future water shortage-induced outages at thermal power plants (*Source* WRI)

2. We construct drought indexes for each power plant using the projections of water balance and the R SPEI package (Vicente-Serrano et al., 2010).[14]
3. We calculate minimum and maximum values of changes in drought indexes for the periods 2019–2035 and 2036–2050 for each power plant relative to the reference period: 2006–2018. We use minimum and maximum values to capture the full uncertainties of scenario experiments. We use relative changes to a reference period to mitigate the difference between projections generated by different models or model versions, which is consistent with IPCC AR5 (Kirtman et al., 2013).
4. We calculate the impact of future climate change on water shortage-induced outages by multiplying the changes in drought indexes by the regression coefficient and holding all other variables constant.
5. To put the impacts in context and stress test them on companies, we calculate what historical water shortage-induced outages would

[14] The drought indexes included in our regression models are calculated using the same statistical package (Aadhar & Mishra, 2017).

have been if power plants had experienced the worst-case drought conditions.

2.4 Assumptions and Limitations

This chapter analyzes a small subset of Indian thermal power companies. As outlined in Luo et al. (2018), power plant outage data availability precludes us from analyzing hydroelectric and captive thermal plants as well as other types of water-related risks such as water quality.[15] We do not consider efficiency losses caused by water shortages, water temperature, and ambient temperature. The efficiency loss calculation requires detailed design information of each power plant that we cannot obtain or reliably estimate. We focus strictly on financial impacts to companies' revenue and EBITDA, which are two of the most important financial metrics used by investors to evaluate companies (FINRA, 2015). Multiple internal and external factors affect stock prices or analyst valuations. As a result, we do not consider how outages might have affected companies but focus on the two fundamental metrics. There are other limitations and assumptions in our analysis that we wish to highlight.

2.4.1 Historical Analysis

Key assumptions and limitations from our historical analysis:

- **Sample size**: Our sample of publicly listed companies limits our ability to derive definitive conclusions for the Indian and broader thermal power sectors.
- **CEA outage reasons**: We rely on outage reasons published by the CEA to calculate water shortage-induced outages. However, CEA data, as currently structured, does not require companies to report the underlying causes of specific outage reasons, so we cannot definitively determine that all water shortage-induced outages occurred

[15] A captive power plant is a facility that is dedicated to providing electricity for a single user. It is usually used and managed by an industrial or commercial energy user for its own energy consumption.

due to drought.[16] Furthermore, as these outage reasons were self-reported by companies, it is possible that water shortage-induced outages could be underreported.[17] Lastly, water shortage-induced outages might also have occurred that we were not able to identify based on the CEA published outage reasons.[18]

- **CEA rate of sale data**: In India, the rate of sale of electricity, referred to by the CEA as "Generation Tariff," is regulated at the power plant level, based on provisions in The Electricity Act, 2003. We use CEA unit- and plant-level rate-of-sale data from FY 2014–2017 to calculate potential revenue losses.
- **Plant capacity**: To calculate water shortage-induced outages, we assume that each power plant would have operated at full capacity for 24 hours a day for each day that it experienced a water shortage-induced outage. We make this assumption based on the fact that thermal power plants were usually used as base load power sources and would operate continuously to meet the minimum level of power demand 24 hours a day if they were not forced to shut down.

[16] For example, "Raw Water Not Available/Low Intake Canal" could be due to drought conditions, but this is not clarified in the data. According to company disclosures and news reports, however, drought conditions may have caused many of the water shortage-induced outages at the five companies in our historical analysis (Adani Power Limited, 2016; Das, 2016). We include the following outage reasons as water shortages: "raw water problem," "raw water not available/low intake canal," "D.M. water problem," "raw water supply misc.," "raw water shortage," "water shortage due to drought." This aligns with the methodology outlined in Luo et al. (2018).

[17] According to CEA (2016), outage reasons were self-reported by companies. In instances where outages could be attributed to multiple factors, companies might not report outages due to water shortages. Instead, they might attribute the outage to another factor, like a coal shortage, to potentially recoup losses from the party at blame for the shortage (Anonymous, interviewed by the authors, World Resources Institute, Washington, DC, October 9, 2018). In some outage cases, companies might first report outages due to water shortages but could attribute later outages to other reasons (e.g., reserve shutdown) even if the power plant unit continued to be affected by water shortages (Dharmadhikary & Sandbhor, 2017). Reserve shutdowns include outages caused by threats to grid security, low demand, transmission congestion, and other anticipated reasons (Luo et al., 2018).

[18] For example, CEA-reported outage reasons such as "generator cooling system failure" might have been caused by water shortages. The outage reason, "generator cooling system failure," was reported as "gen. cooling system failure" by CEA. We determined that the abbreviation, gen, is for generator based on other outage reasons that also include gen, including "gen. transformer tripping/damaged" and "gen. transformer protection relay."

- **Variable costs**: We consider fuel and power purchased costs variable in our analysis, and all other costs are fixed.[19] We also use annual average prices for fuel and power purchased costs in our calculation as daily prices were not available.

2.4.2 Forward-Looking Analysis

The forward-looking analysis has two major limitations. First, the regression results are applicable only to India. However, the regression method and model specifications are universal and flexible and could be applied to other countries or regions if comparable data are available. Second, the regression analysis only includes power plants that had water shortage-induced outages during FY 2014–2017.[20] Therefore, the regression results are only applicable to the power plants that had water shortage-induced outages. This means that the analysis may underestimate the total potential losses from future water shortages to companies if they extend to areas beyond where they are historically present.

The forward-looking analysis also relies on numerous assumptions. The main assumption is that we held all variables constant in calculating the impact of future climate change on water shortage-induced outages. This assumption does not allow for considerations of broader power market dynamics and socioeconomic changes at the local level. We also assume that power plants would source their cooling water within 0.05° (approximately 3.4 miles), but some of them might source water beyond this distance.

Climate projections bring additional uncertainties to the forward-looking analysis. Climate models are extremely sophisticated computer programs that simulate the complex interactions among ocean, atmosphere, land, etc. Every complexity in the models introduces sources

[19] Fixed costs in our analysis generally include employee benefits expenses, general and administrative expenses, and cost of materials consumed, among others.

[20] We do not try to define the factors that might induce a power plant to have water shortage-induced outages in the first place but only the factors that decreased or increased the number of days of outages, given that the plant experienced outages. Therefore, we exclude power plants that did not have water shortage-induced outages in this period for the regression analysis. In addition, it would have resulted in low-confidence (high standard errors) coefficient estimations by including power plants that did not have power outages in this period as a large number of dependent variables would have been zero and had zero variability.

of possible error and leads to model uncertainty. The complexity and uncertainty partially cause a wide range of projections.

3 Analysis

3.1 Historical Analysis

Water shortages caused estimated potential negative impacts as high as 17.4% of EBITDA for one company during one quarter but did not cause persistent financial impacts at the five companies in our historical financial analysis.[21] Adani Power and Reliance Power experienced three quarters with water shortage-induced outages while NTPC experienced five. Only NTPC experienced water shortage-induced outages in contiguous quarters (second quarter of 2016 through the first quarter of 2017). Tata Power and Torrent Power are excluded from the results of the financial analysis because they did not experience any water shortage-induced outages in the FY 2014–2017. Figure 3 outlines the results of the financial analysis.

Three of the five companies in our historical financial analysis experienced outages of varying magnitudes and durations that affected a total of six of their plants in different magnitude. The three companies in our financial analysis that experienced water shortage-induced outages collectively owned 35 thermal power generation plants as of FY 2017 (Fig. 4). Figure 5 illustrates the frequency, magnitude, and duration of outages at the different plants. Two out of the three longest stretches of water shortage-induced outages occurred during the first quarter of FY 2017 (April–June 2016). Adani's Tiroda plant experienced outages for 51 days from May to June 2016 while NTPC's Farakka plant experienced outages for 12 days during the first half of April 2016. These stretches during the first quarter of FY 2017 resulted in the largest potential losses in EBITDA on a quarterly basis for both Adani and NTPC, respectively. Reliance Power experienced its largest potential loss in EBITDA during the third quarter of FY 2017 (October–December 2016) when its Sasan plant shut down for five days during December 2016.

[21] Impacts to EBITDA are more severe, on a percentage basis, than impacts to revenue. As we assume that certain costs are fixed and adjust certain costs that we assume are variable, logic holds that potential EBITDA losses as a percentage of EBITDA are slightly higher than potential revenue losses as a percentage of total revenue. In effect, the margin on potential revenue losses is higher than the overall operating margins of the companies.

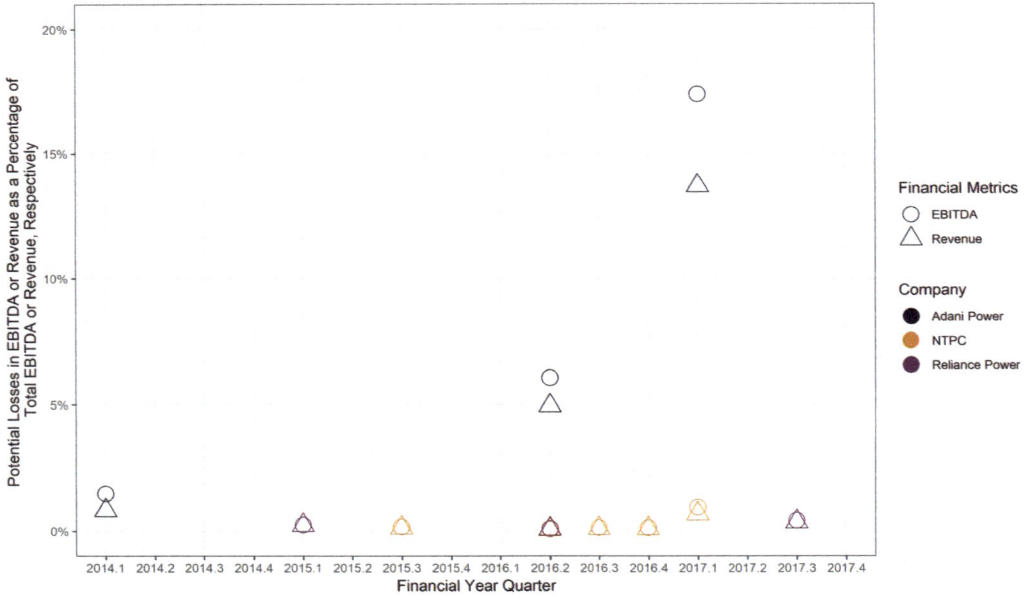

Fig. 3 Potential water shortage-induced losses in revenue and EBITDA as a percentage of total revenue and EBITDA, respectively (*Notes* The chart only includes companies during quarters in which they experienced water shortage-induced outages. Tata Power and Torrent Power, while part of the historical financial analysis sample, are not shown in the figure because they did not experience any water shortage-induced outages within this timeframe. India's Financial Year is from April 1 through March 31, so the first quarter of FY 2014 is April 1, 2013, through June 30, 2013. *Source* WRI, using company financial and CEA outage data)

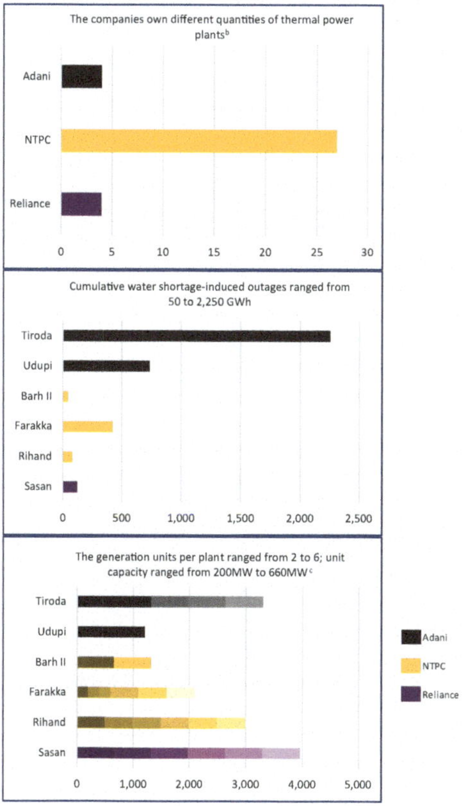

Fig. 4 Context underlying water shortage-induced outages of three Indian companies, FY 2014–2017[a] (*Notes* [a]Tata Power and Torrent Power, while part of the historical financial analysis sample, are not shown in the figure because they did not experience any water shortage-induced outages within this time-frame. India's Financial Year is from April 1 through March 31, so FY 2014 is April 1, 2013, through March 31, 2014; [b]Includes thermal power plants owned and operated as of FY 2017. Plants that were sold or decommissioned during FY 2014–2017 are not included; [c]Different shades represent the different capacities of each generation unit at each plant. Barh II refers to the second stage of the Barh Super Thermal Power Station, which consists of two 660 MW units. The first stage consists of three additional 660 MW units but was not yet completed as of December 2018. *Source* WRI, using company financial and CEA outage data)

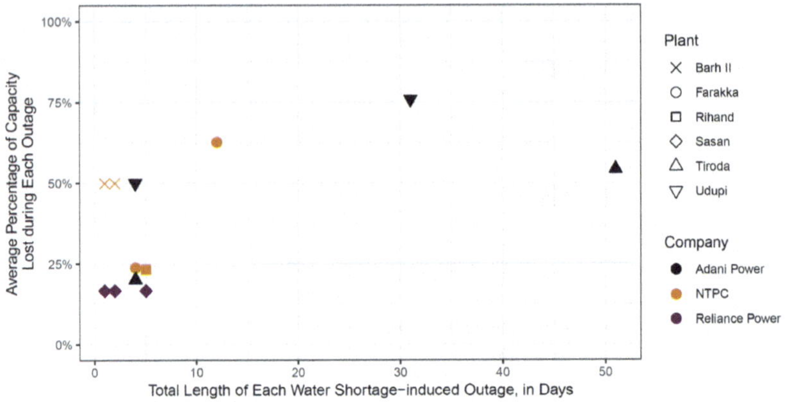

Fig. 5 Frequency, magnitude, and duration of water shortage-induced outages at three Indian companies, FY 2014–2017 (*Note* The percentage of total generation capacity lost is averaged across the duration of the shutdown. Udupi was the only plant to completely shut down; it lost both of its 600 MW units for 16 days of a 31 consecutive-day shutdown during August 2015 (Q2'16). *Source* WRI, using company financial and CEA outage data)

3.2 Forward-Looking Analysis

3.2.1 The Relationship Between Drought and Water Shortage-Induced Outages

We find consistent and robust evidence from historical data that an increase in drought severity is associated with increased water shortage-induced outages at the thermal power plants in India. Our regression analysis finds that an increase in drought severity at a power plant (a unit decrease in the drought index) was associated with 71% more days of water shortage-induced outages, while holding other variables constant (see Table 4).[22] We are 95% confident that the increase in the days of water shortage-induced outages was between 35 and 117 percent greater. We obtain very similar results in five additional model specifications, demonstrating the robustness of the result. The dependent variable, 12-month SPEI, is statistically significant at 0.001 level in all six regressions, and the estimations only vary slightly (from −0.537 to −0.491).

22 $exp^{[(-0.537)\times(-1)]} - 1 = 0.71$.

Table 4 Estimated effects of the drought index on the number of days of water shortage-induced outages by Poisson models

	All observation (Unbalanced)			*Complete observation only (balanced)*		
	Fixed effects (Main model)	*Random effects*	*Pooled*	*Fixed effects*	*Random Effects*	*Pooled*
12-month SPEI	−0.537***	−0.532***	−0.515***	−0.525***	−0.519***	−0.491***
	(0.122)	(0.123)	(0.119)	(0.131)	(0.133)	(0.133)
Age of power plant	0.226*	0.199	0.0217**	0.205	0.186	0.0239**
	(0.107)	(0.108)	(0.00814)	(0.107)	(0.107)	(0.00773)
Installed capacity		0.00657	0.00179		0.00680	0.00342
		(0.00505)	(0.00290)		(0.00508)	(0.00257)
$\log_e(population)$		−1.571	−5.964***		−2.281	−7.152***
		(1.902)	(1.375)		(2.130)	(1.599)
Once-Through Cooling Technology (Baseline)		0	0		0	0
		(.)	(.)		(.)	(.)
Recirculating Cooling Technology		0.673	−7.481**		−0.627	−9.456***
		(1.817)	(2.303)		(2.740)	(2.710)
Company Fixed Effects						
Adani Power (Baseline)		0	0		0	0
		(.)	(.)		(.)	(.)
Chattisgarh State Power		−6.179***	−3.040**		−5.448**	−2.103*
		(1.796)	(1.080)		(1.699)	(0.881)
Durgapur Projects LTD		4.053	10.07***			
		(4.180)	(3.050)			

(continued)

Table 4 (continued)

	All observation (Unbalanced)			*Complete observation only (balanced)*		
	Fixed effects (Main model)	*Random effects*	*Pooled*	*Fixed effects*	*Random Effects*	*Pooled*
Essar Power Ltd		0.976	3.575***		1.670	4.639***
		(1.290)	(0.995)		(1.439)	(1.153)
GMR Group		3.411	4.103**			
		(2.175)	(1.446)			
Gujarat State Elec Corp Ltd		1.787	2.063			
		(2.464)	(1.552)			
Karnataka Power Corp Ltd		−1.117	2.218		0.227	3.713**
		(1.506)	(1.379)		(1.494)	(1.368)
Maharashtra State Power Gen Co.		1.914	7.660***		2.523	9.516***
		(2.136)	(1.708)		(2.340)	(1.869)
MP Power Generating Co Ltd		0.185	0.281			
		(2.193)	(0.999)			
NTPC Ltd		−1.273	1.039		−0.860	1.877*
		(2.847)	(0.879)		(2.725)	(0.885)
Rajasthan RV Utpadan Nigam		0.759	-5.472**		1.282	-5.912**
		(2.589)	(2.041)		(2.785)	(2.152)
Reliance Power Ltd		−2.716***	-2.959***			
		(0.750)	(0.598)			

	All observation (Unbalanced)			*Complete observation only (balanced)*		
	Fixed effects (Main model)	*Random effects*	*Pooled*	*Fixed effects*	*Random Effects*	*Pooled*
Simhapuri Energy Pvt Ltd		2.674	4.118		−2.559	−5.103***
		(3.101)	(2.114)		(2.094)	(1.356)
Vedanta Resources PLC		−1.442	−4.247***		−.859	−6.909***
		(1.436)	(1.198)		(2.423)	(1.790)
Tamil Nadu Gen & Dist Corp Ltd		−3.801*	−6.307***		−2.559	−5.103***
		(1.567)	(1.687)		(2.094)	(1.356)
Telangana State Power Gen Corp		−0.401	3.796*			
		(2.006)	(1.488)			
Uttar Pradesh Rajya Vidyut		−0.183	−5.632***		−0.370	−5.591***
		(2.411)	(1.297)		(2.459)	(1.033)
Month Fixed Effect						
January (Baseline)	0	0	0	0	0	0
	(.)	(.)	(.)	(.)	(.)	(.)
February	0.0286	0.0219	0.0317	0.0244	0.0248	0.0282
	(0.0428)	(0.0432)	(0.0428)	(0.0420)	(0.0426)	(0.0423)
March	0.121	0.124	0.139	0.127	0.129	0.147
	(0.154)	(0.155)	(0.155)	(0.154)	(0.156)	(0.158)
April	0.857**	0.833**	0.668**	0.767**	0.749**	0.599**
	(0.305)	(0.305)	(0.217)	(0.286)	(0.287)	(0.199)
May	0.943**	0.920**	0.760**	0.648*	0.632*	0.488**
	(0.335)	(0.355)	(0.254)	(0.281)	(0.283)	(0.188)
June	0.805*	0.785*	0.639*	0.634*	0.620*	0.491*
	(0.335)	(0.336)	(0.254)	(0.292)	(0.294)	(0.210)
July	0.974**	0.948**	0.782***	0.922**	0.904**	0.752***
	(0.304)	(0.306)	(0.205)	(0.298)	(0.301)	(0.199)

(continued)

Table 4 (continued)

	All observation (Unbalanced)			*Complete observation only (balanced)*		
	Fixed effects (Main model)	*Random effects*	*Pooled*	*Fixed effects*	*Random Effects*	*Pooled*
August	1.218***	1.192***	1.027***	1.144***	1.126***	0.973***
	(0.341)	(0.342)	(0.242)	(0.330)	(0.333)	(0.230)
September	0.927**	0.902**	0.737***	0.836**	0.819**	0.668***
	(0.297)	(0.298)	(0.209)	(0.277)	(0.280)	(0.189)
October	0.647**	0.619**	0.438**	0.493*	0.474*	0.308*
	(0.238)	(0.237)	(0.170)	(0.194)	(0.195)	(0.138)
November	0.415	0.388	0.211	0.286	0.266	0.103
	(0.266)	(0.268)	(0.179)	(0.231)	(0.234)	(0.144)
December	0.262	0.234	0.0592	0.218	0.198	0.0359
	(0.142)	(0.141)	(0.0611)	(0.131)	(0.132)	(0.0521)
Constant		12.48	86.07***		23.43	103.4***
		(27.36)	(20.94)		(31.54)	(24.19)
Observations	2287			1776		

Notes Robust standard errors in parentheses
*$p<0.05$
**$p<0.01$
***$p<0.001$
Source WRI

We find that power plants experienced more days of water shortage-induced outages between April and October, which is consistent with the findings in Luo et al. (2018). Although the coefficient estimations for the April-to-October dummy variables vary, all of them are positive and statistically significant at the 0.05 level in all six regressions, suggesting more days of water shortage-induced outages in these months when compared to January. Coefficient estimations for February, March, November, and December suggest that these months are not statistically significant from January at the 0.05 level (see Table 4).[23] According to Luo et al. (2018), outages in April to October were largely driven by low water availability in the summer and delayed monsoons.

3.2.2 *Estimating Future Impacts*

Future drought projections from climate models under RCP scenarios are highly uncertain, resulting in a wide range of possible futures. We construct future drought indexes for each power plant to project future water shortage-induced outages using geographically specific climate data from IPCC scenario experiments and find significant uncertainties.[24] These uncertainties mainly arise from the random nature of the climate system and climate models' lack of understanding of the climate system. The projections do not agree on the sign of change and have wide ranges for each power plant. Some climate models suggest that the future could be dryer, and power plants could have more days of water shortage-induced outages. Others suggest that the future could be wetter, and power plants could have fewer water shortage-induced outages.

Thermal power plants could experience uneven changes in drought conditions under different RCP scenarios. We use the ranges (minimum and maximum values) from multiple scenario experiments recommended by Rose and Scott (2018) to calculate future changes in drought indexes relative to the reference period 2006–2018 for each power plant and

[23] January is the base category for the monthly dummy variables; thus, the value is zero for January.

[24] Spatial resolution for climate models assessed in the IPCC AR5 varies from model to model (European Network for Earth System Modelling, 2016). The finest resolution is around 70 km, but the average is about 200 km (Climate Change in Australia, 2016).

each scenario.[25] Figure 6 illustrates the unevenness. For example, NTPC's Rihand power plant could have far worse drought conditions under the BAU Scenario (RCP8.5) than the other three scenarios in both periods 2019–2035 and 2036–2050. We also find that the nearer-term projections (e.g., 2019–2035) are generally less sensitive to differences between future scenarios than longer-term projections (e.g., 2036–2050). This finding is consistent with IPCC AR5 (Kirtman et al., 2013).

4 Conclusion

We undertook an analysis of a small subset of Indian thermal power companies to show how water shortages affected financial performances of electric utility companies using real historical data and fundamental financial analysis. Other power generation companies outside of our sample, particularly in drought-prone areas, may also be at risk. In fact, many other Indian thermal power companies also experienced water shortage-induced outages between the calendar year 2013 and 2016 (see Fig. 7), and water scarcity has already caused significant generation capacity reductions and outages in the global hydropower sector (Wang et al., 2017).

In a warming world, droughts are expected to increase in frequency and/or intensity in some regions, including the Mediterranean, Central America, Brazil, South Africa, and Australia, which could pose heightened risks for the energy sector (IPCC, 2014). It is important for investors and companies to understand the financial implications of these risks and take steps to manage them. Our analysis shows that, in recent years, water shortage risks caused occasional significant financial impacts for thermal power companies in India. However, if drought conditions worsen, water shortage risks could become more significant in the future, causing more severe outages and greater financial impacts in Indian thermal power companies, and potentially in other regions or sectors. By combining asset-level data, geospatial physical risk data, and financial information, our methodology presents a comprehensive and robust way for investors

[25] We use these values, rather than summary statistics (e.g., means, medians, percentiles), to interpret scenario experiments, which are not distributions and represent only arbitrary variation in modeling approaches (Rose & Scott, 2018).

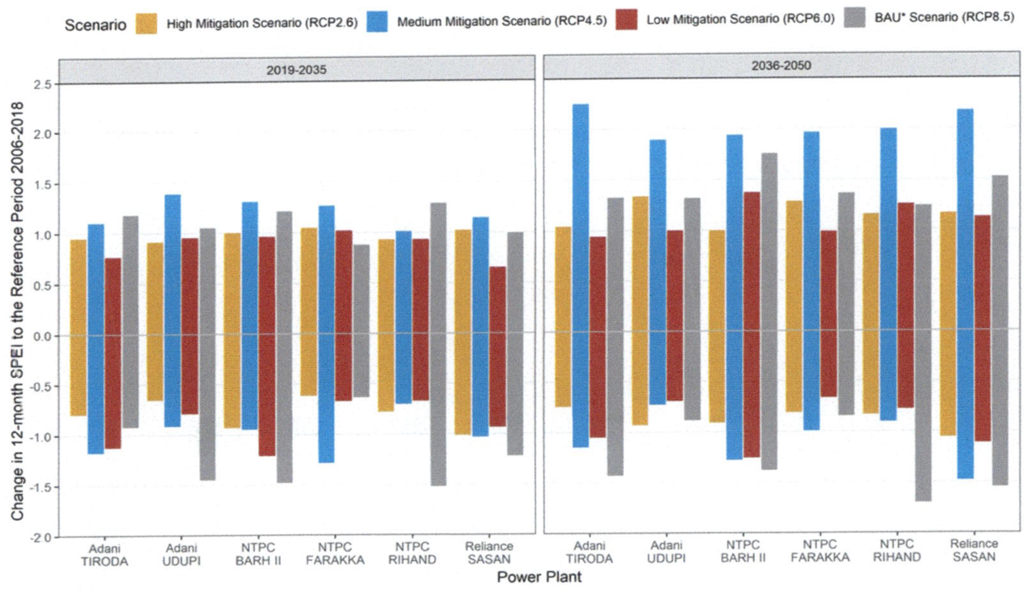

Fig. 6 Minimum and maximum values of changes in the drought index (12-month SPEI) projections in periods 2019–2035 and 2036–2050 relative to the reference period 2006–2018, by power plant (*Note* *BAU: business-as-usual. Negative values mean dryer conditions, and positive values mean wetter conditions. This figure only includes power plants that had water shortage-induced outages in FY 2014–2017 from the five companies covered in our historical financial analysis. *Source* WRI, calculated based on Knutti [2013a, 2013b, 2013c, 2013d])

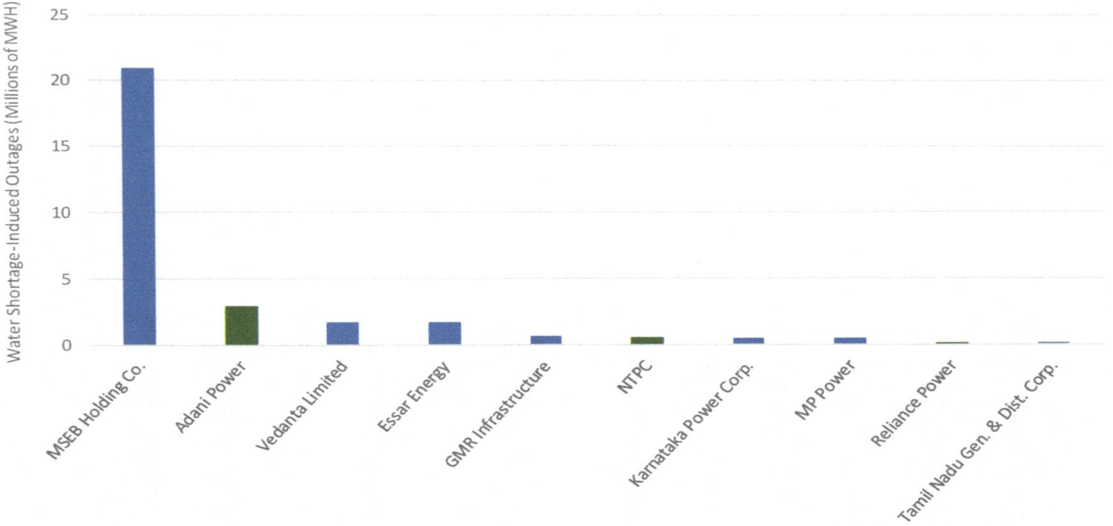

Fig. 7 Water shortage-induced outages during the calendar years 2013–2016 (*Note* Adani Power, NTPC, and Reliance Power are in green to highlight that they were part of our historical financial analysis. For more details on our sample, please see Appendix A. *Source* WRI Luo et al. [2018])

and companies to analyze climate-related risks and conduct scenario analysis based on the TCFD's recommendations and can be applied to other geographical regions when data are available.

Better company disclosures, more comparable and consistent data and improved climate modeling would make it easier to conduct and expand the analysis in this paper. We encountered several major challenges when conducting our analysis:

- Power plant geolocations, water withdrawal and consumption intensities (or cooling technologies), and water shortage-induced generation reductions and outages are important in understanding water-related risks, but they are not consistently disclosed by companies nor available in commercial databases.
- Climate simulation and projection data used in IPCC AR5 only included physical changes in water supply. They did not include localized socioeconomic scenarios and local water demand.
- IPCC AR5 data were not in a format familiar to investors and were difficult to download (World Data Center for Climate, 2017).
- Lastly, climate models project an unambiguous increase in temperatures across the world in the coming decades, but there is a lack of data and methodologies to incorporate temperature risks into our analysis, including potential efficiency losses and potential impacts from temperature regulations on cooling water discharges (Kirtman et al., 2013).

References

Aadhar, S., & Mishra, V. (2017). High-resolution near real-time drought monitoring in South Asia. *Scientific Data, 4*. https://doi.org/10.1038/sdata.2017.145.

Adani Power Limited. (2016). *Updates*. Adani Power Limited. https://nseindia.com/corporate/DisclosuretoStEx_16052016131604_731.zip.

CEA. (2016). *Minutes of the meeting for furnishing of daily/monthly online data on electricity generation held on 9th February, 2016 at 3.00 PM by Chairman of CEA*. CEA. http://www.cea.nic.in/reports/others/god/opm/mom.pdf.

Climate Change in Australia. (2016). *Climate model resolution*. https://www.climatechangeinaustralia.gov.au/en/climate-campus/modelling-and-projections/climate-models/resolution/.

Das, D. (2016). *NTPC plant shutdown hits power supply in 5 states*. https://www.thehindubusinessline.com/companies/ntpc-plant-shutdown-hits-power-supply-in-5-states/article8352253.ece.

Dharmadhikary, S., & Sandbhor, J. (2017). *Loss of 7 billion units power generation due to raw water shortage at coal based thermal power stations in India from June 2016 to April 2017*. Manthan Adhyayan Kendra. http://www.manthan-india.org/wp-content/uploads/2017/05/Report-raw-water-problem-18-May-17.pdf.

European Network for Earth System Modelling. (2016, August 23). *CMIP5 models and grid resolution*. https://portal.enes.org/data/enes-model-data/cmip5/resolution.

EY. (2018). *Climate risk disclosure barometer 2018*. EY. https://www.ey.com/Publication/vwLUAssets/ey-carbon-risk-disclosure-barometer-2018/$File/ey-carbon-risk-disclosure-barometer-2018.pdf.

Feeley, T. J., & Ramezan, M. (2003). Electric utilities and water: Emerging issues and R&D needs. *Proceedings of the Water Environment Federation, 2003*(2), 436–441. https://doi.org/10.2175/193864703784343686.

FINRA. (2015, August 14). *Six financial performance metrics every investor should know*. https://www.finra.org/investors/insights/six-financial-performance-metrics-every-investor-should-know.

IPCC. (2014). *Climate change 2013: The physical science basis: Working group I contribution to the fifth assessment report of the intergovernmental panel on climate change* (T. Stocker, Ed.). Cambridge University Press.

Kirtman, B., Power, S. B., Adedoyin, A. J., Boer, G. J., Bojariu, R., Camilloni, I., Doblas-Reyes, F., Fiore, A. M., Kimoto, M., & Meehl, G. (2013). Near-term climate change: Projections and predictability. In *Climate change 2013: The physical science basis.* Contribution of Working Group I to the Fifth Assessment Report of the Intergovernmental Panel on Climate Change. Cambridge University Press.

Knutti, R. (2013a). *IPCC working group I AR5 snapshot: The rcp26 experiment* [Data set]. In https://cera-www.dkrz.de/WDCC/ui/cerasearch/entry?acronym=ETHr2; https://doi.org/10.1594/WDCC/ETHr2.

Knutti, R. (2013b). *IPCC working group I AR5 snapshot: The rcp45 experiment* [Data set]. In https://cera-www.dkrz.de/WDCC/ui/cerasearch/entry?acronym=ETHr4; https://doi.org/10.1594/WDCC/ETHr4.

Knutti, R. (2013c). *IPCC working group I AR5 snapshot: The rcp60 experiment* [Data set]. In https://cera-www.dkrz.de/WDCC/ui/cerasearch/entry?acronym=ETHr6; https://doi.org/10.1594/WDCC/ETHr6.

Knutti, R. (2013d). *IPCC working group I AR5 snapshot: The rcp85 experiment* [Data set]. In https://cera-www.dkrz.de/WDCC/ui/cerasearch/entry?acronym=ETHr8. https://doi.org/10.1594/WDCC/ETHr8

Luo, T., Krishnan, D., & Sen, S. (2018). *Parched power: Water demands, risks, and opportunities for India's power sector* (p. 40). World Resources Institute. http://www.wri.org/sites/default/files/parched-power-india.pdf.

Occupational Safety & Health Administration. (2002). *Electric power eTool: Illustrated glossary—Power generation plants*. https://www.osha.gov/SLTC/etools/electric_power/illustrated_glossary/power_generation_plant.html.

Pepler, A. (2015). *An introduction to drought indices*. https://www.wmo.int/pages/prog/wcp/ccl/opace/opace4/meetings/documents/Barbados2016/presentations/barbados_Drought_Pepler.pdf.

Platts. (2018). *UDI world electric power plants data base*.

Rose, S., & Scott, M. (2018). *Grounding decisions: A scientific foundation for companies considering global climate scenarios and greenhouse gas goals* (p. 96). EPRI.

TCFD. (2017). *Recommendations of the task force on climate-related financial disclosures* (p. 74). TCFD. https://www.fsb-tcfd.org/wp-content/uploads/2017/06/FINAL-TCFD-Report-062817.pdf.

Verriest, A., Bouwens, J., & de Kok, T. (2018). *The prevalence and validity of EBITDA as a performance measure* (SSRN Scholarly Paper ID 3171131). Social Science Research Network. https://papers.ssrn.com/abstract=3171131.

Vicente-Serrano, S. M., Beguería, S., & López-Moreno, J. I. (2010). A multiscalar drought index sensitive to global warming: The standardized precipitation evapotranspiration index. *Journal of Climate, 23*(7), 1696–1718. https://doi.org/10.1175/2009JCLI2909.1.

Wang, J., Schleifer, L., & Zhong, L. (2017). *No water, no power*. https://www.wri.org/blog/2017/06/no-water-no-power.

Wooldridge, J. M. (2010). *Econometric analysis of cross section and panel data* (2nd ed). MIT Press.

World Data Center for Climate. (2017). *CERA—Jblob documentation*. https://cera-www.dkrz.de/WDCC/ui/cerasearch/info?site=jblob.

World Meteorological Organization. (2012). *Standardized precipitation index user guide*.

Water Insecurity and Climate Risk: Investment Impact of Floods and Droughts

Quintin Rayer, Karsten Haustein, and Pete Walton

1 Introduction

From today's level of approximately 1.2 °C (Gillett et al., 2021), a figure that represents the overall temperature increase on Earth from pre-industrial levels attributable to human factors, anthropogenic global warming (AGW) is increasing at a rate of around 0.2 °C per decade on average; and this rate may be accelerating (Haustein et al., 2017).

Q. Rayer (✉)
P1 Investment Management Ltd, Exeter, UK
e-mail: quintinrayer@p1-im.co.uk

K. Haustein
Climate Service Center Germany (GERICS), Helmholtz-Zentrum Geesthacht, Hamburg, Germany
e-mail: karsten.haustein@hzg.de

P. Walton
UK Climate Impacts Programme (UKCIP), University of Oxford, Oxford, UK
e-mail: peter.walton@ouce.ox.ac.uk

T. Walker et al. (eds.), *Water Risk and Its Impact on the Financial Markets and Society*, Palgrave Studies in Sustainable Business In Association with Future Earth,
https://doi.org/10.1007/978-3-030-77650-3_6

If we follow the current emissions' trajectory, the increase may exceed the 1.5 °C Paris Agreement threshold between the years 2030 and 2050 (Leach et al., 2018). The UN Framework Convention on Climate Change's (UNFCCC) Paris Agreement acknowledges that even in the case of stabilizing global temperatures below the target of 1.5 °C and well below 2.0 °C above pre-industrial levels, the planet will continue to experience the impact of climate change, including more frequent extreme weather events (UN FCCC, 2015).

A warmer atmosphere has a higher water-holding capacity, which leads to an intensified water cycle, and is associated with precipitation and evaporation increases of 2–4% per warming degree (Allen & Ingram, 2002; Durack et al., 2012; Held & Soden, 2006; Hwang et al., 2011). As atmospheric moisture increases, so too does the likelihood of more intense downpours, in addition to the risk of daily extreme temperatures and extended warm spells. Related risks, such as water scarcity and water stress, are dependent on secondary societal and sectoral conditions (associated with river flooding, flash floods, droughts, and heatwaves) which are more difficult to quantify in a climate change context, and act as a threat multiplier.

Water security is a chief concern of environmentally focused investors when considering investment possibilities and risks (Porritt, 2001; Stern, 2006). Despite this, the potential liabilities of extreme floods and droughts (events intrinsically linked to global warming), may not be reflected in share prices (Krosinsky et al., 2012). While some economic impacts from extreme weather events are beginning to be incorporated into risk assessments, many may not be reflected in relevant companies' market valuations (Dietz et al., 2016; Krosinsky et al., 2012).

Recent studies introduce the concept of a hypothetical climate liability (HCL) regime. Under an HCL regime, the hypothetical damage caused by high-emitting companies' historical emissions is estimated, where "damages" refer to insurance payouts, successful legal cases, carbon taxation, or payment as a result of societal pressures leading to voluntary contributions (Rayer & Millar, 2018a, 2018b, 2018c; Rayer et al., 2019, 2020).

Following the approach of Rayer et al. (2019, 2020), we estimate that under an HCL regime, global flood and drought damages for the period of 2012 to 2016 amount to around 2–3% of the market capitalizations of the top nine carbon-emitting companies. This figure is based on the firms' share of historical emissions, with allowance made for baseline flood and drought events (factual versus counterfactual Earth) and estimated producer versus user share of responsibility for emissions. Given the nine considered firms' more recent market capitalizations and following the recent stock market correction due to the COVID-19 pandemic, the estimated damages under the HCL regime rise to approximately 5% of market capitalization.

Despite the current, and already substantial, potential financial implications, future changes are projected to be even more significant. Recent research implies that possible share-price impacts for high-emitting firms could be greater (National Academies of Sciences, Engineering, and Medicine, 2016). The present work, and that of Rayer et al. (2020), only estimates a tiny fraction of attributable climate-related damages.

2 Economic Impact of Floods and Droughts

Global warming is linked to damaging extreme weather-related events, including hurricanes (tropical cyclones), precipitation, floods, droughts, and heatwaves. Such weather events trigger a multitude of detrimental side effects, all of which will become increasingly difficult to adapt to if society fails to introduce effective mitigation measures. Although unlikely, if we continue on our current emission trajectory into the next century, climate change could lead to a "hothouse earth" scenario (Steffen et al., 2018). As self-reinforcing feedbacks could push the Earth climate system toward a threshold that, if crossed, could prevent stabilization of the climate and cause continued warming.

A rise in global mean surface temperature (GMST) increases the risk of daily extreme temperatures as well as the risk of extended warm spells. It also enhances evapotranspiration, which may result in a drying effect in already arid areas. It is therefore plausible to simultaneously experience increased floods and more frequent drought occurrences, while average rainfall amounts may be unchanged overall.

Whereas previous work on damaging extreme weather-related events focuses on tropical cyclones (Rayer et al., 2020), our study considers floods, droughts, and heatwaves. Due to the impact that temperature has on the water cycle, and therefore water security and risk, we find it important to account for heatwave risks concurrently with our analysis of drought and flood risks.

2.1 *Heatwaves*

There is consensus that the intensity, frequency, and duration of heatwaves have increased, both regionally and globally (Perkins-Kirkpatrick & Lewis, 2020). While average heatwave intensity displays little change, cumulative heatwave intensity increases at a similar rate to heatwave frequency. Moreover, the most intense heatwave seasons, as defined by cumulative intensity, have occurred post-2000 (Perkins-Kirkpatrick & Lewis, 2020). At least one extra heatwave day has occurred each decade over most regions between 1950 and 2017, and heatwaves have increased in length. Perkins-Kirkpatrick and Lewis (2020) use a cumulative heat metric for impact and demonstrate a substantial increase in the overall extreme heat experienced during heatwaves across almost every region. The vast majority of the world's worst heatwave seasons have taken place since the 1980s, with many occurring since 2000. Trends are most significant in northern South America, the Middle East, the Maritime Continent, and much of Africa. Globally, cumulative heat increased from 2.84 (in 1950) to 7.46 °C/decade (in 1990). Thus, 62% of cumulative heat is of recent origin, or in other words, since 1950. If data were available for the period before 1950, it is reasonable to assume that this figure would be larger.

For example, in the summer of 2010, Western Russia experienced an extraordinary heatwave, with the region recording the warmest July since at least 1880, and numerous locations setting all-time maximum temperature records (Dole et al., 2011). Analyses found that the heatwave severity was within the bounds of natural interannual variability (Dole et al., 2011), but that climate change had made the event 80% more likely to occur (Rahmstorf & Coumou, 2011). Initial studies looked at different issues, including severity or frequency (Schiermeier, 2018). Further analyses indicate that the average return period for a 2010-magnitude heatwave reduced from 99 years in the 1960s to 33 years in the 2000s; the decrease of 66 years marks a three-fold increase in frequency (Otto et al., 2012).

The recent 2020 Siberian heatwave from January to June included a record-breaking temperature of 38 °C in the town of Verkhoyansk, within the Arctic Circle, on June 20 (Ciavarella et al., 2020). This six-month heatwave contributed to raising the average world temperature to the second hottest on record between January and May. Ciavarella et al. (2020) conclude that this event was made at least 600 times more likely because of AGW and would have effectively been impossible otherwise.

The 2020 Siberian heatwave resulted in a range of disasters leading to the declaration of a state of emergency. Wildfires in Siberian forests caused the release of 56 megatons of CO_2 in June 2020. Trees in areas not affected by the fires were impacted by pests making them more vulnerable to fire. Fish migrated, seeking cooler water, thus affecting catches. Local people are reporting increased physical and mental conditions, including headaches and skin conditions, because of the heat (Ciavarella et al., 2020). Melting permafrost has caused infrastructure to collapse, leading to fuel leakages and further environmental disasters. Further, the melting leads to the release of trapped methane into the atmosphere (Ciavarella et al., 2020).

2.2 *Floods*

A hotter atmosphere has a more energetic water cycle and warmer air holds more moisture; both of these factors are associated with the increased likelihood of more intense downpours (Durack et al., 2012). The warmer atmosphere's higher water-holding capacity results in an intensification of precipitation extremes, thereby increasing the global risk of river flooding and flash floods.

Precipitation extremes (both wet and dry) impact society and ecosystems (Lehmann et al., 2018). For example, flash flood events during an extended dry spell are potentially extremely damaging. Such floods are a highly likely outcome of climate change, with strong evidence that these are already occurring (Lehmann et al., 2018). In addition to thermodynamic effects, changes in atmospheric circulation patterns can alter the intensity of extreme events regionally.

Flooding associated with tropical cyclones adds to the list of common-cause risk factors. For example, rainfall intensity during Hurricane Harvey was estimated to have increased by about 15% due to global warming up to 2017 (van Oldenborgh et al., 2017). Generally, there is high confidence that water vapor increase leads to heavier hurricane rainfall in response to warming (even without increases in hurricane frequency), but globally the trend is not detectable yet due to limited data availability (Patricola & Wehner, 2018).

In 2011, Thailand suffered significant flooding, causing estimated economic damages of US$30 billion (Gale & Saunders, 2013). Thailand's 2011 annual rainfall was the highest in the country's 61-year precipitation record (Gale & Saunders, 2013). For a 2011 scale event, the return period was estimated to be six to ten years based on precipitation, or ten to twenty years using peak river discharge rates (Gale & Saunders, 2013). Based on monsoon (May to October) precipitation, van Oldenborgh et al. (2012) conclude that climate change did not play a role in the 2011 flooding. However, Promchote et al. (2016) identify a climate change role in pre-monsoon (January to April) rainfall trends over the period of 1980 to 2013. Over this period, both the upper and lower basins of Thailand's Chao Phraya River experienced statistically significant trends of increased rainfall. In the upper basin, rainfall increased from around 0.5 mm/day in 1980 to 1.2 mm/day by 2013, while the trend increases in the lower basin was from approximately 0.6 mm/day to 1.4 mm/day over the same period.

2.3 *Droughts*

As well as heatwaves, climate change causes drought and wildfire hazards to increase due to enhanced evaporation and circulatory feedbacks (Schiermeier, 2018; Zhou et al., 2019), all of which are exacerbated by human influences. Contrasting earlier periods less affected by climate change with later periods, Padrón et al. (2020) conclude that there are multiple lines of evidence indicating that changes in dry season water availability between the first half of the twentieth century (1902–1950) and recent times (1985–2014) are attributable to human-induced climate change. Droughts are particularly damaging, as their effects are felt over several years, leading not just to crop failure, but also to the erasure of critical ecosystem components that are not replaced easily.

From 2015 to 2017, the Western Cape province of South Africa suffered from three consecutive years of below-average precipitation, causing the worst drought since 1904. The prolonged lack of rain caused an acute water shortage in Cape Town leading to the possibility of "day zero" in March 2018, which would see the city's water supply run dry (Otto et al., 2018). A meteorological analysis of the event indicated that a strong rainfall anomaly in the shoulder seasons caused the below-average total rainfall (March to May and August to October), despite near normal rainfall during the core winter season (June and July). A combination of observations and computer simulations points to the conclusion that while such an event is rare, with a return time of more than a hundred years, climate change had increased the likelihood by a factor of 3.3 (Otto et al., 2018).

The financial burden of mitigating such stresses is significant; only adequately prepared regions can avoid societal upheaval. The recent drought in Syria (2006–2011), for example, arguably increased the risk of conflict due to its prolonged nature. However, the low signal-to-noise ratio makes the attribution of this particular drought difficult. Statistical analysis alone cannot provide a conclusive explanation, although physical insight helps to interpret the empirical evidence. Yet, the latter is often overlooked, which explains the conflicting results in the scientific literature (e.g., Gleick [2014]; Selby et al. [2017]). Regardless of the individual outcome, some regions will always be particularly affected by climate change-related risks due to political instability and civil unrest.

3 Global Economic Impact

Both floods and droughts increase the risk of water scarcity and water stress in vulnerable regions. Between 1900 and 2016, floods and droughts caused an estimated 19 million deaths and over US$900 billion in damages globally (EM-DAT: The International Disaster Database, 2020). Despite this, the potential liability of extreme floods and droughts, events intrinsically linked to global warming, may not be reflected in share prices (Krosinsky et al., 2012).

To account for the reality that significant floods and droughts do not necessarily occur every year, Fig. 1 presents the rolling five-year totals for flood- and drought-related deaths and damages over the period of 1900 to 2016. We present values by year at the end of each five-year rolling period, thus damages over the five-year period of 1900 to 1904 are dated 1904.

Earlier data may not benefit from the influence of modern communications, potentially making it less comprehensive (i.e., undervaluing damages) while recent data may be more impacted by climate change. Thus, our focus is on the recent five-year period from 2012 to 2016 inclusive, when flood and drought damages totaled $255.97 billion. This data appears in Fig. 1 as year 2016. Per Fig. 1, higher total flood and drought damages ($287.06 billion) occurred during the period of 2010 to 2014.

Given that the damages occurred over a five-year period, values may be subject to inflation. Inflation-adjusted damages are thus estimated in 2016 US dollars, using global inflation data from the World Bank (The World Bank, 2020). To make estimates conservative, damages were assumed to

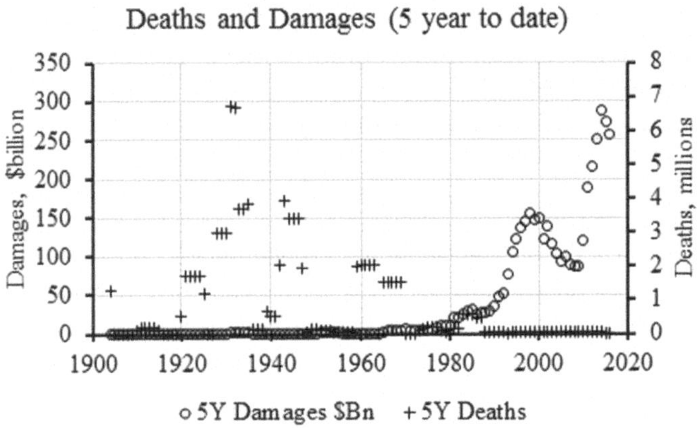

Fig. 1 Rolling five-year total damages (left axis) and deaths (right axis) associated with floods and droughts. Data from EM-DAT: The International Disaster Database (2020). *Note* Over rolling five-year periods, peak damages totaled $287 billion (the five-years leading up to 2014), while peak deaths totaled 6.7 million (the five-years leading up to 1931)

occur at the end of each year. As inflation rates are generally positive, this means that for damages occurring in, say, January of a given year, any uplift in values between January and the end of December for that year are ignored. We then apply inflation-adjustment to subsequent years. For example, over 2013 the global inflation rate was 2.613% (The World Bank, 2020). Thus, damages of, say, $100 million during 2012 might be regarded as all occurring at the end of 2012, to be conservative. These figures are calculated as having an inflation-adjusted value of $102.613 million at the end of 2013. As a result, the $256 billion in damages for the period of 2012 to 2016 based on the "year when occurred" are inflation-adjusted to $265 billion in 2016 USD terms. While inflation over the 2012 to 2016 period has been relatively modest, this should result in a more accurate estimate for the following analysis. The reason for working with damages over a recent finite period is to better capture the contribution made by climate change as global warming has increased.

4 The Emitters

Since the extent and scope of the destruction caused by such events are significant, we investigate whether companies or sectors should be held liable, at least partially, for their activities around emissions. Recent analyses quantify increases to global warming using the historical contributions of individual nations and companies, including changes in extreme event frequencies (Ekwurzel et al., 2017; Otto et al., 2017; Skeie et al., 2017). However, from a stockholder perspective, investments are made in companies rather than into specific countries, thus sidestepping some national (and political) considerations.

Atmospheric CO_2 rose by about 98 parts per million (ppm) from 1880 (290 ppm) to 2010 (388 ppm), reaching 410 ppm in 2018 (IPCC, 2013; Keeling et al., 2001; Scripps Institution of Oceanography, 2019). Industrial-level climate modelling links emissions from 90 major carbon producers and shows that their emissions produce about 57% of the observed rise in atmospheric CO_2, approximately 42–50% of the rise in global mean surface temperatures, and around 26–32% of global sea level rise over the 1880 to 2010 period (Ekwurzel et al., 2017). A significant fraction of this increase took place since 1980 (Table 1). Around three quarters of the increase in atmospheric CO_2, and over two-thirds (0.69–0.70) of the rise in global mean surface, is of recent origin, attributable to accumulated CO_2 emissions between 1980 and 2010.

Estimating global warming from a range of different greenhouse gases, including CO_2, CH_4, and N_2O, is complex, as determining the climate forcing from various greenhouse gases with different atmospheric lifespans is not straightforward (Skeie et al., 2017). Focusing primarily on

Table 1 Proportion of climate change phenomena linked to the emissions of 90 major industrial carbon producers (as per Ekwurzel et al., 2017)

	1880–2010	*1980–2010*	*Fraction of recent origin (i.e., 1980–2010)*
Proportion of observed rise in atmospheric CO_2	57%	43%	43/57 = 0.75
Proportion of observed rise in global mean surface temperatures	42–50%	29–35%	29/42 = 0.69 35/50 = 0.70
Proportion of observed global sea level rise	26–32%	11–14%	11/26 = 0.42 14/32 = 0.44

CO_2 emissions allows useful progress to be made, as cumulative CO_2 emissions are the primary cause of global climate system changes (Allen, 2016). This allows for a relatively simple assignment of historical responsibility. Our first estimate, introduced below, does not account for the starting date of emissions, and baseline flood or drought levels, though we discuss this in the Critique section.

The fossil fuel industry and its products accounted for 91% of global industrial greenhouse gas emissions in 2015 and about 70% of all anthropogenic emissions (Griffin, 2017). If the trend in fossil fuel extraction continues in the same way over the next 28 years as it has over the previous 28, global average temperatures would be on course to rise by around 4 °C (Griffin, 2017). Between 1751 and 2017, 30 entities (both companies and state producers) accounted for 30% of global industrial emissions (Heede, 2019). Since 1988, only 25 entities (both companies and state producers) accounted for 51% of global industrial emissions in 2015 (Griffin, 2017). Seven of these top 25 emitters are publicly owned companies (see Table 2), collectively accounting for 9.5% of Scope 1 and 3 emissions between 1988 and 2015, or 12.9% of Scope 1 and 3 emissions between 1751–2017 (Griffin, 2017; Heede, 2019). Between 1751 and 2017, nine of the top 20 emitters were publicly owned (Table 2), collectively accounting for 14.5% of Scope 1 and 3 emissions (Heede, 2019). Scope 1 emissions are direct emissions that originate from sources owned and controlled by a company, for example, fuel used in company vehicles. Scope 2 emissions are indirect, resulting from energy used by a company, such as electricity, steam, heating, and cooling. This could include, for example, electricity purchased by a company from an external supplier. Scope 3 emissions cover all indirect emissions arising due to company activities. Scope 3 emissions include upstream and downstream value chain emissions. Scope 3 emissions include those of suppliers and customers using companies' products.

Heede's (2019) methodology covers the period from 1751 to 2017 and includes the emissions of companies that are no longer extant but have become part of another company through mergers and acquisitions. For example, Royal Dutch Shell acquired BG in 2016; BG's historical production of oil and gas from 1988 to 2015 was added to Shell's oil and gas (and coal) data from 1892 to the present (Heede, 2019). This approach also extends the dataset to earlier years (e.g., RWE prior to 1965 from its founding in 1898, Phillips gas production prior to 1937, Wintershall prior to 1998).

Table 2 Cumulative industrial greenhouse gas emissions from Scopes 1 + 3 in percent: 1988–2015 (Griffin, 2017), 1751–2017 (Heede, 2019)

		Cumulative emissions from Scopes 1 + 3 (%)				
Producer	*Type (see note)*	*1751–2017 (Heede, 2019)*			*1988–2015 (Griffin, 2017)*	
Chevron Corp	Oil & gas	1	3.13	3.13	4	1.3
ExxonMobil Corp	Oil & gas	2	2.98	2.98	1	2.0
BP PLC	Oil & gas	3	2.29	2.29	3	1.5
Royal Dutch Shell PLC	Oil & gas	4	2.18	2.18	2	1.7
ConocoPhillips	Oil & gas	5	1.08	–	–	–
Peabody Energy Corp	Coal	6	0.92	0.92	5	1.2
Total SA	Oil & gas	7	0.83	0.83	6	0.9
BHP Billiton Ltd	Coal, oil & gas	8	0.57	0.57	7	0.9
CNX Resources (CONSOL)	Coal	9	0.56	–	–	–
Totals			14.5	12.9		9.5

Note Firms are listed in descending order of emissions from largest to smallest, from 1751 to 2017 (Heede, 2019). The ranking for 1988–2015 (Griffin, 2017) is included for reference. Historically, the following companies also produced coal: BP, Chevron, ExxonMobil, and Royal Dutch Shell (Heede, 2019)

Heede's production and emissions data for crude oil, natural gas, and coal include estimates of fossil fuel emissions from international sources that document global emissions of carbon dioxide from 1751 (for coal emissions), from 1870 (oil), from 1885 (gas), from 1928 (cement), and from 1950 (flaring) (Boden et al., 2012; Global Carbon Project, 2018; Le Quéré et al., 2018).

The selection of Scope 1 and 3 emissions is based on the carbon producers' primary business: the extraction of crude oil, natural gas liquids, natural gas, and coal. This selection focuses on each company's extracted carbon, although most vertically integrated multi-national oil and gas companies sell more carbon fuels than they extract (Heede, 2019). Heede's approach is to focus on tracing carbon from the lithosphere to the atmosphere by way of each company's net production of primary carbon fuels.

5 Attributable Financial Contribution to Climate Damages

In this chapter, we extend the concept of climate liability regime to estimate market capitalization (share price) impacts (expressed in US dollars throughout) related to global flood and drought damages by nine major carbon-emitting firms. Rayer et al. (2020) report that the combined market capitalization of the seven companies that Table 2 lists, in August 2018, was approximately $1303 billion. On August 6, 2018, the combined market capitalization for the same seven firms was $1271 billion (Macrotrends, 2020). We use this figure in the analysis below. For the nine firms that Table 2 lists, the combined market capitalization was $1358 billion on August 6, 2018.

During the period of January 1 to May 31, 2020, which includes the onset of the COVID-19 pandemic, stock prices fell significantly before recovering. During this period, for the nine firms that Table 2 lists, their minimum market capitalizations occurred between March 12 and 26, 2020. On March 18, 2020, the combined market capitalization for the seven firms that Table 2 lists was $512 billion, and $538 billion for the nine firms (Macrotrends, 2020). We refer to these as the COVID-19 market capitalizations below. By June 1, 2020, the combined market capitalization recovered to $785 billion for the seven firms, and $834 billion for the nine firms (Macrotrends, 2020). The relevant combined market capitalizations are included in Table 3.

The combined market capitalization for the seven companies in November 2017 was estimated to be at around $1230 billion (Rayer & Millar, 2018a, 2018b, 2018c). As mentioned above, Rayer et al. (2019, 2020) used a combined market capitalization from August 2018 of $1303 billion. The changed market capitalization figures mean our analysis differs from previous analyses in this respect.

Consider the 9.5% cumulative emissions for the seven firms that Table 2 lists and the $265 billion (in 2016 inflation-adjusted USD) damages estimate for floods and droughts over the period 2012 to 2016. If, hypothetically, the seven firms contributed 9.5% of these damages, this would equal $25.2 billion [= $265bn × 9.5%]. This could be shared between the companies based on their respective emissions (see Table 2), say, $3.4 billion [= $265bn × 1.3%] for Chevron and so on. Unless companies chose to contribute voluntarily, decisions about how hypothetical payments might be made would be a matter for government

Table 3 Hypothetical climate liabilities for floods, droughts, and heatwaves over the period of 2012 to 2016 for the seven and nine companies that Table 2 lists based on emissions over the periods of 1988 to 2015 and of 1751 to 2017

	Seven companies 1988–2015	*Seven companies 1751–2017*	*Nine companies 1751–2017*
Hypothetical climate liability %	9.5%	12.9%	14.5%
Hypothetical climate liability $	$25.2 billion	$34.2 billion	$38.4 billion
Combined market capitalizations 6th August 2018	$1271 billion	$1271 billion	$1358 billion
Combined market capitalizations COVID-19	$512 billion	$512 billion	$538 billion
Combined market capitalizations 1st June 2020	$785 billion	$785 billion	$834 billion
Detriment to market cap or share price 6th August 2018	25.2/1271 = 2.0%	34.2/1271 = 2.7%	38.4/1358 = 2.8%
Detriment to market cap or share price COVID-19	25.2/512 = 4.9%	34.2/512 = 6.7%	38.4/538 = 7.1%
Detriment to market cap or share price 1st June 2020	25.2/785 = 3.2%	34.2/785 = 4.4%	38.4/834 = 4.6%

Note Detriments are presented as a percentage of combined market capitalizations (or share price terms), based on market capitalizations during the COVID-19 pandemic (18th March 2020) and following a degree of recovery (1st June 2020)

policy, the courts, or other (currently undefined) mechanisms. Based on the August 2018 combined market capitalizations for the seven firms, this amount represents a 2.0% [= $25.2 billion/$1271 billion] detriment to their market capitalizations (or share price). This is a significant sum, particularly considering that similar contributions might be requested respecting other past and future extreme weather events.

Over the longer period of 1751 to 2017, the seven companies' cumulative emissions were 12.9%. A hypothetical 12.9% contribution would be $34.2 billion [= $265bn × 12.9%], or 2.7% [= $34.2 billion/ $1271 billion] in August 2018 share-price terms. Based on the June 2020 capitalizations of $785 billion, this would be 4.4% [= $34.2 billion/$785 billion] in share-price terms. Table 2 shows the total for the nine listed companies. Table 3 lists the resulting HCLs. While these figures include historical emissions over the period of 1751 to 2017, they omit other extreme weather events and consequences, such as sea level rise.

We define hypothetical climate liabilities as the percentage of market capitalization for the firms in question. Thus, the HCL indicates the share-price impact that might be expected for the liabilities in question. Over five-year periods, these figures show HCLs of 2.0–2.8% based on August 2018 market capitalizations (i.e., 2.0–2.8% of market capitalizations or a similar detriment in share-price terms). During the 2020 COVID-19 pandemic, due to the sudden cut in demand for fossil fuels, reduced share prices resulted in the HCLs rising to 4.9–7.1% of market capitalizations. While the firms' market capitalizations have since partially recovered, based on prices on June 1, 2020, the HCLs would amount to 3.2–4.6% in share-price terms. Based on the June 1, 2020 figures, the five-year period HCLs would amount to 0.6–0.9% on an annual basis. Thus, if the five-year HCLs are representative, climate-induced flood and drought damages would be equivalent to around 0.6–0.9% in share-price terms, annually.

The collapse in market capitalizations for the firms that Table 2 lists during the COVID-19 pandemic was attributed to a sudden drop in demand for fossil fuels and related products. This may be interpreted as a stress-test for market reaction, should the widespread decarbonization of economies occur. If investors made a wholesale move away from fossil fuels or were to disinvest—prompted by fear that companies' fossil reserves may prove significantly over-valued—this may result in fossil reserves becoming so-called stranded assets, as they suffer from premature write-downs, devaluations, or conversion to liabilities. Increasingly, risk factors related to the environment may strand assets, potentially profoundly altering asset values across a wide range of sectors (Caldecott et al., 2015). In this case, the HCL detriment figures that Table 3 lists for the market capitalizations seen during the COVID-19 pandemic may prove a useful indicator. Based on these figures, hypothetical liabilities during the five-year period amount to 1.0–1.4% on an annual basis.

In a world, where global warming causes more intense flood and drought events that result in greater financial losses, if an HCL regime were to develop, damage contributions of around 0.6–1.4% of these high-emitting companies' market capitalizations (and share prices) might be anticipated more frequently with each annual wet or dry season.

Furthermore, these figures ignore other global warming impacts, such as hurricanes and sea level rise, and could easily reach much larger sums if they were also included. These figures also neglect the likelihood of increasingly powerful climate responses as global warming intensifies (Rayer et al., 2020).

6 Critique

In terms of the scope of extreme weather events explored, this chapter builds on the extant analysis by Rayer et al. (2020) that examines the 2017 North Atlantic hurricane season to consider floods and droughts. This analysis is an initial exploration of how global warming-induced extreme flood and drought events may be attributed to specific companies. Estimates of the costs of those events can be expressed in terms of the decrease in the share price for the firms affected. At this stage, the arguments are incomplete, even though the broad conclusions above appear scientifically robust. Although the scale of the estimates is uncertain, the potential for impact on the share prices of individual companies is clear. The results presented here only relate to flood and drought damages over the period of 2012 to 2016.

As our analysis neglects other effects of global warming, such as sea level rise, which has the capacity to cause huge damage to affected areas, the risks are only partially quantified for investors in those companies. In terms of hurricanes, intense storms with strong winds and high precipitation occur in many parts of the world apart from the North Atlantic, and Rayer et al. (2020) only consider a single year. Therefore, the global damages associated with extreme weather events are expected to be a significant multiple of the estimated costs below. Despite the significant costs mentioned in the following section, climate change-induced extreme weather events are expected to rise as AGW intensifies. Thus, although the analysis presented here does not cover all climate risks, it quantifies some in share-price terms and outlines a method that can be followed by other workers in this area.

To complete the above analysis, two additional aspects are required. Firstly, the proportion of the above damages that should be attributed to global warming must be highlighted (after all, floods and droughts occurred before the onset of climate change), so that only changes due to AGW (anthropogenic global warming) are considered. Secondly, the split between producer and user responsibility must be identified. To what extent are fossil fuel extraction companies responsible compared with users who burn the fuel? These two aspects are explored in Sect. 6.1 (the pre-industrial, pre-AGW baseline for climate change induced extreme weather events) and Sect. 6.2, which explores producer-versus-user responsibility.

6.1 *Pre-industrial Baseline*

A major concern is the use of appropriate baseline levels. In other words, what is the pre-industrial baseline against which it is appropriate to compare current observed climate changes? While it may be difficult to estimate what the pre-industrial baseline should be, it is problematic to assume that the frequency or intensity of extreme events would be linearly dependent on atmospheric CO_2 concentrations, or on GMST (global mean surface temperature) rise, as there could easily be cliff-edge or non-linear effects. For example, melting permafrost could lead to the release of trapped methane into the atmosphere, which, as a potent greenhouse gas, would lead to further global warming (Ciavarella et al., 2020). Thus, a small increase in CO_2 or GMST could generate a disproportionately large extreme weather response. For example, the recently observed rainfall deficit during mid-Spring in Central Europe could be a harbinger of lasting circulation changes in some regions (Ionita et al., 2020). As floods and droughts occurred before the onset of AGW, we posit that only the costs linked to additional frequency, or additional intensity of floods and droughts, should be considered, as it is human emissions that generate AGW.

Per Rayer et al. (2020), we estimate the fraction of the impact of extreme weather events that can be attributed to AGW, which we denote using F_{AGW}. Thus, if some damaging extreme event were to occur six times as often because of global warming, then there would be an increase in frequency from one to six per unit time, resulting in $F_{AGW} = (6 - 1)/6 = 0.83$. This calculation implies that 83% of recent occurrences of the event in question are due to AGW, so 83% of damages would be attributable to human emissions. Similarly, if another extreme event were to occur no more frequently, but with double the intensity or volume, this would imply an increase of intensity from 1 to 2 units, so $F_{AGW} = (2 - 1)/2 = 0.50$. Thus, half the recent impact would be due to AGW.

To estimate the pre-industrial baseline, we draw from evidence based on the case studies that Sect. 2 summarizes. As these are an incomplete sample, our results can only be estimates. Adopting a precautionary approach can cut either way, in corporate liability terms or in climate damage terms. In financial terms, we can treat these as potential maximum risks, thus suggesting an upper limit to financial damages for the firms in question. In climate terms, it is possible that climate impacts may be far greater than our estimates, making these a lower limit for financial damages. Table 4 shows the resulting estimates of F_{AGW}.

Based on the data above, an indicative range for F_{AGW} would be 0.575–0.998, which we interpret as meaning that between 0.575 and 0.998 of damages from these events are caused by AGW. We take the mid-point of $F_{AGW} = 0.79$ for our analysis below. Figure 2 shows the range of F_{AGW} values for the events above. Considering Table 3, and to allow for the fraction of extreme flood and drought events that occurred before the onset of human-induced global warming, the HCL for the seven firms listed over the period of 1751 to 2017 of \$34.2 billion is reduced to \$127.0 billion [= \$34.2bn × 0.79]. This amounts to 2.1% [= \$27.0 billion/ \$1271 billion] in August 2018 share-price terms.

6.2 *Producer-Versus-User Responsibility*

A further question arises: to what extent should fossil fuel companies be held responsible? Humanity's role in CO_2 emissions that results in global warming is well established. In 1857, Eunice Foote reported the atmospheric heating effect of carbonic acid gas (carbon dioxide, CO_2) (Foote, 1857). Between 1859 and 1861, John Tyndall demonstrated the absorption of radiant heat by CO_2, water vapor, and hydrocarbon vapors, as well as their effect on climate (Jackson, 2020; Tyndall, 1859, 1861). In 1938, Guy Stewart Callendar demonstrated that the Earth's land surface was warming, suggesting that the production of CO_2 by the combustion of fossil fuels was responsible for much of this change in climate (Callendar, 1938; Hawkins & Jones, 2013). Callendar stated "[b]y fuel combustion man has added about 150,000 million tons of CO_2 to the air during the past half century. The author estimates ... that approximately three quarters of this has remained in the atmosphere" (Callendar, 1938, p. 223).

Shue (2017) explores the question of moral responsibility. He notes that society classifies responsibilities into positive and negative, general

Table 4 Proportion of impact of extreme weather events attributable to human-induced global warming

Type of event	*Source*	*Location*	*Comments*	*Evaluation period*	F_{AGW}
Hurricane	Emanuel (2017), Rayer et al. (2020)	Houston, USA (2017)	Event frequency increased six-fold	1981–2000 versus 2017	$F_{AGW} = (6 - 1)/6 = 0.83$
Flooding	Promchote et al. (2016)	Thailand (2011)	2011 rainfall in the upper and lower river basins increased from 0.5–1.2 mm/day and 0.6–1.4 mm/day respectively	1980–2013	Upper: $F_{AGW} = (1.2 - 0.5)/1.2 = 0.58$, Lower: $F_{AGW} = (1.4 - 0.6)/1.4 = 0.57$ Average: $F_{AGW} = 0.575$
Drought	Otto et al. (2018)	South Africa (2015–2017)	Event likelihood p increased by a factor 3.3	Counterfactual (simulated world without AGW) versus 2017	$F_{AGW} = (3.3p - p)/3.3p = 0.70$
Heatwave	Otto et al. (2012)	Russia (2010)	Three-fold increase in frequency	1960s versus 2010	$F_{AGW} = (3 - 1)/3 = 0.67$
Heatwave	Ciavarella et al. (2020)	Siberia (2020)	Event more than 600 times more likely	Start of twentieth century versus 2000	$F_{AGW} = (600p - p)/600p = 0.998$
Heatwave	Perkins-Kirkpatrick and Lewis (2020)	Global (2010s)	Cumulative heat increased from 2.84 to 7.46 °C/decade	1950 versus 1990	$F_{AGW} = (7.46 - 2.84)/7.46 = 0.62$

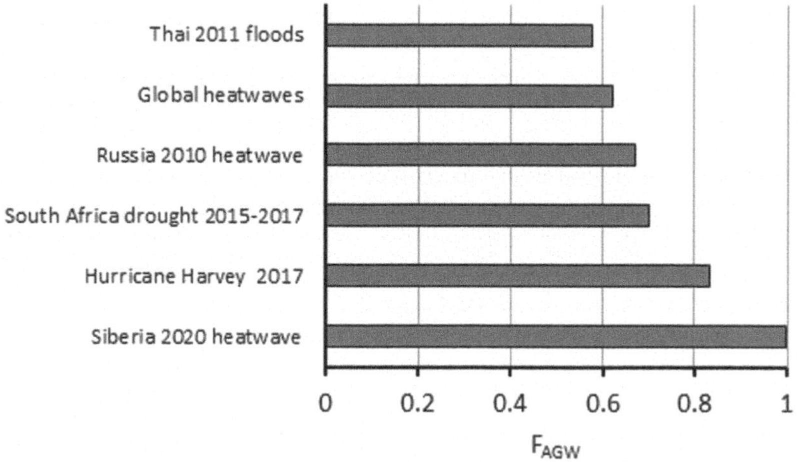

Fig. 2 Range of F_{AGW} values for the fraction of events caused by human-induced global warming for the examples discussed

and special, and backward-looking and forward-looking.[1] By the 1960s, once it became clear that continuing with CO_2 emissions would progressively endanger the climate, major carbon producers could see that they were marketing harmful products.[2] Shue argues that the negative responsibility to "do no harm" required major carbon producers to reduce that

[1] Positive responsibilities are duties to act in certain ways; negative responsibilities are those that ask us to not act in certain ways. A general responsibility may be owed to all human beings, or even all living beings. A special responsibility is only owed to some others, those with which one has a special relationship. Examples might include family, customers, stockholders, creditors, benefactors, or someone to whom a promise was made. A backward-looking responsibility is based on something that has already occurred (e.g., "clean up the mess you made"). By contrast, forward-looking responsibilities rest on opportunity, not blame. A forward-looking responsibility implies a responsibility to act, not because one has caused a problem, or there is a prior connection, but because one is able to improve the situation (Shue, 2017).

[2] In 1965, the US President said in a message to Congress: "this generation has altered the composition of the atmosphere on a global scale through radioactive materials and a steady increase in carbon dioxide from the burning of fossil fuels" (Jamieson, 2014). Later that year, the President's Science Advisory Committee issued a report treating CO_2 as a pollutant, with an appendix on "Atmospheric Carbon Dioxide" (US White House, 1965).

harm rapidly either by modifying the product to capture its dangerous emissions, or by developing safe substitutes, such as carbon-free energy. Harm reduction also includes ending the exploration of additional fossil fuels. Shue sees the half-century of corporate carbon producers' failure to reduce the harm caused by their products as grounds for attributing them with additional responsibility to correct the damage done by decades of neglecting the negative responsibility. If major carbon producers wished to make more than a minimal positive contribution (i.e., do more than only reduce the harm caused by their products), their distinctive political power, wealth, and expertise qualify them to lead the transition to an energy regime which would be safe for future generations to rely on (Shue, 2017).

Beyond the responsibility of the fossil fuel extractors and producers, what is the responsibility of those who burn the fuel, such as motorists in cars, or individuals heating their homes? For motor vehicle emissions, it seems counterproductive to argue that the onus lies with the motorist driving; in practice, their choice is not whether to emit, but whether they require a car. Fossil fuel-free cars are relatively expensive in the present day,[3] and historically either unavailable or not readily available.[4] Therefore, most of the responsibility lies with industrial suppliers to make efficiency gains and alternative technologies that enable fossil-free transport. Additionally, Oreskes and Conway (2010) point to active global warming denial activities by fossil fuel extractors and producers, which would appear to compound their responsibility. For example, Supran & Oreskes (2017) suggest that, although ExxonMobil contributed to advancing science through its scientists' academic publications, it also promoted doubt by emphasizing scientific uncertainties. The authors

[3] According to one study in 2019, the average price of an electric car was 81% higher than conventional cars in the first half of 2019 (Braithwaite-Smith, 2019).

[4] The first small-scale electric cars appeared in Hungary, the Netherlands, and the US around 1828–1835, with the debut of the first electric vehicle in the US in 1889. They become popular around 1900, but were undercut by the Model T Ford, resulting in an almost total decline by 1935. A new generation of electric vehicles started to be developed around 1973, but limited performance and range led interest to fade by 1979. GM released the EV1 in 1996 which gained a cult following, with Toyota introducing the first mass-produced hybrid in 1997 (the Prius). In 2010, GM launched the plug-in hybrid Chevy Volt and Nissan launched the all-electric LEAF. As battery costs started to drop from 2013 gradually, the availability and choice of electric cars has increased (US Department of Energy, n.d.).

conclude that the oil company misled the public, a decision which ExxonMobil rejects.[5]

This does not excuse users, but it highlights the difference in users' level of responsibility versus that of fossil fuel extraction companies. For example, private motorists generate non-industrial emissions. For users to assume some responsibility, the allocation could be based on estimates of the current split between industrial and non-industrial emissions. Based on 2015 global industrial and non-industrial emissions, Rayer et al. (2020) suggest that fossil fuel extraction companies should be responsible for 77% of emissions.

It is necessary to consider the extent to which non-industrial emissions are a consequence of industrial emissions. After all, it is companies that provide most of the greenhouse-gas-emitting products that private individuals use. The analysis also omits Scope 2 emissions resulting from energy used by a company, including electricity, steam, heating, and cooling, although in global terms, these would be caught by the emissions of utility companies providing these services. Per Shue's (2017) arguments, and active climate denial (mentioned above), it seems reasonable to argue that a far greater share than 77% should be attributable to the fossil extraction firms considered, possibly as high as 100%.

A "neutral" estimate might be to split the difference between the two bounds to regard 88.5% as the appropriate split between producer and user damages for the firms in question. Consider the HCL attributable to the nine firms that Table 3 lists over the period of 1751 to 2017 of $38.4 billion. Based on the producer-user share of responsibility, these should perhaps be reduced to $34 billion [= $38.4bn × 0.885], resulting in a reduced—but still substantial—detriment to their share prices of 2.5% [= $34.0bn/$1358bn] based on August 2018 market capitalizations. The higher share of responsibility relative to users represents an increase over the factor of 0.77, per Rayer et al. (2019, 2020), reflecting the greater failure of moral responsibility indicated by Shue (2017).

[5] For a more extended discussion on climate change denial, see Oreskes and Conway (2010), with brief comments in Rayer et al. (2020).

6.3 Combination of Flood and Drought Baseline and Producer Responsibility

Our argument highlights two important aspects: firstly, the fraction of damages that should be attributed to global warming, $F_{AGW} = 0.79$, and secondly, the split between extractor and user responsibility, estimated above at 88.5%. On this basis, the initial damage estimate of $38.4 billion for the nine firms due to their emissions over the period of 1751 to 2017 should be reduced by both factors. Thus, hypothetical climate damages attributable to the nine listed firms should perhaps be reduced to $26.9 billion [= $38.4bn × 0.79 × 0.885]. As a percentage of the market capitalizations of the nine firms that Table 2 lists, this figure amounts to a share-price detriment of 2.0% [= $26.9bn/$1358bn]. Considered together, all these figures still fit within the 2–3% share-price detriment mentioned above and do not alter the overall conclusion. Table 5a summarizes these results.

As previously discussed, the COVID-19 pandemic had a significant effect on the market capitalizations of the firms, with partial recovery by June 1, 2020. Reduced market capitalizations result in higher detriments in share-price terms. Table 5b shows the resulting HCLs in share-price terms.

7 How Should Investors React?

It is difficult to determine whether or how quickly investors should respond to the possibility of companies having, or deciding, to make contributions for damages associated with global warming caused by their past emissions. As market reaction to the COVID-19 pandemic demonstrated, a global shift in perceptions (perhaps relating to the need for the low-carbon transition), can cause stock markets to reprice shares rapidly. Should investors react at all? Cautious investors might be concerned, particularly if they are uncertain about how much these risks are priced into the relevant companies' shares. Undoubtedly, potential AGW damages increase uncertainties, which financial risk analysts would wish to factor into their calculations. Market participants seek to anticipate investment opportunities and avoid risks ahead of competitors, making markets forward-looking. Investors must, therefore, make judgments and forecasts about the economy and other developments to gauge investment outcomes in the face of incomplete information. Incomplete or fragmented information results in a wide range of views, which cannot, of course, all be correct. However, as other investors seek to anticipate trends, market participants do not have the luxury of waiting for certainty to develop, since, by then, markets will have already responded and share prices will have adjusted (Rayer, 2018).

The impact of legal cases should also be considered. Rayer et al. (2020) review these more extensively. For example, on January 9, 2018, the City of New York filed a climate change lawsuit against BP, Chevron, ConocoPhillips, ExxonMobil, and Royal Dutch Shell (Mooney & Crookes, 2018). The case was dismissed on July 19, 2018, with the judge saying that climate change should be addressed through federal regulation and foreign policy (Pierson, 2018; van Voris, 2018). While unsuccessful, future similar cases may weaken the illusion that the oil industry is too big to take on. Although oil companies maintain that the issue of climate change does not belong in court, but rather with the legislative authorities of the country, repeated litigation may undermine their moral position (Slav, 2018). Despite the legal outcome, the City of New York stated that it would start seeking ways to divest its $189 billion pension schemes out of fossil fuel companies in a responsible way, consistent with fiduciary obligations (Mooney & Crookes, 2018). While legal cases may not have made progress to date, the change in attitude among large institutional investors does point to ethical and sustainable investing as a means of promoting change.

Thus, any movement toward an active liability regime could risk shares in such companies becoming stranded assets, with other investors being reluctant to buy them, except at a significant discount, a fact of which oil companies are aware (Supran & Oreskes, 2017). Given mounting evidence, investors may question whether these risks are appropriately priced into high-CO_2-emitting companies' shares.

8 Conclusions

This chapter extends the climate liability regime concept proposed by Rayer and Millar (2018a, 2018b, 2018c) and Rayer et al. (2019, 2020) relating to the 2017 North Atlantic hurricane season, applying it to global flood and drought damages arising over the period of 2012 to 2016. Hypothetical climate liability is expressed in terms of the market capitalizations (or share prices) of nine high carbon-emitting firms.

Inflation-adjusted, global flood and drought damages over the period of 2012 to 2016 were \$265 billion in 2016 USD terms. The climate science reviewed in this chapter progressively links the increased frequency or intensity of extreme weather events to human-induced global warming. We review the literature with specific emphasis on the atmospheric water cycle, floods, droughts, and heatwaves. We also discuss water scarcity and water stress in the context of increasingly robust attribution results.

As global warming is primarily caused by cumulative CO_2 emissions (Allen, 2016), this allows historical responsibility to be allocated relatively simply. Between 1751 and 2017, nine high-emitting publicly owned firms collectively accounted for 14.5% of Scope 1 and 3 emissions (Heede, 2019). Based on the historical emissions of the nine carbon-intensive firms that Table 2 lists between 1751 and 2017, this figure implies a potential hypothetical climate liability of US\$38.4 billion with respect to global flood and drought damages arising during the period of 2012 to 2016. The potential HCL amounts to between 2.8–7.1% in share-price terms. The variation arises from expressing the potential HCL as a fraction of company market capitalizations. During the COVID-19 pandemic, the nine firms' market capitalizations declined significantly. Thus, the potential HCL of 7.1% of share price relates to March 18, 2020. Thereafter, share prices somewhat recovered, and by June 1, 2020, the HCL was 4.6% of the nine firms' market capitalizations. Throughout the pandemic, the demand for fossil fuels reduced considerably, therefore the market response could be interpreted as a stress-test of those companies' valuations should there be a significant move toward decarbonizing the global economy.

Extreme weather occurred before the onset of AGW, so only the costs linked to additional frequency or intensity of floods, droughts, and heatwaves should be considered. Based on the examples given, we estimate the fraction of the impact of these events caused by AGW, denoted F_{AGW}. The resulting range for F_{AGW} was 0.575–0.998, with a mid-point of 0.79, which we used in the subsequent analysis.

We also consider the extent to which fossil fuel companies should be held responsible, from a historical and moral perspective. As it became clear in the 1960s, that CO_2 emissions were progressively altering Earth's climate in dangerous ways, Shue (2017) argues that major carbon producers knew they were marketing harmful products and should have acted to reduce their carbon footprint or alter their product portfolio. This would have included ending exploration for additional fossil fuels.

The producers' failure to do so places additional responsibility upon them to correct the externalized damage brought about by decades of their neglect.

Much responsibility to provide efficiency gains and alternative technologies lies with industrial suppliers, as it is companies that create most of the greenhouse-gas-emitting products that private individuals use. Additionally, active global warming denial activities by fossil fuel extractors and producers (see, e.g., Oreskes and Conway [2010]) appear to magnify their responsibility. This does not excuse users, but the extent to which users or fossil fuel extraction companies should be held accountable is a value-based judgment, and although not straightforward, this chapter attempts to progress the debate on user accountability. Based on 2015 global industrial and non-industrial emissions, Rayer et al. (2020) suggest that fossil fuel extraction companies should be responsible for 77% of emissions. We argue that a greater share should be attributable to the fossil extraction firms that our analysis considers. Our "neutral" estimate is to propose 88.5% as the appropriate share of damages. The higher share of responsibility relative to users represents an increase over the factor of 0.77 used in Rayer et al. (2019, 2020), and reflects a greater failure of moral responsibility.

Accounting for both the occurrence of floods, droughts, and heatwaves before AGW and the producer-user share of responsibility, the potential HCL is reduced to US$26.9 billion for the nine high-emitting firms under consideration. For the different market capitalizations considered, this amounts to a HCL of a 2.0–5.0% detriment in share-price terms (Tables 5a and 5b). Since our analysis neglects other effects of global warming, such as sea level rise, we expect figures to be significantly greater if those are included. Future research would allow for a more comprehensive overview of global damages associated with other extreme weather events.

Given the mounting evidence, studies such as ours may allow investors to question whether these risks are appropriately priced into high-CO_2-emitting companies' shares. Investors may be concerned, particularly if they are uncertain about how much these risks are priced into the relevant companies' shares. Some pension fund managers are seeking ways to divest in a way consistent with fiduciary obligations. This may signal that a movement toward an active liability regime could risk shares in such companies becoming stranded assets, a fact that oil companies appreciate

Table 5a Estimated flood and drought hypothetical climate damage contributions for the nine firms, and as a percentage of their combined firm market capitalizations in August 2018 (see main text)

Nine firms, 1751–2017 August 2018 market capitalizations	*No allowance for increase above estimated flood and drought baseline levels*	*Allowance for increase above estimated flood and drought baseline levels,* $F_{AGW} = 0.79$
No allowance for producer-versus-user responsibility	\$38.4 billion 2.8%	\$30.4 billion 2.2%
Allowance for producer-versus-user responsibility at 88.5%	\$34.0 billion 2.5%	\$26.9 billion 2.0%

Note Hypothetical climate damage contributions are presented with and without allowance for estimated flood and drought baseline levels and producer-versus-user responsibility

Table 5b Estimated flood and drought hypothetical climate damages for the nine firms expressed as a percentage of their combined firm market capitalizations during the 2020 COVID-19 pandemic and on June 1, 2020 (see main text)

Nine firms, 1751–2017 Share price detriments	*No allowance for increase above estimated flood and drought baseline levels*	*Allowance for increase above estimated flood and drought baseline levels,* $F_{AGW} = 0.79$
No allowance for producer-versus-user responsibility	COVID: 7.1% 1st June 2020: 4.6%	COVID: 5.6% 1st June 2020: 3.6%
Allowance for producer-versus-user responsibility at 88.5%	COVID: 6.3% 1st June 2020: 4.1%	COVID: 5.0% 1st June: 3.2%

Note Hypothetical climate damage contributions are presented with and without allowance for estimated flood and drought baseline levels and producer-versus-user responsibility

(Supran & Oreskes, 2017). Although legal cases to date may not have made progress, the change in attitude among large institutional investors does point to ethical and sustainable investing as a means of promoting change.

References

Allen, M. R. (2016). Drivers of peak warming in a consumption-maximizing world. *Nature Climate Change, 4*. https://doi.org/10.1038/NCLIMATE2977.

Allen, M. R., & Ingram, W. J. (2002). Constraints on future changes in climate and the hydrologic cycle. *Nature, 419*(6903), 224–232. https://doi.org/10.1038/nature01092.

Boden, T. A., Andres, R. J., & Marland, G. (2012). Global, regional, and national CO_2 emissions 1751–2009. *Carbon Dioxide Information Analysis Center*, Oak Ridge National Laboratory. U.S. DOE. https://doi.org/10.3334/CDIAC/00001_V2012.

Braithwaite-Smith, G. (2019, October 14). *Electric cars still twice as expensive to buy as normal ones*. https://www.motoringresearch.com/car-news/electric-cars-twice-expensive-buy-normal-ones/.

Caldecott, B., Dericks, G., & Mitchell, J. (2015, March). *Stranded assets and subcritical coal, the risk to companies and investors*. Smith School of Enterprise and the Environment, University of Oxford.

Callendar, G. S. (1938). The artificial production of carbon dioxide and its influence on temperature. *Quarterly Journal of the Royal Meteorological Society, 64*, 223–240. https://doi.org/10.1002/qj.49706427503.

Ciavarella, A., Cotterill, D., Stott, P., Kew, S., Philip, S., van Oldenborgh, G. J., Skålevåg, A., Lorenz, P., Robin, Y., Otto, F., Hauser, M., Seneviratne, S. I., Lehner, F., & Zolina, O. (2020, July 15). *Prolonged Siberian heat of 2020*. World Weather Attribution. https://www.worldweatherattribution.org/wp-content/uploads/WWA-Prolonged-heat-Siberia-2020.pdf.

Dietz, S., Bowen, A., Dixon, C., & Gradwell, P. (2016). Climate value at risk of global financial assets. *Nature Climate Change, 6*, 676–679. https://doi.org/10.1038/nclimate2972.

Dole, R., Hoerling, M., Perlwitz, J., Eischeid, J., Pegion, P., Zhang, T., Quan, X-W., & Murray, D. (2011). Was there a basis for anticipating the 2010 Russian heat wave? *Geophysical Research Letters, 38*. https://doi.org/10.1029/2010GL046582.

Durack, P. J., Wijffels, S. E., & Matear, R. J. (2012). Ocean salinities reveal strong global water cycle intensification during 1950 to 2000. *Science, 336*, 455–458. https://doi.org/10.1126/science.1212222.

Ekwurzel, B., Boneham, J., Dalton, M. W., Heede, R., Mera, R. J., Allen, M. R., et al. (2017). The rise in global atmospheric CO_2, surface temperature, and sea level from emissions traced to major carbon producers. *Climatic Change, 144*, 579–590.

Emanuel, K. (2017). Assessing the present and future probability of Hurricane Harvey's rainfall. *Proceedings of the National Academy of Sciences of the USA*. https://doi.org/10.1073/pnas.1716222114.

EM-DAT: The International Disaster Database. (2020). C. f. CRED, Producer, & Universite catholique de Louvain (UCL), Brussels, Belgium. Retrieved March 24, 2020, from www.emdat.be.

Foote, E. (1857). Circumstances affecting the heat of the sun's rays. *American Journal of Science*, *66*, 382–383.

Gale, E. L., & Saunders, M. A. (2013, September). The 2011 Thailand flood: Climate causes and return periods. *Weather*, *68*(9), 233–237.

Gillett, N. P., Kirchmeier-Young, M., Ribes, A., Shiogama, H., Hegerl, G. C., Knutti, R., Gastineau, G., John, J. G., Li, L., Nazarenko, L., Rosenbloom, N., Seland, Ø., Wu, T., Yukimoto, S., & Ziehn, T. (2021). Constraining human contributions to observed warming since the pre-industrial period. *Nature Climate Change*. https://doi.org/10.1038/s41558-020-00965-9.

Gleick, P. H. (2014). Water, drought, climate change, and conflict in Syria. *Weather, Climate, and Society*, *6*(3), 331–340. https://doi.org/10.1175/WCAS-D-13-00059.1.

Global Carbon Project. (2018, December 5). *Carbon budget and trends 2018*. www.globalcarbonproject.org/carbonbudget.

Griffin, P. (2017). *The carbon majors database*. CDP Carbon Majors Report 2017. CDP (Carbon Disclosure Project). https://www.cdp.net/en/articles/media/new-report-shows-just-100-companies-are-source-of-over-70-of-emissions.

Haustein, K., Allen, M. R., Forster, P. M., Otto, F. E., Mitchell, D. M., Matthews, H. D., & Frame, D. J. (2017). A real-time global warming index. *Scientific Reports*, *7*(15417). https://doi.org/10.1038/s41598-017-14828-5.

Hawkins, E., & Jones, P. D. (2013). Notes and correspondence on increasing global temperatures: 75 years after callendar. *Quarterly Journal of the Royal Meteorological Society*, *139*, 1961–1963. https://doi.org/10.1002/qj.2178.

Heede, R. (2019). *Carbon majors: Updating activity data, adding entities, & calculating emissions: A training manual*. Snowmass, CO: Climate Accountability Institute. https://climateaccountability.org/pdf/TrainingManual%20CAI%2030Sep19hires.pdf.

Held, I. M., & Soden, B. J. (2006). Robust responses of the hydrological cycle to global warming. *Journal of Climate*, *19*, 5686–5699. https://doi.org/10.1175/JCLI3990.1.

Hwang, Y.-T., Soden, B. J., & Held, I. M. (2011). Corrigendum for Held and Soden (2006). *Journal of Climate*, *24*, 1559–1560. https://doi.org/10.1175/2010JCLI4045.1.

Ionita, M., Nagavciuc, V., Kumar, R., & Rakovec, O. (2020). On the curious case of the recent decade, mid-spring precipitation deficit in central Europe. *npj Climate and Atmospheric Science*, *3*(49). https://doi.org/10.1038/s41612-020-00153-8.

IPCC. (2013). *Climate change 2013: The physical science basis*. Contribution of Working Group I to the Fifth Assessment Report of the Intergovernmental Panel on Climate Change 2013. Cambridge, UK and New York, NY, USA: Cambridge University Press. www.ipcc.ch.

Jackson, R. (2020). Eunice Foote, John Tyndall and a question of priority. *Notes and Records of the Royal Institution, 74*, 105–118. https://doi.org/10.1098/rsnr.2018.0066.

Jamieson, D. (2014). *Reason in a dark time: why the struggle against climate change failed–And what it means for our future*. Oxford University Press.

Keeling, C. D., Piper, S. C., Bacastow, R. B., Wahlen, M., Whorf, T. P., Heimann, M., & Meijer, H. A. (2001). *Exchanges of atmospheric* CO_2 *and* $13CO_2$ *with the terrestrial biosphere and oceans from 1978 to 2000* (I. Global aspects, SIO Reference Series, No. 01-06, 88 pp.), Scripps Institution of Oceanography, San Diego.

Krosinsky, C., Robins, N., & Viederman, S. (2012). *Evolutions in sustainable investing: Strategies, funds and thought leadership*. John Wiley & Sons.

Le Quéré, C., Andrew, R. M., Friedlingstein, P., Sitch, S., Hauck, J., Pongratz, J., Pickers, P. A., Korsbakken, J. I., Peters, G. P., Canadell, J. G., Arneth, A., Arora, V. K., Barbero, L., Bastos, A., Bopp, L., Chevallier, F., Chini, L. P., Ciais, P., Doney, S. C., ... Zheng, B. (2018). Global carbon budget 2018. *Earth System Science Data, 10*, 2141–2194. https://doi.org/10.5194/essd-10-2141-2018.

Leach, N. J., Millar, R. J., Haustein, K., Jenkins, S., Graham, E., & Allen, M. R. (2018). Current level and rate of warming determine emissions budgets under ambitious mitigation. *Nature Geoscience, 11*, 574–579.

Lehmann, J., Mempel, F., & Coumou, D. (2018). Increased occurrence of record-wet and record-dry months reflect changes in mean rainfall. *Geophysical Research Letters, 45*(24) https://doi.org/10.1029/2018gl079439.

Macrotrends. (2020). *Market cap*. Retrieved June 11, 2020, from https://www.macrotrends.net/; https://www.macrotrends.net/stocks/charts/XOM/exxon/market-cap.

Mooney, A., & Crookes, E. (2018, January 10). New York sues big oil companies over climate change. *Financial Times*. Retrieved August 1, 2018, from https://www.ft.com/content/4de8e4fc-f62b-11e7-88f7-5465a6ce1a00.

National Academies of Sciences, Engineering, and Medicine. (2016). *Attribution of extreme weather events in the context of climate change*. The National Academies Press.

Oreskes, N., & Conway, E. M. (2010). *Merchants of doubt*. Bloomsbury Press.

Otto, F. E., Massey, N., van Oldenborgh, G. J., Jones, R. G., & Allen, M. R. (2012). Reconciling two approaches to attribution of the 2010 Russian heat wave. *Geophysical Research Letters, 39*. https://doi.org/10.1029/2011GL050422.

Otto, F. E., Skeie, R. B., Fuglestvedt, J. S., Berntsen, T., & Allen, M. R. (2017). Assigning historic responsibility for extreme weather events. *Nature Climate Change, 7,* 757–759.

Otto, F., Wolski, P., Lehner, F., Tebaldi, C., van Oldenborgh, G. J., Hogesteeger, S., Singh, R., Holden, P., Fučkar, N. S., Odoulami, R. C., & New, M. (2018, July 13). *Likelihood of Cape Town water crisis tripled by climate change*. World Weather Attribution. https://www.worldweatherattribution.org/the-role-of-climate-change-in-the-2015-2017-drought-in-the-western-cape-of-south-africa/.

Padrón, R. S., Gudmundsson, L., Decharme, B., Ducharne, A., Lawrence, D. M., Mao, J., et al. (2020). Observed changes in dry-season water availability attributed to human-induced climate change. *Nature Geoscience, 13,* 477–481. https://doi.org/10.1038/s41561-020-0594-1.

Patricola, C. M., & Wehner, M. F. (2018). Anthropogenic influences on major tropical cyclone events. *Nature, 563,* 339–346. https://doi.org/10.1038/s41586-018-0673-2.

Perkins-Kirkpatrick, S. E., & Lewis, S. C. (2020). Increasing trends in regional heatwaves. *Nature Communications, 11*(3357). https://doi.org/10.1038/s41467-020-16970-7.

Pierson, B. (2018, July 19). Oil majors win dismissal of New York City climate lawsuit. *Reuters*. Retrieved August 1, 2018, from https://www.reuters.com/article/us-new-york-climatechange-lawsuit/oil-majors-win-dismissal-of-new-york-city-climate-lawsuit-idUSKBN1K931T.

Porritt, J. (2001). *The world in context: Beyond the business case for sustainable development*. Cambridge: HRH The Prince of Wales' Business and the Environment Programme, Cambridge Programme for Industry.

Promchote, P., Wang, S.-Y. S., & Johnson, P. G. (2016). The 2011 great flood in Thailand: Climate diagnostics and implications from climate change. *Journal of Climate, 29,* 367–379. https://doi.org/10.1175/JCLI-D-15-0310.1.

Rahmstorf, S., & Coumou, D. (2011). Increase of extreme events in a warming world. *PNAS, 108*(44), 17905–17909. www.pnas.org/cgi/doi/10.1073/pnas.1101766108.

Rayer, Q. G. (2018). Market crises: Should we discuss them more—Including with our clients? *The Review of Financial Markets, Q1*(16), 66–69.

Rayer, Q. G., & Millar, R. J. (2018a, February 8). Investing in Extreme Weather Conditions. *Citywire Wealth Manager®*, (429), 36.

Rayer, Q. G., & Millar, R. J. (2018b, June 11). Hurricanes hit company share prices. *Citywire New Model Adviser®*, (596), 18.

Rayer, Q. G., & Millar, R. J. (2018c, August). Investing in the climate. *Physics World, 31*(8), 17.

Rayer, Q. G., Pfleiderer, P., & Haustein, K. (2019, September 20–21). *Global warming and extreme weather investment risks (abstract)* [Conference session].

Emerging Risks in Finance: Challenges and Opportunities for Investors, Institutions and Regulators, Conference and Joint Academic/Practitioner Workshop, Concordia University, John Molson School of Business. Montreal, Canada.

Rayer, Q., Pfleiderer, P., & Haustein, K. (2020). Global warming and extreme weather investment risks. In T. Walker, D. Gramlich, M. Bitar, & P. Fardnia (Eds.), *Ecological, societal, and technological risks and the financial sector* (Vol. Palgrave Studies in Sustainable Business in Association with Future Earth, pp. 39–68). Palgrave Macmillan. https://doi.org/10.1007/978-3-030-38858-4_3.

Schiermeier, Q. (2018, August 2). Climate as culprit. *Nature*, 560, 20–22.

Scripps Institution of Oceanography. (2019). *The Keeling Curve*. Retrieved February 4, 2019, from https://scripps.ucsd.edu/programs/keelingcurve/.

Selby, J., Dahi, O. S., Frohlich, C., & Hulme, M. (2017). Climate change and the Syrian civil war revisited. *Political Geography, 60*, 232–244. https://doi.org/10.1016/j.polgeo.2017.05.007.

Shue, H. (2017). Responsible for what? Carbon producer CO_2 contributions and the energy transition. *Climatic Change, 144*, 591–596. https://doi.org/10.1007/s10584-017-2042-9.

Skeie, R. B., Fuglestvedt, J., Berntsen, T., Peters, G. P., Andrew, R., Allen, M., & Kallbekken, S. (2017). Perspective has a strong effect on the calculation of historical contributions to global warming. *IOP, Environmental Research Letters, 12*(2) 024022. Retrieved June 28, 2021, from https://iopscience.iop.org/article/10.1088/1748-9326/aa5b0a/pdf.

Slav, I. (2018, July 3). *Rhode Island launches anti-oil lawsuit*. Oilprice.com Retrieved July 4, 2018, from https://oilprice.com/Latest-Energy-News/World-News/Rhode-Island-Launches-Anti-Oil-Lawsuite.html.

Steffen, W., Rockström, J., Richardson, K., Lenton, T. M., Folke, C., Liverman, D., et al. (2018). Trajectories of the Earth System in the Anthropocene. *PNAS, 115*(33), 8252–8259. https://doi.org/10.1073/pnas.1810141115.

Stern, N. (2006). *Stern Review executive summary*. New Economics Foundation. https://webarchive.nationalarchives.gov.uk/+/http:/www.hm-treasury.gov.uk/media/4/3/executive_summary.pdf.

Supran, G., & Oreskes, N. (2017). Assessing ExxonMobil's climate change communications (1977–2014). *Environmental Research. Letter, 12*. 084019. Retrieved June 28, 2021, from https://iopscience.iop.org/article/10.1088/1748-9326/aa815f/pdf.

The World Bank. (2020). *Inflation, consumer prices (annual %)*. Retrieved June 11, 2020, from https://data.worldbank.org/indicator/FP.CPI.TOTL.ZG?end=2019&start=1960&view=chart.

Tyndall, J. (1859). Note on the transmission of radiant heat through gaseous bodies. *Proceedings of the Royal Society of London, 10*, 37–39.

Tyndall, J. (1861). The Bakerian Lecture: On the absorption and radiation of heat by gases and vapours, and on the physical connexion of radiation, absorption, and conduction. *Philosophical Transactions of the Royal Society of London, 151*, 1–36. http://www.jstor.org/stable/108724.

UN FCCC. (2015). Adoption of the Paris Agreement. United Nations Framework Convention on Climate Change. https://unfccc.int/resource/docs/2015/cop21/eng/l09r01.pdf.

US Department of Energy. (n.d.). *Timeline: History of the electric car*. Retrieved January 13, 2021, from https://www.energy.gov/timeline/timeline-history-electric-car.

US, White House. (1965). *Restoring the quality of our environment*. Report of the Environmental Pollution Panel, President's Science Advisory Committee Washington. Washington: US Government Printing Office. https://carnegiedge.s3.amazonaws.com/downloads/caldeira/PSAC,%201965,%20Restoring%20the%20Quality%20of%20Our%20Environment.pdf.

van Oldenborgh, G. J., van der Wiel, K., Sebastian, A., Singh, R., Arrighi, J., Otto, F., Haustein, K., Li, S., Vecchi, G., & Cullen, H. (2017). Attribution of extreme rainfall from Hurricane Harvey, August 2017. *IOP Environmental Research Letters, 12*. 124009. Retrieved June 28, 2021, from http://iopscience.iop.org/article/10.1088/1748-9326/aa9ef2.

van Oldenborgh, G. J., van Urk, A., & Allen, M. (2012). The absence of a role of climate change in the 2011 Thailand floods. *Bull Amer Meteor Soc, 93*, 1047–1049. https://doi.org/10.1175/BAMS-D-12-00021.1.

van Voris, B. (2018, June 13). Oil companies ask to throw out New York's climate change lawsuit. *Bloomberg*. Retrieved August 1, 2018, from https://www.bloomberg.com/news/articles/2018-06-13/oil-companies-ask-to-throw-out-new-york-s-climate-change-lawsuit.

Zhou, S., Williams, A. P., Berg, A. M., Cook, B. I., Zhang, Y., Hagemann, S., Lorenz, R., Seneviratne, S. I., & Gentine, P. (2019, September 17). Land–atmosphere feedbacks exacerbate concurrent soil drought and atmospheric aridity. *PNAS, 116*(38), 18848–18853. www.pnas.org/cgi/doi/10.1073/pnas.1904955116.

Chronic Coastal Water Threats Warrant a Valuation Re-Think

Dharisha Mirando and Debra Tan

1 Introduction: New Risk Landscape Will Cause Reassessment of Valuations

Rising seas are being locked-in today, and they will continue to rise. This will permanently erode valuable and densely populated coastlines. In addition, we will likely reach 1.5 °C by 2030, 70 years earlier than the original Paris Agreement target of 2100. Therefore, sea level rise (SLR) is more imminent than many realize; three meters (m) of SLR is thus a plausible reality by 2100 (CWR, 2020b; IPCC, 2018).

A study in 2020 points out that on our current climate path (RCP8.5), without any coastal protection or adaptation, coastal flooding could affect 48% of the world's land area and 52% of the global population by 2100. It could also damage 46% of global assets which could be worth as much as

D. Mirando (✉) · D. Tan
China Water Risk (CWR), Hong Kong, China
e-mail: dmi@chinawaterrisk.org

D. Tan
e-mail: dt@chinawaterrisk.org

T. Walker et al. (eds.), *Water Risk and Its Impact on the Financial Markets and Society*, Palgrave Studies in Sustainable Business In Association with Future Earth,
https://doi.org/10.1007/978-3-030-77650-3_7

US$14.2 trillion, approximately 20% of the global gross domestic product (GDP) (Kirezci et al., 2020).

The Asia-Pacific region (APAC) is particularly at risk; five of the top 10 countries most affected by "weather-related loss events" from 1998 to 2017 were in Asia (Germanwatch, 2019). In addition, with many of the APAC's capitals and financial, commercial, and transport hubs found in low lying regions, coastal threats will be significant (CWR, 2020a).

Central banks and regulators have recognized sea level rise as a significant financial risk with the need to factor this into finance (Network for Greening the Financial System, 2020). Yet, the valuation of the risks lags due to its long-term nature and the expectation that we have time. However, the same way freehold assets are afforded a higher valuation compared to a leasehold property, long-term SLR will have a clear downward impact on valuations as it will shorten the lifespan of all coastal areas. This means that discount rates, terminal values, capital adequacy ratios, and the cost of capital should all be re-assessed for assets in vulnerable locations.

Not factoring in these long-term coastal threats has led to (1) a negative finance feedback loop where capital continues to flow to carbon intensive sectors and high-risk locations which increase risks (Fig. 1); (2) governments are not focused on adaptation to protect vulnerable cities; and (3) the financial sector is not assessing if portfolios are clustered

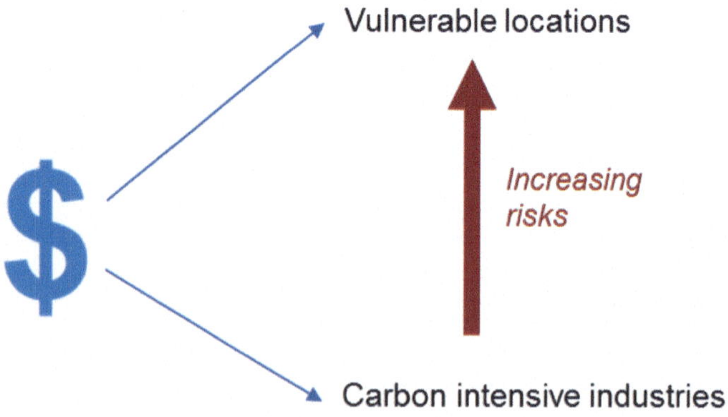

Fig. 1 Negative finance feedback loop. *Note* This figure has been reproduced for this chapter with the kind permission of CWR (*Source* [CWR, 2020b])

in vulnerable locations which could lead to economic collapse (CWR, 2020b).

The financial sector, businesses, and governments must factor in these tail risks from chronic coastal threats to avoid this. With central banks and regulators starting to recognize the risks, valuation adjustments are inevitable.

This chapter and the charts have been summarized with permission from China Water Risk (CWR) from its five-report series "CWR Coastal Capital Threat Series" (2020), which benchmarks risks for 20 coastal capitals and economic hubs in APAC from Tokyo to Sydney. The five reports in the series include an overview of the latest science on rising seas; a ground-breaking CWR APACCT 20 Index to benchmark risks; physical threat levels of 20 APAC cities; what governments are doing or not doing to protect from existential threats; analyses of sovereign and clustered financial risks from GDP and trade to bank loans; as well as next steps and to-do lists for multiple stakeholders to waterproof APAC.

2 Changing Physical and Regulatory Landscapes

CWR's analyses and research show that there is an outlying chance that we will see multi-meter SLR by 2100 (CWR, 2020b). Plus, our actions today are locking in significantly more SLR. With central banks and regulators starting to take notice of these risks, the financial sector will have to act sooner rather than later; as Mark Carney, the then governor of the Bank of England says "once climate change becomes a defining issue for financial stability, it may already be too late" (Carney, M., 2015, p. 4).

2.1 Sea Level Rise Is Accelerating

Sea levels are rising faster than we think:

- Reaching 1.5 °C 70 years early will bring forward SLR: The planet has already warmed by at least 1.1 °C with a high of 1.2 °C in 2016 (WMO, 2020). The World Meteorological Organization (WMO) also warned in 2020 of a 20% chance that one of the next five years (2020–2024) will be at least 1.5 °C warmer. The harsh reality is that we will hit 1.5 °C by 2030 (IPCC, 2018). Reaching 1.5 °C seventy years ahead of our original target in the Paris Agreement of 2100 will no doubt bring forward SLR impacts.

- Oceans have warmed at an alarming rate leading to SLR: So far 90% of the earth's excess heat has been absorbed by the ocean. A recent paper shows that in the last 25 years, the oceans have warmed at an unprecedented rate of 450% compared to the period before (Cheng, 2020). According to the lead author of the research, "the amount of heat we have put in the world's oceans in the past 25 years equals to 3.6 billion Hiroshima atom-bomb explosions" (Kottasová, 2020).
- Grim news points to polar regions warming (atmospheric and ocean) at a faster rate: Average temperatures are projected to rise by 5–12 °C on land and 12–28 °C over the Arctic Ocean by 2100, which will only accelerate SLR (Landrum & Holland, 2020). Already, Greenland and Antarctica are losing ice six times faster than they were in the 1990s (NASA News, 2020).

Clearly, the combination of the factors above will accelerate rapid SLR. The Intergovernmental Panel on Climate Change's (IPCC) high-end SLR estimate is only 1.1 m by 2100 under its worst-case scenario RCP8.5, which is our current climate pathway (IPCC, 2019). However, experts warn it can be multi-meter; the "very likely" range SLR high-end for 2100 in an SLR expert survey in 2020 (H20 Expert Survey) was 1.65 m (Horton et al., 2020) and a Structured Expert Judgement (SEJ) focusing on ice sheets had a high-end "very likely" range of 2.38 m while the top end was 3.29 m (Bamber et al., 2019). The late and renowned scientist Konrad Steffen worried that current observations point to 5 m of SLR in 50–100 years (Bruggen, 2018).

The difference between IPCC estimates and other experts is because the IPCC is intentionally cautious: it shows consensus estimates in the "likely range" (greater than a 66% probability) rather than the "very likely" range (greater than a 90% probability). This means that a large chunk of estimates is "hidden" from the public. Indeed, ice experts warn that IPCC SLR estimates "may be misleading and will likely lead to a poor evaluation of the true risks" (Bamber et al., 2019, p. 11199).

Governments, corporations, and the financial sector must remain dynamic as climate science is evolving. We currently do not know how feedback loops can trigger rapid SLR. Thus, governments and corporations just relying on the IPCC to assess worst-case exposure for adaptation planning will likely leave their countries or assets exposed.

Seas also do not stop rising in 2100; the last time our world faced similar warming (0.5 °C–1 °C), our seas were 6 m–9.3 m higher

(DeConto & Pollard, 2016; Meehl et al., 2020). So, the question is not if this SLR will happen, but *when*—on our current 4 °C climate path we could see 3 m of SLR by 2100 and lock in at least 8.9 m of SLR (CWR, 2020b; Strauss et al., 2015).

2.2 Central Banks and Regulators Starting to Recognize Risks

Rising seas will shorten the lifespan of all assets located in vulnerable regions. Chronic locked-in tail risks from such coastal threats should be factored into sovereign and corporate credit risk ratings as well as equity and project valuations, but they have yet to. A group of now 80+ central banks and supervisors under the Network for Greening the Financial System (NGFS) has recognized that physical climate risks have microeconomic and macroeconomic impacts that translate into credit, market, underwriting, operational, and liquidity risk for all financial institutions (Network for Greening the Financial System, 2020). With regards to coastal threats, they fall into two distinct, but interlinked, types of physical risks:

- Acute: These are event-driven risks. Storm surge risks accompanying typhoons, storms, and hurricanes fall into this category. As a recent report by the NGFS on environmental risk analysis (ERA) notes, historically acute impacts "were considered transient but this will change with increased global warming." They warned that "these events can increase underwriting risks for insurers and impart asset values" (Network for Greening the Financial System, 2020, p. 6); and
- Chronic: These are underlying risks. SLR and subsidence fall into this category. The NGFS warns that chronic impacts will affect capital productivity and require "a significant level of investment and adaptation from companies, households and governments" (Network for Greening the Financial System, 2020, p. 6).

However, as the financial sector has prioritized carbon transition risk scenarios so far, some have factored in acute storm surge risks, but most have not explored the downside risk of locked-in chronic SLR scenarios due to the longer timeframe. Even though NGFS has classified SLR as a chronic risk it has yet to delve into how this would transmit to the financial sector. This is worrying given the severity and speed of rising seas as explained previously. However, it is only a matter of time before regulators push the financial sector to assess and disclose these risks.

3 A Case for Accounting for Chronic Risks Today

Millions of people live in coastal locations, where expensive real estate and economically important trade activities tend to occur. Yet, all of this is at risk from rising seas. The APAC is particularly vulnerable due to the coastal location of its capitals and economic hubs. In 2020 CWR explored the risks to the APAC in five reports in "The CWR Coastal Capital Threat Series," which delve into how 20 APAC cities[1] in 14 countries or territories will be affected by the absolute and relative chronic tail risks from coastal threats.

These cities generate US$5.7 trillion of GDP annually and account for 22% of GDP of their respective 14 countries or territories presenting significant clustered risk as shown in Fig. 2. People and assets are clustered in vulnerable locations, yet governments are not doing enough, and the financial sector needs to factor these in to break the negative finance feedback loop.

3.1 Clustered Assets and People Lead to Increased Risks

CWR's analysis shows that at 2.9 m of SLR, which will be locked-in when we reach 1.5 °C, 28 million people from just 20 capitals and cities in APAC will lose their homes, and urban real estate equivalent to 22 Singapores will be underwater, as will 20 ports and 12 airports as shown in Fig. 3 (CWR, 2020a).

On our current 4 °C climate path, these impacts could happen as early as 2100. In addition, 4 °C will also commit swathes of APAC to permanent submersion as we will lock in 8.9 m of SLR, a plausible reality by 2200. CWR estimates that 102 million people from the 20 APAC cities will be displaced, and urban real estate equivalent to 59 Singapores will be

[1] Aichi/ Nagoya, Auckland, Bangkok, Guangzhou, Ho Chi Minh, Hong Kong, Jakarta, Macao, Manila, Osaka, Seoul, Shanghai, Shenzhen, Singapore, Suzhou, Sydney, Taipei, Tianjin, Tokyo & Yangon.

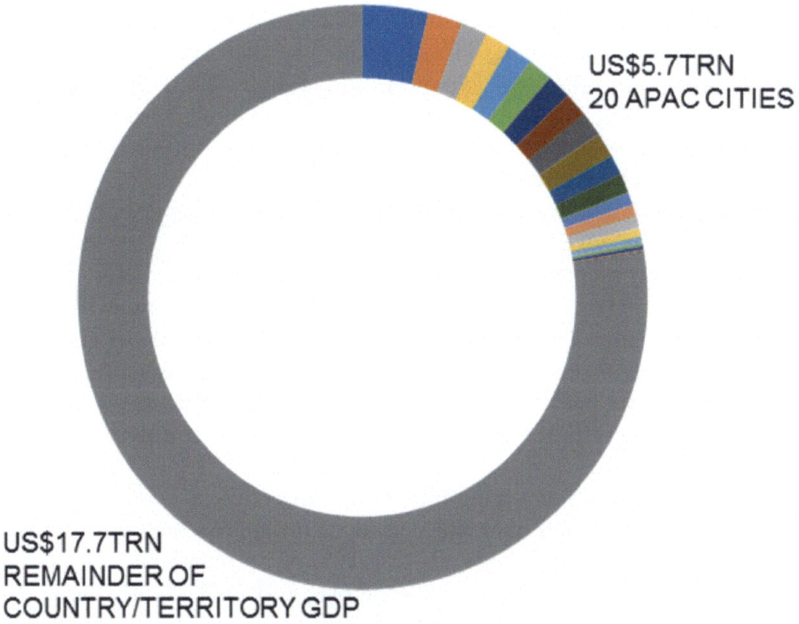

Fig. 2 CWR APACCT 20 Index—20 APAC capitals & cities drive 22% of GDP of 14 countries/territories. *Note* This figure has been reproduced for this chapter with the kind permission of CWR (*Source* [CWR, 2020c])

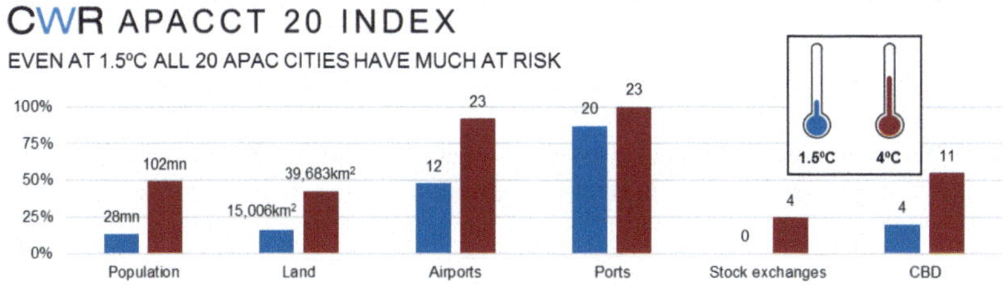

Fig. 3 CWR APACCT 20 Index. *Note* This figure has been reproduced for this chapter with the kind permission of CWR (Source [CWR, 2020c])

underwater including the 11 Central Business Districts (CBDs) of Auckland, Bangkok, Guangzhou, Hong Kong, Ho Chi Minh City, Macao, Osaka, Shanghai, Singapore, Suzhou, and Tianjin. Only 2 out of the 25 airports serving the 20 cities will be operational; all 23 ports will be permanently submerged, which will be damaging for trade reliant APAC economies (CWR, 2020a).

The clustered risk exposure of APAC could lead to economic collapse unless action is taken to decarbonize and adapt for the locked-in impacts. Governments must take significant action to protect its citizens and economies. However, due to sometimes short time horizons and planning, not all governments are acting as expected. In addition, many governments just rely on dated information or conservative IPCC SLR estimates to assess worst-case exposure for adaptation planning as explained previously, but this will likely leave their countries exposed (CWR, 2020b).

Some governments are doing much more than others—CWR's analysis shows that Japan overall has a fairly laissez-faire approach to government adaptation action even though at least 70% of its population is located in large coastal cities; Tokyo, Aichi/Nagoya, and Osaka which make up 33% of GDP are vulnerable to coastal threats, and they face high typhoon risks (CWR, 2020c). On the other hand, Singapore, which is a low-lying island outside the path of typhoons, is raising critical infrastructure by five meters to protect its residents (The Meteorological Service Singapore, 2015).

Clearly governments that are investing in adaptation will protect more of their people and assets; therefore, the assessment of coastal threats must include what governments are doing to adapt as that will directly impact each city or country. Mitigation action is also important but must be evaluated at a global level as it requires collective action to change the risks facing any city.

3.2 *Negative Finance Feedback Loop Will Continue if Valuations Do Not Include Such Risks*

When it comes to climate change, the financial sector has traditionally prioritized assessing carbon transition risks over chronic risks. However, assessing one set of risks but not the other will provide an incomplete picture of the risk landscape. Because of this, wrong investment decisions are being made that will add to risks and perpetuate bad decisions.

On our current climate path of 4 °C, carbon transition risks will be negligible—estimated at US$16 billion in Asia (CWR, Manulife Asset Management & Asia Investor Group on Climate Change, 2019)—but physical impact risks are high, especially from chronic SLR risks. As the sector is not fully accounting for the rising tail risks from the latter set of risks, it continues to support business-as-usual as it believes it can reap the benefits in the short term.

As Fig. 1 shows, not only are carbon intensive industries being supported, but locations that are highly vulnerable to future chronic tail risks due to physical risks and/or the lack of government adaptation action continue to see capital inflows. Unfortunately, allocating capital to business-as-usual and carbon intensive sectors only increases physical risks by reaffirming commitment to a 4 °C pathway (CWR, 2020b). This negative finance feedback loop will continue until chronic physical risks are assessed and the sector can make the proper risk comparisons.

As explained by CWR, to break this negative loop, the financial sector must factor in tail risks from coastal threats as it will realize that the costs of 4 °C chronic and acute physical risks are dramatically larger than the costs of 4 °C carbon transition risks.

In fact, the costs of 4 °C chronic and acute physical risks will be larger than the costs of 1.5 °C carbon transition risks. According to Carbon Delta, the carbon transition cost of a 1.5 °C scenario is US$221 billion in Asia (CWR, Manulife Asset Management & Asia Investor Group on Climate Change, 2019), which is negligible considering that just the 20 cities in APAC analyzed by CWR generate US$5.7 trillion of GDP annually and will be severely impacted at 4 °C.

3.3 *Adjust the Terminal Value to Incorporate Chronic Coastal Threats into Valuations*

Generally, investments have time horizons of three-to-five years at the most. But a discounted cash flow model (DCF) used to value companies and assets takes into account the terminal value (TV), which is typically the value beyond the three-to-five-year forecast period. According to the 2° Investing Initiative and The Generation Foundation, TV accounts for over 82% of a company's fair value (2° Investing Initiative & The Generation Foundation, 2017).

As chronic SLR will create a permanent overhang where everything in vulnerable areas will be changed from "freehold to leasehold" status, and since leaseholds never command the same valuation, adjustments to TV must be made.

We recommend TV to be discounted, rather than wiped out by carrying out sensitivity analysis by doing the following dependent on model type:

- Perpetual growth: Reduce the cash flow perpetual growth rate;
- Exit multiple: Reduce the expected exit multiple (for the chosen financial statistic) below the current average market rate; and
- Either method: Use two different discount rates for the TV and initial forecasted period with the TV discount being higher due to the uncertainty of the value of the company or project beyond the forecasted period.

These valuation adjustments have yet to be made but we must start now; since the magnitude of SLR and storm risks are predicted to rise with time, valuation gaps will only widen.

4 How to Assess Coastal Threats

4.1 A Coastal Risk Index for Risk Assessment

Many see coastal threat assessments as complex and daunting as they are geo-locational. It is important to not only understand the absolute risks per location, but also the relative risks as this informs risk spread across the region. This can then be factored into sovereign and corporate credit risk ratings as well as equity or project valuations to inform re-allocation (if necessary) and future capital allocation decisions.

Ascertaining this is not easy. To plug this gap and to complement the tools that already exist, CWR created a new tool to show the actual physical threat exposure of APAC from coastal threats by also taking into account government adaptation action and benchmarking cities.

CWR engaged over 100 finance professionals from chairs or directors of bank boards to research analysts as well as financial industry associations, asset owners, and financial regulators, where two-thirds were from mainstream finance and were also focused on APAC.

This led to the creation of the CWR APAC Coastal Threat Index for 20 cities (CWR APACCT 20 Index) to assess the absolute and relative coastal threats facing 20 APAC capitals and economic hubs. The index reflects mapped stacked SLR risks for four climate scenarios (1.5 °C, 2 °C, 3 °C, and 4 °C) across key indicators (population, land area, and key infrastructure assets) for each city. Storm surge threats from typhoons, tropical cyclones, or hurricanes, subsidence plus government adaptation actions to reduce physical risks have also been assessed and benchmarked in the index. Government adaptation action was included through multiple indicators in four main categories of research and planning; adaptation investments; acknowledgment and buy-in; and commitments. Finance input was used for the design and development, which physical threats and government actions to include, proxy indicators to use plus index weightings (CWR, 2020a).

CWR shows results and rankings for two consensus scenarios:

- Best-case: 1.5 °C CWR APACCT 20 Index reflects the coastal threat landscape we will likely lock-in by 2030; and
- Base case: 4 °C CWR APACCT 20 Index reflects the coastal threat landscape of the current climate path of 4 °C by 2100.

The index can also be adjusted to show the worst-case scenario to reflect deep uncertainties from polar ice sheet melt linked to SLR as well as thermal expansion of our oceans that have yet to be included in the IPCC SLR estimates. This can be done by shifting timelines forward as the deep uncertainties could bring forward impacts: 1.5 °C CWR APACCT 20 Index's locked-in SLR of 2.9 m may be a reality by 2100; and 4 °C CWR APACCT 20 Index's locked-in SLR of 8.9 m may be a reality by 2200.

4.2 *Absolute and Relative Risk Assessment*

It is important to understand the absolute risks cities face because those that face low risks and contribute a negligible amount of GDP to its country or territory need not be worried about. The 20 APAC cities evaluated by CWR, however, account for a sizeable share of their country/territory's GDP—around 25% (Sydney, Seoul, Taipei, and Yangon), 38% (Auckland and Manila), and up to 100% (Hong Kong, Macao, and Singapore) (CWR, 2020c).

In addition, each city faces different physical risks in terms of what SLR will affect—for example, at 1.5 °C Macao's port, airport, and CBD are impacted as well as 10% of its populations and 44% of landmass, which increases to 40% and 59% respectively at 4 °C (CWR, 2020d) (Fig. 4).

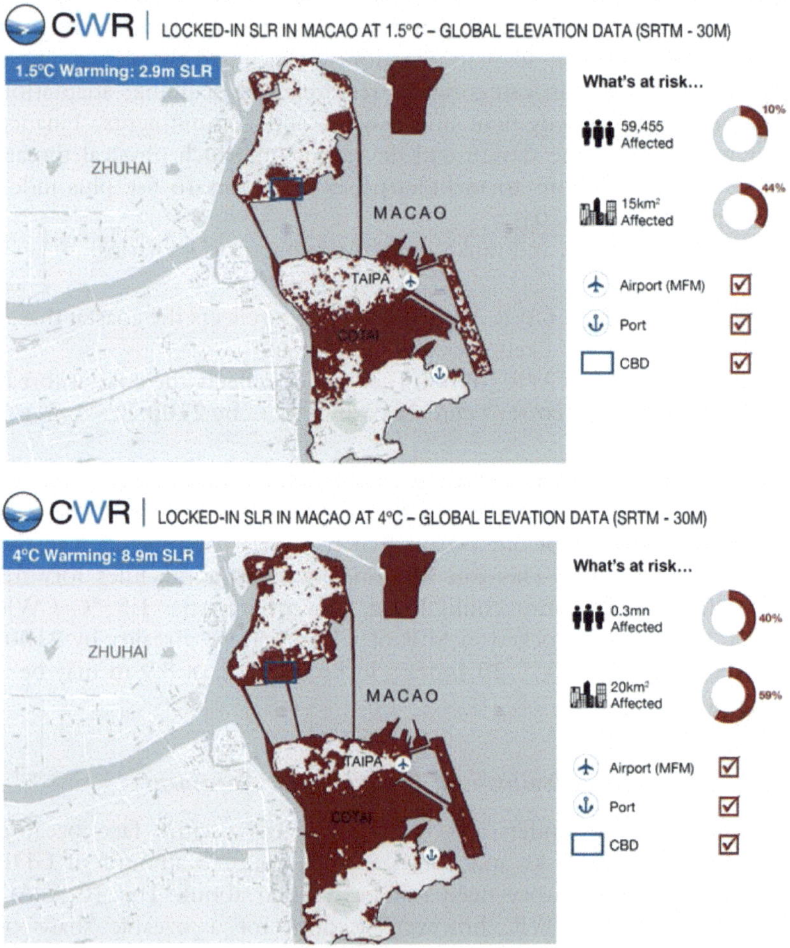

Fig. 4 Locked-in SLR in Macao at 1.5 °C and 4 °C. *Note* This figure has been reproduced for this chapter with the kind permission of CWR (*Source* [CWR, 2020d])

Another example is Ho Chi Minh City where at 1.5 °C only its port is affected as well as 19% of its population and 38% of land but again this all increases to include the airport, CBD, and stock exchange as well as 68% of population and 76% of land at 4 °C (CWR, 2020d).

The CWR APACCT 20 Index can also be used to assess relative risks of the 20 APAC cities with and without government action (Fig. 5).

Key points to note: 1.5 °C and 4 °C CWR APACCT 20 Index

- Cities vulnerable to typhoons and tropical cyclones face more threats and have higher risk scores. Out of the 20 cities in the index, 13 are in the path of typhoons and tropical cyclones and have already experienced the strongest level of typhoon recently.
- At 1.5 °C the most vulnerable cities in Quartile 4 are Taipei (#20), Macao (#19), Aichi/Nagoya (#18), Tokyo (#17), and Hong Kong (#16).
- Some cities become relatively more/less risky as rankings change at 4 °C vs. 1.5 °C, but scores increase for all cities as they all face higher risks as temperatures rise.
- At 4 °C Taipei, Macao, Aichi/Nagoya, and Hong Kong remain the most vulnerable cities whereas Suzhou replaces Tokyo.

While the CWR APACCT 20 Index factors in government adaption action, these were carved out from the index so that we can see just the physical threats faced by all cities at 1.5 °C and 4 °C in an index that excludes government action, which is the CWR APACCT 20 Index—ex. Govt Action (Fig. 6).

Key points to note: 1.5 °C and 4 °C CWR APACCT 20 Index—ex. Govt Action

- Accounting for physical risks alone, Manila is most at risk at #20 at 1.5 °C as it faces material subsidence as well as storm surges. At 4 °C it moves up at #12 as other cities are more at risk.
- At 4 °C, Shanghai is the most at-risk city at #20 accounting for physical risks alone.
- Singapore, Sydney, and Auckland are the three least at-risk at both 1.5 °C and 4 °C.

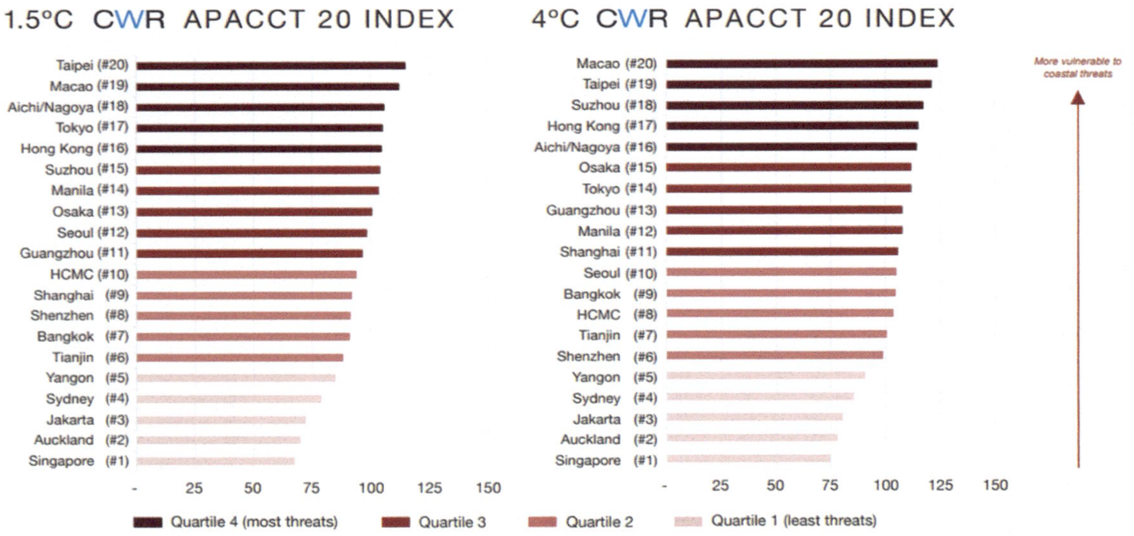

Fig. 5 CWR APACCT 20 Index at 1.5 °C and 4 °C. *Note* This figure has been reproduced for this chapter with the kind permission of CWR (*Source* CWR, 2020a)

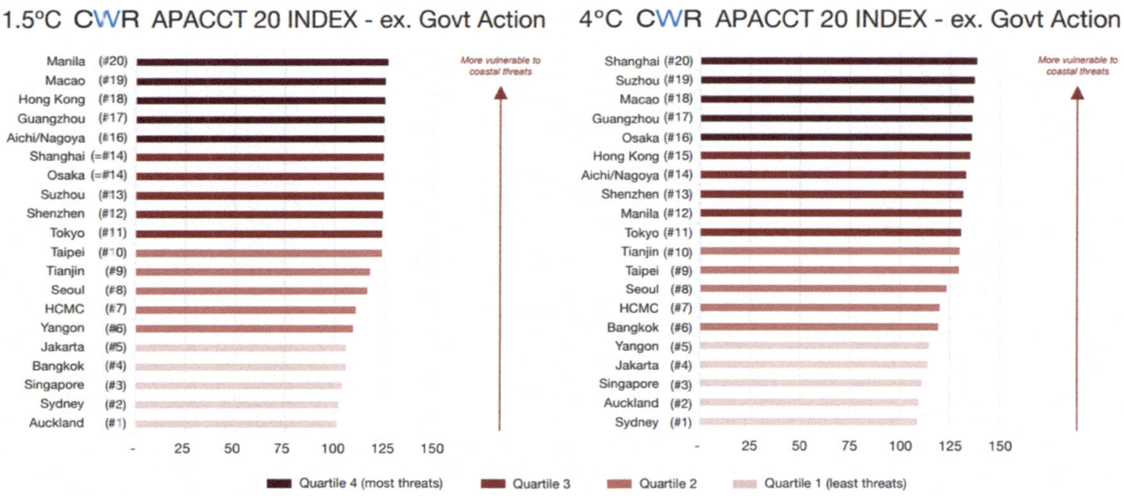

Fig. 6 CWR APACCT 20 Index—ex. Govt Action at 1.5 °C and 4 °C. *Note* This figure has been reproduced for this chapter with the kind permission of CWR (*Source* CWR [2020a])

4.3 *Understanding Government Action*

When it comes to tackling climate change, governments can (1) push for a reduction in carbon emissions so that the worst risks are avoided (mitigation) and (2) invest in protecting people and assets from the expected physical threats due to a warming planet (adaptation).

Finance has been focused predominantly on assessing the former, but mitigation action will only alter future climate risks if efforts are carried out globally with multigovernmental cooperation. Conversely a government's adaptation efforts will directly help reduce risks facing its residents and economy. Government adaptation action can protect cities from the worst of coastal threats to a degree, which will also help sovereigns to be financially resilient in the long term, leading to lower costs.

The CWR APACCT 20 Index only factors in government adaptation actions, not mitigation, as the consensus from the 100+ finance experts indicated they wanted an index that would reflect actual physical threats ahead facing the cities. By carving out government action from the indices, we see which governments are doing more than others, and conversely which are doing the least and so are more at risk.

At 4 °C, Shanghai is the most at risk from coastal threats (#20) in the CWR APACCT 20 Index—ex. Govt Action as 97% of the city's land area will be permanently submerged as well as the homes of 87% of the population, both airports, the port, CBD, and stock exchange (CWR, 2020d). However, due to significant government adaptation action, Shanghai moves up to #11 in the CWR APACCT 20 Index as shown in Fig. 7.

In contrast, at 4 °C, Taipei is only at #10 in the CWR APACCT 20 Index—ex. Govt Action as only 5% of the city will be permanently submerged. Yet, 16% of the population will lose their homes, and the port and one airport will be impacted (CWR, 2020d). However, due to relatively less government adaptation action compared to the other 19 cities, Taipei moves down to #20 in the CWR APACCT 20 Index.

Not only can this information be used to assess risks, but it can also be used to engage corporations based in these locations as well as to lobby governments to do more and learn from best practices elsewhere.

In addition to using the index with and without government action to gauge risks, mitigation action should also be assessed to ensure that government climate strategies are aligned. Governments which have mitigation strategies that head to 4 °C, especially if they are adapting for 1.5 °C, should be further penalized as this will lead to failure.

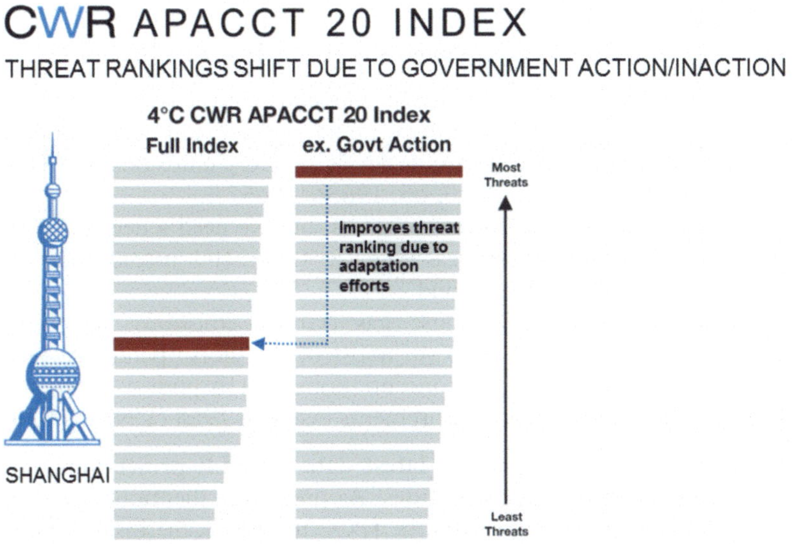

Fig. 7 CWR APACCT 20 Index *Note* This figure has been reproduced for this chapter with the kind permission of CWR (*Source* [CWR, 2020d])

4.4 Dealing with Hotspots

The analysis above will help identify the hotspots, i.e., locations that are vulnerable and contain a significant percentage of a portfolio. For these areas, we recommend a more in-depth analysis to understand the risks. Below are five tips on how to do this:

1. **Use realistic climate scenarios:** In 2020 the WMO Secretary-General warned of a warming path of 3 °C–5 °C (WMO, 2020). In addition, financial consensus from the CWR survey shows that 69% of those surveyed thought that a 3 °C–4 °C was "very likely" and "likely" (CWR, 2020a).

   Therefore, we recommend the use of RCP8.5 as the base case. Some say that this may be too severe as emissions will likely be falling before the end of the century, especially with China's recent pledge toward carbon neutrality by 2060 (Ministry of Foreign Affairs of the People's Republic of China, 2020) and with the United States (US) rejoining the Paris Agreement (The White House, 2021). However, we believe that RCP8.5 should still be the base case because:

- RCP8.5 tracks our current emissions, and it is also "the best match out to midcentury under current and stated policies with still highly plausible levels of CO_2 emissions in 2100" (Schwalm et al., 2020, pp. 19656–19657);
- COVID-19 did lower emissions, but we need a decade of it if we are to push out reaching 1.5 °C to 2100 instead 2030 (Le Quéré et al., 2020); and
- Our global greenhouse gas (GHG) emissions have been rising—in April 2020 (in midst of the global pandemic lockdown) record CO_2 concentrations of 416 ppm were logged, the highest since measurements began in 1958 (UNEP, 2020).

2. **Use local SLR, adjust for latest data for SLR projections, and factor in subsidence and tides**: As previously mentioned, relying only on SLR estimates from the IPCC to build scenarios is discouraged as they have a conservative bias and this can impede risk management. In addition, local SLR data can differ from the set available globally so use local data.

   Worse still, some cities are sinking, which brings forward SLR impacts. For example, North Jakarta's sinking means it will likely face 2.37 m of SLR as early as 2050 (Deltares, 2018; Kopp et al., 2014) and projected SLR in Bangkok could be as high as 1.94 m by 2050; and 3.9 m by 2100 (Duangyiwa, 2020). In addition, local tides must be considered; for example, astronomical high tide in Hong Kong is 2.5 m (Hong Kong Observatory, 2018), which should be added to SLR to gauge impacts.

   The resulting SLR will differ depending on which scenario is used. For example, CWR estimates that under the current climate path, Hong Kong will be affected by SLR of 4.32–5.55 m during astronomical high tide by 2100—for detailed steps on how this was calculated please see (CWR, 2020b).
3. **Test sensitivities by looking at the extremes, not just the averages or medians.** Once the base case for the current climate pathway has been chosen, it is possible to decide which part of the impact range to use, because for each temperature scenario, there is a range of impacts. As can be seen from the case of Hong Kong below, the differences between the low/median/high end of the range of

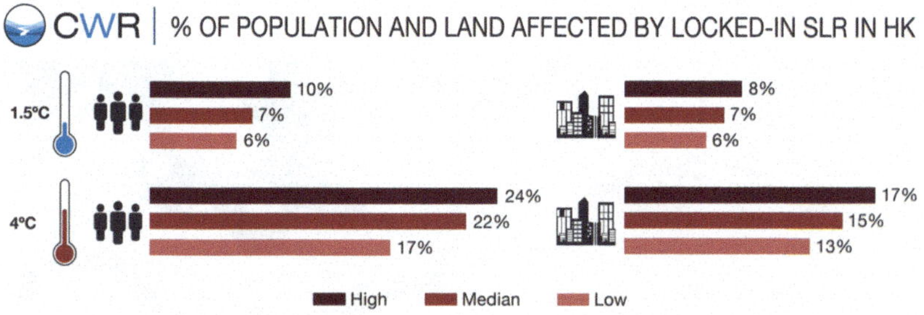

Fig. 8 CWR APACCT 20 Index. *Note* This figure has been reproduced for this chapter with the kind permission of CWR (*Source* [CWR, 2020b])

population and land affected by locked-in SLR in Hong Kong can be significant (Fig. 8):

4. **The correct underlying elevation data is important.** Coastal threats are geo-locational risks, so assets must be mapped against the flood risk. The accuracy of elevation data in the underlying map is thus important. The best and most accurate maps are those that use top-accuracy airborne laser-based (LIDAR) bare earth elevation data. However, this is only available for coastal US, much of coastal Australia, and parts of Europe and Asia, as they are expensive to produce.

   The CWR APACCT 20 Index used NASA's SRTM-30 m elevation data to ensure consistency and comparability in the results. However, impacts will differ if higher granularity analysis was performed; the maps below show flooding from locked-in SLR in Hong Kong at 4 °C using the global elevation map—NASA's SRTM-30 m (left) against the results using Hong Kong's own Digital Terrain Model (DTM) with a 5 m grid (right). Clearly more land is flooded, but also the share of people impacted doubles at 4 °C from 1.5 million using the 30 m map vs. 3.1 million using the more granular 5 m map (CWR, 2020a) (Fig. 9).

   Our recommendation is to use the SRTM 30 m to benchmark various areas as this continues to be the best available elevation data for global comparison in densely populated areas. It is also free. For more in-depth analysis of the risks, higher granularity bare earth digital elevation data should be used if available.

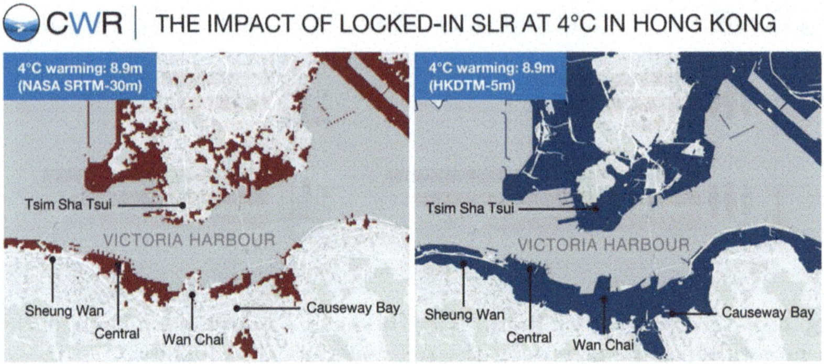

Fig. 9 CWR APACCT 20 Index. *Note* This figure has been reproduced for this chapter with the kind permission of CWR (*Sourc*e [CWR, 2020a])

5. **Factor in impacts of storm tides as these bring forward and exacerbate inundation impacts.** Solely analyzing the impact of SLR underestimates the risks as multiple cities in Asia, such as Hong Kong, Tokyo, and Shanghai, are in the path of typhoons, hurricanes, or tropical cyclones. According to research by The Japan Agency for Marine-Earth Science and Technology and the University of Tokyo published in 2017, the number of strong tropical cyclones will increase by 6.6% by the end of this century compared to 1979–2008 (Japan for Sustainability, 2018).

   As typhoons have had devastating consequences and are only expected to get worse, scenario analysis should be used to gauge storm tide impacts. Costly modeling of "hard to predict accurately" magnitude and frequency of storms, in addition to the exact levels of storm tide, means that models must accurately predict the exact angle and time of arrival as well; this is near impossible at the moment.

   But the impacts of extreme storm tides can still be assessed by estimating extreme storm tide levels and then applying it to the city or region to see impacts. As shown below, the storm tide can be estimated by adding the astronomical tide at the time of the storm, sea level rise projections as well as the storm surge estimate of the storm to chart datum. Here, the chart datum is the base water level to which tide heights are referenced (Fig. 10).

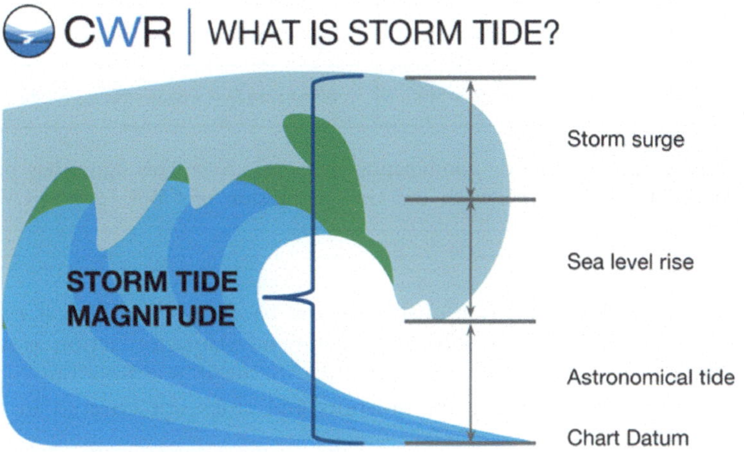

Fig. 10 CWR APACCT 20 Index. *Note* This figure has been reproduced for this chapter with the kind permission of CWR (*Source* [CWR, 2020b]

CWR's report "Changing Risk Landscapes: Coastal Threats to Central Banks" provides step-by-step detail on how to build storm tide scenarios. In summary, assess the potential storm tide today by (1) selecting the largest recent typhoon and isolating the storm surge for the chosen area, (2) adjusting for astronomical high tide, and (3) adjusting for a maximum hit. Determining extreme storm tide in the future is more complicated and would require further discussion with local meteorological departments, but it can still be done by determining extreme storm tide without SLR, and then determining SLR and storm tide combination scenarios. Using this data, base and worst-case storm surge combination scenarios can be estimated as illustrated in Fig. 11; these can be used for stress testing—an example of this for Hong Kong can be found in CWR's report (CWR, 2020b).

To assess impacts, the above combinations of SLR and storm tides must be applied to the highest granularity maps. Only if these scenarios show dire impacts on assets should storm and storm surge modeling for key hotspots be commissioned.

Base case SLR + Base case storm tide	Base case SLR + Worst-case storm tide
Worst-case SLR + Base case storm tide	Worst-case SLR + Base case storm tide

Fig. 11 SLR & storm tide combination scenarios. *Note* This figure has been reproduced for this chapter with the kind permission of CWR (*Source* CWR, [2020b])

The above discussion demonstrates how to assess the physical risks; however, it is important to look at government adaptation action as well since they can reduce risks. The index can be used as a base but more analysis might be necessary for hotspots and cities not covered in the index.

5 Challenges

Data problems continue to be a challenge, but we must start assessing these risks today so that the data can be improved as waiting for the perfect set may be leaving it until it is too late. CWR faced five key challenges in building a consistent methodology to benchmark the 20 APAC cities in the CWR APACCT 20 Index:

1. Determining SLR exposure—which level of locked-in SLR to use?
2. Flooding methodology—bathtub or hydrological?
3. Mapping granularities—trade-off granularity for consistency?
4. Storm surge projections—simple metrics vs. expensive modeling?
5. Government actions are inconsistent and unique—what proxy indicators to use?

The CWR report "Avoiding Atlantis: CWR APACCT 20 Index – Benchmarking coastal threats for 20 APAC cities with finance sector input" goes into detail on how these challenges were overcome, taking into account advice from the financial professionals surveyed. Given that 94% wanted the index despite its known imperfections shows that it is better to start with something than not have anything at all.

The absolute risks may not always be accurate due to data inconsistencies, but benchmarking allows investors to understand which locations are relatively more or less at risk, which has been an extremely useful aspect of the index. In addition, this can be used to find hotspots if a significant part of a portfolio is located in a vulnerable location. Only for these locations do we recommend in-depth analysis as otherwise, it would be expensive to carry out this type of analysis for an entire portfolio in terms of time and expenses.

However, the CWR APACCT 20 Index, as a pilot project, only includes 20 APAC capitals and economic hubs and is solely focused on coastal threats, so not all risks and locations have been evaluated.

6 Conclusion

Assessing coastal threats is possible but there are multiple challenges. Currently the financial sector is focused on carbon transition risks and acute physical risks. Therefore, chronic risks from SLR have been largely ignored as something that can be considered later. However, the risks are being locked-in today and will lead to a permanent loss of homes, expensive real estate, and economically important infrastructure.

With central banks and regulators starting to recognize the risks, it is only a matter of time before banks, asset managers, and asset owners incorporate these risks into valuations.

6.1 Preparing for the Financial Evolution

Central banks will drive chronic risks to be factored in, or else all valuations are off and the negative finance feedback loop will be perpetuated. Pushing for accountability of such chronic risks will result in a recalibration of risks with valuation adjustments across the board. Proper pricing allows for smart allocation of capital away from sectors that increase such risks toward sectors that prevent them and will also reallocate capital away from vulnerable locations, which will break the negative finance feedback loop that otherwise will send us headlong into a dire 5 °C Atlantean future by 2100.

Pricing in chronic risks that are guaranteed in our future will also allow governments and corporations to fast-track both adaptation and decarbonization measures. By doing all of this to ensure financial resiliency, the sector will protect millions of livelihoods and savings.

6.2 Next Steps

As per the CWR report, "Waterproofing APAC To Avoid Atlantis – Executive Summary & Next Steps" (2020) to break the negative finance feedback loop as well as protect our cities and savings from imminent and inevitable chronic dangers ahead all stakeholders must take action:

1. **Assess and reprice risks by applying the "freehold vs leasehold" concept to all valuations.** The index shows which city areas face permanent submergence, so switching assets there from freehold to leasehold status would drastically reduce their value. Applying this concept, all project, equity, or corporate credit ratings and valuations will face downward adjustments unless governments or corporations take action to adapt in addition to pursue significant collective decarbonization action. Financial regulators must ensure that all vulnerable assets are repriced as only this will lead to more proactive action;
2. **Recalibrate sovereign credit ratings for tail risks to ensure financial resilience.** Financial mispricing of the fundamentals has allowed no-sense climate strategies to pervade across governments and all sectors. The CWR APACCT 20 Index can be used to inform credit rating agencies on their re-rating of sovereigns as the 14 countries or territories are clearly exposed through the 20 APAC cities. This does not even include other coastal economic hubs and cities in these countries or territories;
3. **Spread risks by reallocating capital.** Governments should invest in adaptation to protect cities. But if they do not, cities will be left exposed and financial institutions must re-assess their portfolio spread to allocate capital away from vulnerable areas and toward those that are less at risk. The benchmark indices discussed in this report can be used to assess government adaptation action;
4. **Increase planning and investment in adaptation efforts.** Unfortunately, the 1.5 °C climate scenario is unavoidable, so we must plan and invest to adapt using "no-regret scenarios";
5. **Increase planning and investment in mitigation efforts.** Our current climate path of 4 °C will bring dire consequences for the APAC—we must avoid this by reducing emissions;
6. **Actively engage governments to step up adaptation for 1.5 °C and to decarbonize to avoid 4 °C.** Use the potential impact to

push governments as they have a major role to play in ensuring people and assets are protected. Governments must also create a suitable policy environment with appropriate incentives and deterrents to ensure we avoid 4 °C, as the risks rise significantly with temperatures;
7. **Discuss with scientists.** Climate scientists are at the forefront of understanding what we are doing to our planet and the potential impacts, but very few working in finance have come across a cryosphere scientist nor would have likely read scientific research papers. CWR aims to bridge this gap in its Coastal Capital Threat Series, but this must continue as climate science will keep evolving and we must all keep up; and
8. **Fund more research to fill the data gaps.** The current gaps in research make reallocating capital as well as adaptation and mitigation efforts more difficult as there are multiple blind spots. Governments cannot fill all these gaps alone, so the private sector also has an important part to play.

References

2° Investing Initiative, & The Generation Foundation. (2017). *All swans are black in the dark*. https://2degrees-investing.org/resource/all-swans-are-black-in-the-dark/.

Bamber, J., Oppenheimer, M., Kopp, R., Aspinall, W., & Cooke, R. (2019). Ice sheet contributions to future sea-level rise from structured expert judgment. *Proceedings of the National Academy of Sciences of the United States of America, 116*(23), 11195–11200. https://doi.org/10.1073/pnas.1817205116.

Bruggen, J. V. (Director). (2018). *Swiss Polar Institute - Konrad Steffen* [Motion Picture].

Carney, M. (2015, p. 4, September 29). *Speech: Breaking the tragedy of the Horizon—Climate change and financial stability*. https://www.bankofengland.co.uk/-/media/boe/files/speech/2015/breaking-the-tragedy-of-the-horizon-climate-change-and-financial-stability.pdf?la=en&hash=7C67E785651862457D99511147C7424FF5EA0C1A.

Cheng, L. A. (2020). Record-setting ocean warmth continued in 2019. *Advances in Atmospheric Sciences, 37*(2), 137–142.

CWR. (2020a). *Avoiding Atlantis: CWR APACCT 20 Index*. https://www.chinawaterrisk.org/research-reports/avoiding-atlantis-cwr-apacct-20-index-benchmarking-coastal-threats-for-20-apac-cities-with-finance-sector-input/.

CWR. (2020b). *Changing risk landscapes: Coastal threats to central banks.* https://www.chinawaterrisk.org/research-reports/changing-risk-landscapes-coastal-threats-to-central-banks-everything-you-need-to-know-about-sea-level-rise-storm-surge-financial-regulations-to-recalibrate-risks/.

CWR. (2020c). *Sovereigns at risk: APAC capital threats.* https://www.chinawaterrisk.org/research-reports/sovereigns-at-risk-apac-capital-threats/.

CWR. (2020d). *The CWR APACCT 20 Index city factsheets.* https://www.chinawaterrisk.org/research-reports/cwr-apacct-20-index-city-factsheets-at-a-glance-coastal-threat-assessment-for-20-apac-cities/.

CWR, Manulife Asset Management & Asia Investor Group on Climate Change. (2019). *Are Asia's pension funds ready for climate change?* https://www.chinawaterrisk.org/wp-content/uploads/2019/04/Are-Asian-Pension-Funds-Ready-For-Climate-Change-CWR-MAM-AIGCC-Report_FINAL2.pdf.

DeConto, R. M., & Pollard, D. (2016). Contribution of Antarctica to past and future sea-level rise. *Nature, 531*(7596), 591–597. https://doi.org/10.1038/nature17145.

Deltares. (2018). *Deltares. Updated subsidence scenarios Jakarta: MODFLOW SUB-CR calculations for Sunter, Daan, Mogot and Marunda.* https://www.deltares.nl/app/uploads/2018/09/11202275-008-BGS-0004-r-Updated-subsidence-scenarios-Jakarta-DEF.pdf.

Duangyiwa, C. (2020, October 20). *The present & future flood risks in Bangkok.* http://www.chinawaterrisk.org/interviews/the-present-future-flood-risks-in-bangkok.

Germanwatch. (2019). *Global climate risk index 2019.* https://germanwatch.org/en/16046.

Hong Kong Observatory. (2018, October 29). *A wake up call from Mangkhut.* https://www.hko.gov.hk/en/blog/00000216.htm#:~:text=Mangkhut%2C%20packing%20extremely%20high%20winds,unforgettable%20to%20many%20of%20us.

Horton, B. P., Khan, N. S., Cahill, N., Lee, J., Shaw, T., Garner, A., Kemp, A. C., Engelhart, S. E., & Rahmstorf, S. (2020). Estimating global mean sea-level rise and its uncertainties by 2100 and 2300 from an expert survey. *Npj Climate and Atmospheric Science, 3(1).* https://doi.org/10.1038/s41612-020-0121-5.

IPCC. (2018). *IPCC special report: Global warming of 1.5 °C.* https://www.ipcc.ch/sr15/.

IPCC. (2019). *Special report on the ocean and cryosphere in a changing climate.* https://www.ipcc.ch/srocc/.

Japan for Sustainability. (2018, June 11). *Global warming likely to expand radius of gale force wind of tropical cyclones.* https://www.japanfs.org/en/news/archives/news_id036092.html.

Kirezci, E., Young, I., Ranasinghe, R., Muis, S., Nicholls, R., Lincke, D., & Hinkel, J. (2020). Projections of global-scale extreme sea levels and resulting episodic coastal flooding over the 21st century. *Scientific Report, 10*(1), 11629. https://doi.org/10.1002/2014EF000239.

Kopp, R. E., Horton, R. M., Little, C. M., Mitrovica, J. X., Oppenheimer, M., Rasmussen, D. J., Strauss, B. H., & Tebaldi, C. (2014). Probablistic 21st and 22nd century sea-level projections at a global network of tide-gauge sites. *Earth's Future, 2*(8), 383–406. https://doi.org/10.1002/2014EF000239.

Kottasová, I. (2020, January 13). *Oceans are warming at the same rate as if five Hiroshima bombs were dropped in every second*. CNN. https://www.cnn.com/2020/01/13/world/climate-change-oceans-heat-intl/index.html.

Landrum, L., & Holland, M. (2020). Extremes become routine in an emerging new Arctic. *Nature Climate Change, 10*(12), 1108–1115. https://doi.org/10.1038/s41558-020-0892-z.

Le Quéré, C., Jackson, R., Jones, M., Smith, A., Abernethy, S., Andrew, R., De-Gol, A., Shan, Y., Canadell, J., Friedlingstein, P., Creutzig, F., & Peters, G. (2020). Temporary reduction in daily global CO2 emissions during the COVID-19 forced confinement. *Nature Climate Change, 10*(7), 647–653. https://doi.org/10.1038/s41558-020-0797-x.

Meehl, G., Senior, C., Flato, G., Eyring, V., Lamarque, J., Stouffer, R., Taylor, K. E., & Schlund, M. (2020). Context for interpreting equilibrium climate sensitivity and transient climate response from the CMIP6 Earth system models. *Science Advances, 6*(26), eaba1981. https://doi.org/10.1126/sciadv.aba1981.

Ministry of Foreign Affairs of the People's Republic of China. (2020, September 22). *Statement by H.E. Xi Jinping President of the People's Republic of China At the General Debate of the 75th Session of The United Nations General Assembly*. https://www.fmprc.gov.cn/mfa_eng/zxxx_662805/t1817098.shtml

NASA News. (2020, March 16). *Greenland, Antarctica melting six times faster than in the 1990s*. https://climate.nasa.gov/news/2958/greenland-antarctica-melting-six-times-faster-than-in-the-1990s/#:~:text=Observations%20from%2011%20satellite%20missions,they%20were%20in%20the%201990s.&text=All%20total%2C%20Greenland%20and%20Antarctica,of%20ice%20since%20the%201990s.

Network for Greening the Financial System. (2020). *Overview of environmental risk analysis*. https://www.ngfs.net/sites/default/files/medias/documents/overview_of_environmental_risk_analysis_by_financial_institutions.pdf.

Schwalm, C., Glendon, S., & Duffy, P. (2020). Rcp8.5 tracks cumulative CO2 emissions. *Proceedings of the National Academy of Sciences of the United States of America, 117*(33), 19656–19657. https://doi.org/10.1073/pnas.2007117117.

Strauss, B., Kulp, S., & Levermann, A. (2015). Carbon choices determine US cities committed to futures below sea level. *Proceedings of the National Academy of Sciences, 112*(44), 13508–13513. https://doi.org/10.1073/pnas.1511186112.

The Meteorological Service Singapore. (2015). *The Singapore 2nd national climate change study—Phase 1.* https://data.globalchange.gov/report/singapores-second-national-climate-change-study-phase-1.

The White House. (2021, January 20). *Paris climate agreement.* https://www.whitehouse.gov/briefing-room/statements-releases/2021/01/20/paris-climate-agreement/.

UNEP. (2020, May 11). *Record global carbon dioxide concentrations despite COVID-19 crisis.* https://www.unep.org/news-and-stories/story/record-global-carbon-dioxide-concentrations-despite-covid-19-crisis.

WMO. (2020, January 15). *Press release: WMO confirms 2019 as second hottest year on record.* https://public.wmo.int/en/media/press-release/wmo-confirms-2019-second-hottest-year-record.

Managing Water Risk: Investing in the Future

Water Risks, Conflicts, and Sustainable Water Investments: A Case Study of Ontario, Canada

Guneet Sandhu, Olaf Weber, and Michael O. Wood

1 Introduction

Water security is the ability to maintain a sufficient quantity and quality of water resources, which is necessary for social well-being, economic development, ecological integrity, and climate resilience (UN-Water, 2013). Therefore, ensuring water security for current and future generations is the underlying foundation of Sustainable Development Goal (SDG) 6, but much progress needs to be made globally to attain this goal (United Nations, 2018). Water availability (quantity, quality, and access)

G. Sandhu (✉) · O. Weber · M. O. Wood
School of Environment, Enterprise and Development, University of Waterloo, Waterloo, ON, Canada
e-mail: guneet.sandhu@uwaterloo.ca

O. Weber
e-mail: oweber@uwaterloo.ca

M. O. Wood
e-mail: michael.wood@uwaterloo.ca

T. Walker et al. (eds.), *Water Risk and Its Impact on the Financial Markets and Society*, Palgrave Studies in Sustainable Business In Association with Future Earth,
https://doi.org/10.1007/978-3-030-77650-3_8

is continuously being threatened due to increasing population, economic activities, and climatic uncertainty (Hoekstra, 2014). Given the wide social, economic, and ecological implications of water challenges, it becomes necessary to conceptualize, assess, and manage interdisciplinary water-related risks as a means to address various dimensions of water security, including the investment needs (OECD, 2013; United Nations, 2018).

Risks to water resources are multi-faceted due to the duality of water used as a material input for manufacturing, agriculture, power generation, as well as a common good essential for human health, sanitation, recreation, and ecosystem productivity (Hanemann, 2006). While certain uses of water are obvious (e.g., extractive uses for manufacturing, agriculture, municipal water supply, power generation), many in-stream uses (hydroelectricity generation, fisheries, recreation, tourism, and ecosystem services) connected with water resources are equally pertinent (Sandhu et al., 2020). Therefore, adverse water events can have a domino effect across entire watersheds and aquifers that multiplies the impact manifold (Vörösmarty et al., 2010). Moreover, degraded water quality due to chemical or biological contamination events also makes water sources completely unfit for use or highly cost-intensive to treat, thus contributing to water scarcity in the region (Sandhu et al., 2020).

Even though the access to safe drinking water was declared as a human right by the United Nations General Assembly in 2010, 2.2 billion people across the globe are still devoid of this basic human necessity (WHO, 2019). Economic scarcity of drinking water infrastructure is also a major water challenge not only in the Global South, but also in Indigenous communities of high-income countries like Canada (White et al., 2012). As the globe grapples with the recent COVID-19 crisis, equitable access to safe water and wastewater treatment is seen at the forefront of preventing the spread of such viruses and bacteria. This intricate connection of water with public health, sanitation, and hygiene further prioritizes the allocation and provision of safe water for human well-being and re-establishes focus on water security (Sanitation & Water for All, 2020).

Water management is typically under public jurisdiction governed by different levels of national, provincial, and municipal governments and varies considerably across countries. However, sustainable water use and management to attain water security is not only the responsibility of governments, but also of different water-using sectors including the corporate sector. As major users of this economically significant resource, businesses and the financial sector fueling these businesses have a crucial role in designing proactive sustainable water management practices by accounting for various water risks and response strategies.

Even with the increasing academic and practitioner interest in integrating water risk data in operational and investment decisions, there is a gap in understanding and assessing granular interdisciplinary water risks to design holistic decision support tools and guide sustainable water investments (Christ & Burritt, 2018; Hogeboom et al., 2018). The chapter explores different dimensions of water risks, including an overview of the current literature, gaps, and future opportunities in the field. Then using the case study of Ontario, Canada, we gather interdisciplinary water risk evidence at the granular sub-watershed scale and synthesize it into a framework that can guide proactive decision-making in seemingly water-secure regions typically outside the investor or corporate scanner.

2 Background

To categorize water risks and tease out their interdisciplinary nuances, it is essential to understand the theoretical aspects of risk assessment, management, and current water risk management practices in the corporate and financial sectors. The field of risk assessment and management has been spurred by adverse industrial contamination events, natural hazards, or national security threats with severe human and ecological health implications as well as economic implications due to risky financial activities in the past. OECD (2003) and Aven and Renn (2010) define emerging systemic risks as an interdisciplinary and highly networked set of risks that threaten the underlying core systems (e.g., health, infrastructure, environment, energy, transport, etc.) responsible for the functioning of society. Since there is an interconnected nexus of water security with food security, public health, economic development, energy security, ecosystem health, and social well-being, water issues (quantity, quality, and access) are considered as "emerging systemic risks" requiring interdisciplinary assessment and management approaches rooted in risk theory (Aven & Renn, 2010).

2.1 Theoretical Background on Risk Assessment and Management

Aven and Renn (2009, p. 6) define risk as "uncertainty about and severity of the events and consequences (or outcomes) of an activity with respect to something that humans value." Uncertainty is a distinct concept that is discussed in tandem with risk; it is the possibility of different consequences after a hazardous event or the unpredictability of risks due to a lack of information or knowledge (Aven & Renn, 2009; Gough, 1997). Risk assessment involves verifiable data, theory, probabilistic analysis, and modeling using objective methods (Cienfuegos Spikin, 2013). Moreover, risk is rooted in society, and the perception of risk is contingent on values, norms, beliefs, personal experiences, attitudes, and cognitive knowledge (Slovic, 1987). Therefore, in conjunction with the objective or physical measurement of risk, there is an inherent subjectivity in how risk and severity of the outcome is observed and how tolerability of risk is judged, leading to public concern or outrage (Klinke & Renn, 2002; Renn, 1998).

Environmental, fiscal, and organizational policies are largely based on quantitative risk assessments, but the journey from data to an implementable decision-making framework is not devoid of judgment (Aven & Renn, 2009). From an implementation perspective, the field of risk management informs policies and formulation of regulations while operating within given social, economic, and political constraints (Renn, 1998). In practice, environmental protection policies are designed to prioritize foremost the concerns of the public constituents and contain contextual nuances (Russell & Gruber, 1987). In certain cases, public perception and concerns about the potential impacts of a technology or industrial activity may catalyze policy actions triggering risk assessments pre-emptively (Renn, 1998; Russell & Gruber, 1987). Therefore, political, social, economic, and cultural aspects intersect for resources like water with multiple stakeholders, along with hydrological evidence for risk assessments (Gough, 1997; OECD, 2003).

Risk management in totality involves assessment, comparison, and design of strategies of avoiding, reducing, transferring, or adapting to identified risks. Since environmental risk includes ecological, social, cultural, and economic risks, existing risk management approaches can provide a comprehensive foundation where interdisciplinary data can be merged to make sustainable decisions for current and future generations (Gough, 1997; OECD, 2003). Three typical approaches used are ***Technical Risk Assessment*** (engineering, health, environment) that uses purely scientific assessments for decision-making, ***Decision Analytic Approach*** that includes subjective value judgement of decision-makers about possible outcomes based on statistical estimates of an event, and ***Comparative Risk Assessment*** that includes ranking the issues and assigning priorities for response items (Gough, 1997). While the risks (social, ecological, human health) may be assessed from different data sources with varying reliability, the emphasis is on prioritizing issues to design action strategies (Gough, 1997).

Exposure to water risks is also closely related to society in general, hence the tolerability or acceptability of risks is defined by public perception shaped by technical, cultural, and economic contexts (Wyrwoll et al., 2018). In order to account for the contextuality and variability of water issues, the use of granular temporal and spatial scales also becomes warranted for holistic impact and risk assessment affecting a range of stakeholders (Loucks, 2000). A risk-based framework is often recommended in the literature for public water management where the source of risks, corresponding actions, and stakeholders are identified. This process often entails detailed hydrological studies, stakeholder engagement, and impact assessments leading to developing policy responses through the use of regulatory, economic, or stewardship instruments (OECD, 2013; Wyrwoll et al., 2018). Similarly, the field of corporate water accounting, disclosure, and reporting is a means of assessing and managing water-related risks. In modern risk management, risks are often not only a cost or regulatory liability, but also an avenue to identify opportunities for innovation, increased efficiency, and sustainability (Cienfuegos Spikin, 2013).

2.2 Corporate Water Risk Assessment and Management

Most industrial sectors not only require a certain quality and quantity of water for production operations and waste assimilation, but also rely on water-intensive supply chains (Christ & Burritt, 2017b). Given the complexity of water as a shared public, economic, and environmental resource, water risk has been articulated in corporate management literature as physical, regulatory, and reputational risks that can potentially culminate into adverse financial impacts for corporations and their investors (detailed extensively in Barton, 2010; Christ & Burritt, 2017a; Freyman et al., 2015; Morrison & Schulte, 2010; Schulte et al., 2012). ***Physical or operational risks*** arise from a direct scarcity of intake water due to decreased hydrological flows or contamination of supply, or indirectly through water stress-related disruptions in supply chains or electricity inputs (Barton, 2010). ***Regulatory risks*** arise from changing public policies and legislation regarding the allocation and use of water resources resulting in reduced allowed extraction volume, increased water extraction fees, pollution fines, consultation, restrictions, monitoring, and reporting requirements (Signori & Bodino, 2013). ***Reputational risks*** arise from conflicts between competing water use sectors during, for example, drought conditions or existing conflicts with local communities over inequitable allocation or access to water resources that can negatively impact stakeholder perceptions and potential public lawsuits, and lead to regulatory actions like moratoriums or a loss of "social license to operate" (Schulte et al., 2012).

These three categories of water-related risks ultimately culminate in increased operational costs, loss of reputation, potential loss of investors, and increased scrutiny by internal and external stakeholders (e.g., governments, NGOs, suppliers, customers, investors, employees, local communities). Therefore, from a corporate perspective, sustainable water management by managing identified risks is essential to safeguard financial viability, maintain investor interests, and dissipate external stakeholder pressure (Freyman et al., 2015; Hogeboom et al., 2018). Nonetheless, the degree of water risk exposure of each organization is highly variable and contingent on operational factors, i.e., type of sector (water intensity, supply chain, quality of water needed, waste generation) and use of water-efficient technologies. Moreover, many external factors at the watershed or basin level like hydrological conditions, surrounding land use, socio-economic conditions, political and institutional conditions, legislations on water allocation, and use and discharge need to be accounted in such assessments (Signori & Bodino, 2013).

From the perspective of the private sector, water management is inherently complex due to the multidimensionality of water risks and the range of stakeholders involved (Christ & Burritt, 2017b). Nonetheless, the corporate water management framework employed to reduce, mitigate, and adapt to water risks involves firstly the acquisition of global, national, and local data on watershed trends; water accounting data at the facility level on the extraction, discharge and impact on local water resources; and regulatory and reputational water risk data (Christ & Burritt, 2017a; Signori & Bodino, 2013). A plethora of methods and tools have been used by businesses to collect data on watershed conditions, operational water use, the impact on watersheds (availability and quality) as well as regulatory and reputational risks (WBCSD-IUCN, 2012). While some tools like WWF Water Risk Filter and WRI Aqueduct provide hydrological risks based on different global hydrological simulation models, they do not capture the regulatory and reputational risks reflecting public concerns, regulatory context, or local watershed conditions at disaggregated sub-watershed scales (Jorisch et al., 2018; Morgan et al., 2020).

The practice of corporate disclosure of water risks and opportunities is seen as a means to develop accountability especially with rising demands by the public and institutional investors where third party rating agencies are involved in assessing companies (Freyman et al., 2015; Signori & Bodino, 2013; Vörösmarty et al., 2018). While water accounting is done internally to inform strategic decisions at the organizational level, disclosure is done for both internal and external stakeholders in a prescribed format using available best practices or guidance. Global Reporting Initiative (GRI) 303: Water and Effluents, CDP Water Questionnaire, Sustainability Accounting Standards Board (SASB), Alliance for Water Stewardship, UN CEO Water Mandate Corporate Water Disclosure Guidelines, and European Union Non-Financial Reporting Directive are some of the prominent disclosure mechanisms adopted voluntarily by firms to increase transparency and define reporting protocols (Freyman et al., 2015; WBCSD-IUCN, 2012).

While disclosure can facilitate stakeholder dialogue and better water management practices, there is widespread criticism of current corporate water reporting practices. These reports generally lack granular basin-level data, contextual information, stakeholder engagement, and supply chain risks that can help assess the financial and operational risks more accurately (Money, 2014a; Signori & Bodino, 2013). There is stagnancy in water reporting since firms react to issues that can improve operational efficiency within their boundaries rather than engaging with stakeholders on a watershed basis (Christ & Burritt, 2018; Signori & Bodino, 2013). The gaps found in water accounting and disclosure highlight the scarcity of appropriate sub-watershed water security risk data needed by stakeholders to make proactive assessments. The tools designed are fragmented, seldom encompassing, and the management frameworks lack the collaborative interdisciplinary approaches necessary for different types of risks (Christ & Burritt, 2017a, 2018).

2.3 Financial Sector and Water Risks

The flow of financial capital sourced through debt (loans, bonds) or equity is necessary to drive and sustain economic activities in every region and sector. The financial sector comprising of commercial and investment banks, insurance companies, and asset management companies are the backbone of a healthy economy and have employed risk assessment and management techniques to assess borrowers, insurance policyholders, or investees (Alaerts, 2019). Risks to water security pose a significant threat to the financial sector through different asset classes of their portfolios if they culminate in rising operational costs, loss of production, decreased credit ratings, or rising insurance claims due to water-related disasters like floods, hurricanes, or droughts (Hoekstra, 2014; Hogeboom et al., 2018). Reduced cash flows due to a water scarcity event negatively affect the value of assets, repayment of loans, profitability, and share value. Thus, these water-related risks can adversely affect financial viability due to loss of valuation and credit default as well as reputational damage across assets in prime sectors like manufacturing, agriculture, energy, logistics, and real estate (Alaerts, 2019; Jorisch et al., 2018; Money, 2014b). While climate-driven water risks are highly variable across sectors, companies, and locations, investors with long-term investment horizons like pension funds or insurers are most exposed (Alaerts, 2019).

Given the adverse impact of climate change on the financial performance of corporate investments and financial markets, there is increasing emphasis on assessing and disclosing the exposure to climate risks by investors, regulators, and central banks (Alaerts, 2019; Hogeboom et al., 2018; Money, 2014b). Water-related risks are closely associated with climate change, and the management of these risks is foreseen as a leading strategy for climate adaptation leading to a resilient and sustainable economy (Alaerts, 2019). Therefore, the financial sector has a unique position to proactively integrate these risks in their investment policies, risk assessments proposal requirements, credit ratings, or underwriting, and incentivize water security initiatives (Alaerts, 2019; Hoekstra, 2014; Hogeboom et al., 2018). However, with the lack of granular context-based data, regulatory mandate, and complexity of interdisciplinary risk assessment, the integration of water risks in the financial sector has yet to peak (Jorisch et al., 2018; Vörösmarty et al., 2018).

The financial sector at large has been curious about water risks awareness, action, and stakeholder engagement, but the systematic evaluation of local risks using a standardized framework presents a research opportunity (Jorisch et al., 2018). While investor-specific risk assessment and guidance tools like WWF Water Risk Filter, WRI Aqueduct, Ecolab Water Risk Monetizer, Ceres Aquagauge, CEO Water Mandate, and Pacific Institute's Conflict Tool exist, the need for local, reliable, publicly available, and peer-reviewed data persists. Freyman et al. (2015) report that investors have identified the need for collaboration with the scientific community to obtain reliable water security data as well as with regional experts to gain insights about the contextual nuances of water resources to evaluate risks. In the wake of limited information about a company's actual water performance from corporate disclosure, sustainable water investment and practices are expected to benefit from collaboration between investors, academics, and water management and governance experts at granular geographic scale (Vörösmarty et al., 2018).

Nonetheless, the need for water security also reveals investment opportunities and solutions catering to water efficiency, treatment, and reuse in line with SDG 6 (Alaerts, 2019; Hogeboom et al., 2018). The need for investment in sustainable drinking water, wastewater, and stormwater treatment infrastructure and technologies is on the rise in both the Global South and developed nations and estimated at around USD one trillion by 2025 worldwide (Alvarez & Rodriguez, 2015; RobecoSAM, 2015). Moreover, water-related mutual funds are not only socially responsible investments, but also financially perform at par with conventional funds with added diversification value (Alvarez & Rodriguez, 2015). Rising above the notion of socially responsible investments as philanthropic specialty products offered by a few financial institutions, sustainable investing is seen as a valuable proposition where financial, environmental, and social benefits are realized when sustainability metrics are used to guide investments (Weber & Feltmate, 2016). However, institutional investors and banks need realistic assessments, including social and environmental aspects to safeguard the financial viability of projects (Jorisch et al., 2018). Thus, granular water risk assessment in terms of physical, reputational, and regulatory risks is needed for investing in water security solutions and managing water risk exposure in financial portfolios (Alaerts, 2019).

2.4 Gaps and Opportunities in Water Risk Assessment

The extent to which a particular firm or asset is exposed to water risks is highly contingent on the local or granular spatial–temporal hydrological conditions, regulatory environment, type of sector (water extraction, consumption, or potential for water pollution), and competing water users (Hogeboom et al., 2018). Water scarcity is evident in emerging economies like China and India, or in developing countries where population growth, economic growth, and weak regulations have already led to unsustainable resource extraction (Alaerts, 2019; Schulte et al., 2012). Even in developed countries like Australia, Israel, and the USA, the emphasis on sustainable water management is more pronounced in arid regions or in European countries that are densely populated (Morrison et al., 2009). Therefore, rather than using a proactive approach of mitigating emerging water risks, decisions regarding water management are made as a reaction to major water events or focus only on currently water-scarce areas (OECD, 2013). While most water risk assessment tools tend to focus on operational risks for water-intensive sectors based on facility-level data and global hydrological models, the contextual watershed risks including water availability, quality, regulatory risks, and reputational risks are aggregated nationally. Moreover, these tools, based on different models with certain assumptions, may not fully represent water security risks at the disaggregated sub-watershed level that can have adverse financial implications for future businesses and investors (Jorisch et al., 2018; Morgan et al., 2020).

With an evolving understanding of water scarcity in terms of quantity, quality, and access, there is a serious omission of seemingly water-abundant countries, where emerging water risks need to be proactively managed (Sandhu et al., 2020; Wolfe & Brooks, 2003). For instance, Canada is typically considered as a water-rich country with large freshwater resources per capita. However, there is a vast spatial variability of water resources across Canada where 60% of the freshwater resources are available in the sparsely populated North, and the densely populated southern watersheds have been facing temporal water scarcity and quality issues (Mitchell, 2017). In the Canadian context, the assessment and management of flood-related risks and their economic impact are gaining momentum (Thistlethwaite et al., 2020). However, the perception of water abundance masks the stressors contributing to water scarcity with increasing demand amidst diminishing seasonal flows, quality, and equitable access (Mitchell, 2017).

In the recent staff notice published for guidance on "Reporting of Climate Change-related Risks," the Canadian Securities Administrators (CSA) emphasize water availability and quality as material physical risks arising from climate change that can have adverse economic implications for businesses and investors (Canadian Securities Administrators, 2019). Even though water risks are exacerbated manifold by climate change, many cumulative factors exist within Canada like temporal and spatial hydrological conditions, increasing demand, pollution as well as changing legislation pertaining to access, minimum environmental flows, lack of drinking water infrastructure, and social conflicts that contribute to water scarcity (Bakker & Cook, 2011; Curran, 2019). Given these gaps, ripe opportunities prevail to explore and design water risk indicators by harmonizing interdisciplinary quantitative and qualitative data available at the granular spatial and temporal scale (Christ & Burritt, 2018; Jorisch et al., 2018). Therefore, using Ontario as a case study, the ensuing sections aim to tease out multidimensional aspects of water risks intertwined with spatial and temporal resource variability, regulatory environment, and legacy issues that can help design decision support tools for businesses, investors, governments, regulators, and local communities.

3 Case Study: The Province of Ontario, Canada

Canada is a compelling case of a developed resource-rich economy that is relatively water-rich with established water management regulatory mechanisms. From an investment and business perspective, it is a prime location for water-intensive sectors and trade (Sandhu et al., 2020). However, if we consider the regional and local scales, the picture of water security, regulatory stability, and equitable water access changes considerably (Mitchell, 2017). Recently, Environment and Climate Change Canada (2020) has released a report that reveals significant changes in the seasonal availability of freshwater under various carbon emission scenarios. Earlier snowmelts, warmer winters, permanent loss of glacier ice, and declining groundwater tables (reduced baseflow for connected surface water) will contribute to lower summer flows when the demand is at its highest. While the average annual precipitation is expected to increase, summers will be much drier, resulting in temporal water stress (Bush & Lemmen,

2019). These emerging water crises, further intensified by climate change across provinces in Canada, need to be accounted for in corporate and investment decisions (Heinmiller, 2017).

The province of Ontario, Canada, (Fig. 1) is surrounded by the Great Lakes and many surface water and groundwater sources, supporting a growing population, natural ecosystems, as well as agricultural and manufacturing activities. It hosts approximately 40% of the Canadian society and contributes to about 40% of the Canadian GDP (Ontario Ministry

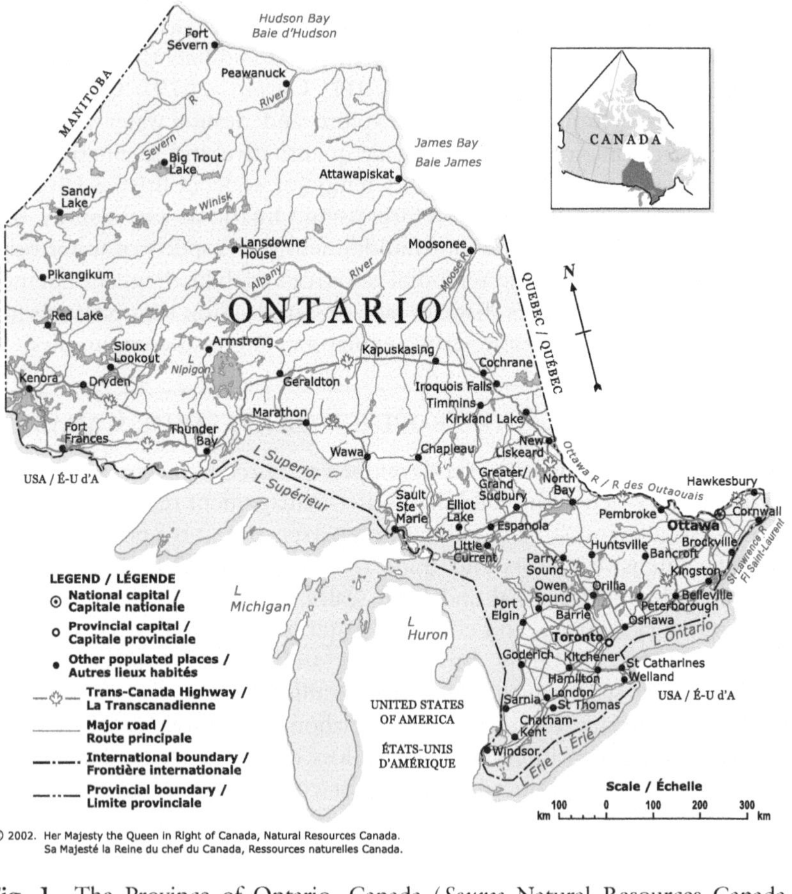

Fig. 1 The Province of Ontario, Canada (*Source* Natural Resources Canada, 2002)

of Finance, 2019). However, this seemingly water-rich province has not been immune to conflicts around water availability, degrading water quality, and lack of drinking water infrastructure in Indigenous communities (Galway, 2016; Mitchell, 2017). These conflicts and changing regulatory landscape have exposed the vulnerability of businesses to water risks and generated a keen interest in sustainable water management (Sandhu et al., 2020). Therefore, the assumption of water security in Ontario needs to be verified, especially from the perspective of corporations and investors looking to integrate water risks in long-term investment and operational decision-making.

3.1 Water Management and Governance Landscape for Ontario

According to the Constitution Act of 1867, the management of water resources is distributed between federal, provincial, municipal, and First Nations governments (Heinmiller, 2017). The constitution delegates ownership and management of water resources to provincial governments who can assign the right to use water through allocation frameworks that vary across different provinces. The federal government retains rights over certain divisions like fisheries, navigation, shipping, international, interprovincial and federal works, rivers, lakes, canal, harbor improvement, reserved aboriginal lands, and internationally shared waters (with the United States) (Brandes & Curran, 2017). Given the colonial history of Canada, treaty rights of Indigenous Peoples (First Nations, Inuit, and Métis) in the form of historic, modern, and claims over ownership of traditional lands, water, and natural resources are yet to be resolved. Therefore, engagement with Indigenous communities is an essential component of water-related decision-making in Canada (Bakker & Cook, 2011; Bradford et al., 2017).

Delving into the water management in Ontario, the province uses a *regulated riparian model* to allocate water to different users under the Ontario Water Resources Act, 1990, where extraction[1] of more than 50,000 L/day of surface or groundwater requires a Permit to Take Water (PTTW) from the Ontario Ministry of Environment, Conservation and Parks (MECP) (Ontario Ministry of Environment, Conservation & Parks, 2020). While the bulk transfer of water is prohibited from individual

[1] Water used for domestic purposes, livestock watering, firefighting, wetland conservation, and construction is exempt from requiring a PTTW.

watersheds, the export of water-intensive products is possible (Jaffee & Case, 2018). MECP uses a risk-based approach in granting these permits and categorizes applicants into three groups based on increasing environmental risk of the proposed water use and location (Ontario Ministry of Environment, Conservation & Parks, 2020). In addition to an administrative fee for PTTW applications contingent on the risk category, an additional "water conservation charge" of CAD 3.71/million liters of water extracted is imposed on highly water consumptive industrial sectors[2] (Ontario Ministry of Environment, 2007). However, post the moratorium on new water bottling permits, the volumetric charge for existing permit holders in the water bottling sector was increased to 503.71/million liters (Sandhu et al., 2020). Moreover, the MECP-PTTW Director has considerable discretion over the fate of the applications that can be refused, cancelled later, or subjected to additional protocol. While this gives enough flexibility for the ministry to adapt water policies, it adds a high level of uncertainty for businesses, which may not have assessed the implications of local regulatory conditions (Kreutzwiser et al., 2004).

In addition to the provincial legislation on water allocation, there is a whole tier of legislation on source water quality, protection and sustainable water use, and management, e.g., Clean Water Act, 2006; Safe Drinking Water Act, 2002; Environmental Protection Act, 1990; Environmental Assessment Act, 1990; Ontario Water Opportunities and Water Conservation Act, 2010; Nutrient Management Act, 2002; Great Lakes Protection Act, 2015; Lakes and River Management Act, 1990; and Sustainable Water and Sewage Systems Act, 2002, among others (Ontario Ministry of Environment, Conservation & Parks, 2017). The concerns of algal blooms, industrial activity, nuclear energy and fracking development, groundwater over-extraction, and climate change have also led to binational Governments (US-Canada) to update their existing transboundary Great Lakes agreements with an emphasis on ecological health and sustainable water use in the basin (Johns, 2017). Moreover, there is also considerable authority with individual municipalities in Ontario pertaining to water use, quality, and wastewater under the Municipal Act, 2001; Planning Act, 1990; and Health Promotion and Protection Act,

[2] Water bottling, beverage manufacturing, fruit/vegetable canning, certain chemical manufacturing where the majority of the water extracted is incorporated in the final product and not returned as wastewater to watershed.

1990, that can intersect with provincial authority (Kreutzwiser & de Loë, 2002; Kreutzwiser et al., 2004).

The province of Ontario has also instituted thirty-six sub-watershed specific agencies called "conservation authorities" under the Conservation Authorities Act, 1990, that are responsible for water and natural resource management interfacing with the provincial government, municipalities, and other stakeholders (Kreutzwiser et al., 2004). Conservation authorities have grassroots expertise in monitoring water levels, quality, and technical assessments of water-related risks, and they lead the Ontario Low Water Response program for watershed-based drought planning and management (Disch et al., 2012). These conservation authorities have undertaken many watershed assessments that are not only useful for policymaking, but also for corporate water risk assessments (Sandhu et al., 2020).

There is a plethora of federal, provincial, and sub-watershed data sources and scientific reports capturing hydrological, regulatory, conflicts, public perception, and concerns that can lend insights to water-related risks from an investor and business perspective. In the ensuing sections, we present local water issues and risks within the physical, regulatory, and reputational risk framework at the sub-watershed scale for Ontario, Canada.

3.2 Sub-Watershed Based Physical Risks

Spatial water scarcity (quantity and quality): The risks stemming from spatial and temporal water scarcity in regions sensitive to declining water quantity and quality (contamination) need to be assessed at the sub-watershed scale. The Ontario Clean Water Act, 2006, instituted after the tragic bacterial contamination of drinking water wells from surrounding farmland in the town of Walkerton, lay the foundation for the "Source Water Protection Plans." Conservation authorities quantify water quality and quantity threats and subsequent risks at the quaternary watershed scale to surface water and groundwater sources used for drinking water (Conservation Ontario, 2018; Sandhu et al., 2020). While the aim is to protect the drinking water sources using regional data and scientific modeling, these water quantity (surface and groundwater) risks and quality threats have been assessed for high use sub-watersheds provincially and nationally. Therefore, this high quality, scientifically reviewed granular data used for provincial and municipal policymaking can be used to account for current and future water demand, environmental flows, seasonal availability, drought scenarios, and aquifer vulnerability (Sandhu et al., 2020).

Temporal Water Scarcity: The Ontario Ministry of Natural Resources and Forestry has also issued the "Ontario Low Water Response Level" maps that depict different severity of drought conditions developing in real-time across sub-watersheds (Disch et al., 2012). The provincial surface water monitoring center, in association with individual conservation authorities, issue these maps based on stream-flows and precipitation monitoring under the Ontario Low Water Response Program (Disch et al., 2012).

Sector Specific Risks: Certain industrial sectors tend to consume more water than others (water intensive sectors), wherein the water extracted is permanently removed (incorporated in the product, evaporated, or diverted) from the watershed. The Province of Ontario uses specific metrics to distinguish high, moderate, and low water consumptive sectors (Ontario Ministry of the Environment, 2007). Therefore, this provincial categorization of water consumptive sectors can be used for more harmonized water risk assessments by the corporate and financial sector.

Legacy Contaminated Sites and Areas of Concern: The federal government has maintained a database of active, suspected, or remediated brownfield sites across each province that have been contaminated due to past industrial activity, e.g., underground petroleum, chemical storage tanks, and landfills, which can have an adverse environmental impact to neighboring water resources and soil (Government of Canada, 2020). Provincially, The Office of the Auditor General of Ontario has also compiled their findings on "Management of Contaminated Sites" in their 2015 report detailing active contaminated sites along with their location, contaminants, and financial liability. In 2004, the report on the "Groundwater Program" also listed contamination groundwater sites where people had to resort to an alternate source of drinking water (Auditor General of Ontario, 2004, 2015). Under the Canada-US Great Lakes Water Quality Agreement, forty-three "Areas of Concern" (twelve in Canada and five binationally shared) were identified in the Great Lakes basin with severely degraded water quality and ecosystem health (Environment & Climate Change Canada, 2020). These sites are currently under the policy scanner for remediation and have been a source of public concern resulting in subsequent stringent regulations (Johns, 2017). From a business or investment context, these sites must be accounted at relatively high risk for source water treatment requirements as well as regulatory and reputational implications.

3.3 Federal, Provincial, and Municipal Regulatory Risks

The fragmented, decentralized legislative and regulatory framework for water management in Canada and Ontario has been discussed in Sect. 2.2 (Bakker & Cook, 2011). Evidently, there is a whole suite of overlapping international, federal, provincial, and municipal laws and regulations pertaining to water resources, especially in the Great Lakes basin that cumulatively makes the regulatory landscape much more complicated especially for businesses and investors trying to assess regulatory risks. However, the current legislative and regulatory frameworks have also been criticized for being insufficient in catering to existing and emerging water issues and the assertion of Indigenous rights to water and land (Bakker & Cook, 2011; Brandes & Curran, 2017). The assessments used decades ago did not consider the dynamic ecological, social, and economic conditions of modern times. Therefore, there is a lot of potential reform and revisions being proposed in water-related legislation and regulation (Curran, 2019).

It is also important to acknowledge that the regulatory actions undertaken by the province are typically reactive to public concerns, especially around the water bottling sector. For instance, following drought conditions in the Grand River watershed in 2016, public criticism regarding groundwater extraction for water bottling by Nestlé in Wellington County spurred the province to issue a moratorium on new water taking permits for water bottlers (Jaffee & Case, 2018; Ministry of Environment, Conservation & Parks, 2020b). As a consequence, a series of policy reviews, groundwater management initiatives, and monitoring and environmental assessments have been undertaken (Sandhu et al., 2020). In response to the policy review in June 2020, the MECP has proposed changes to the water quantity framework, including placing the highest priority of water use for environment and drinking water, followed by irrigation in high-risk locations (Ontario Ministry of Environment, Conservation & Parks, 2020b). Even though the Ministry's review found no significant impact of water bottling operations on the sustainability of water sources, public concerns and perceptions were seen as drivers for proposed actions and regulatory changes for the sector, including the requirement to obtain support from the "host municipality" (Professional Geoscientists Ontario, 2020).

In another instance, in May 2018, the automotive glass manufacturer Xinyi Glass Holding Ltd., proposed to build a CAD 450 million glass manufacturing plant in Guelph Eramosa township in Wellington County, Ontario (Wellington Advisor, 2018). The proposed plant raised public concerns over traffic, emissions, and the company's 1.6 million liters/day groundwater withdrawal, resulting in a motion against the proposal by local citizens and water advocacy groups. Based on the municipal zoning bylaw that categorized the site for "dry-use" industries, the Township's council accepted the motion to reject the proposal on water usage even before the provincial review of water taking (Council of Canadians, 2018). Therefore, there has been an emphasis on integrating stakeholder concerns and perceptions about water security in management decisions and mitigation strategies for all water-using sectors (Wolfe & Brooks, 2017). Normative perception is also a key element in risk assessment theory and needs to be accounted for in water risk assessments (Gough, 1997).

The legislation on minimum environmental flows is also gaining traction across Canada. Given the importance of water-dependent ecosystems and biodiversity, the amount of surface water and groundwater that can be sustainably extracted primarily during low flows needs to be regulated (Curran, 2019). The conservation authorities across Ontario have given due consideration for maintaining minimum flows for the environment in their water quantity risk assessments, and the spatial variability of these ecological conditions have been acknowledged (Shifflett, 2014). The MECP does warrant technical and ecological impact assessments of specific water-taking permit applicants and provides technical guidance on evaluating environmental flows (Ontario Ministry of Environment, Conservation & Parks, 2020a). However, there has been a call for legally mandating minimum environmental flows within provincial allocation regulations and policies that can potentially impact existing and future water users in the sub-watersheds (Brandes & Curran, 2017; Curran, 2019).

Unresolved and outstanding land and water treaties concerning Indigenous Peoples of Canada have also been cited as a driver for change in water law and governance across Canada (Bakker & Cook, 2011). The UN Declaration on the Rights of Indigenous Peoples (UNDRIP) endorsed by the Government of Canada in 2016 can also have an impact on how water resources are allocated and to whom (Black & McBean, 2017; Von der Porten & de Loë, 2013). Moreover, taking a more active role in decision-making and co-management with Indigenous communities will alter the existing legal and social license to operate (Bradford et al., 2017; Brandes & Curran, 2017). The private and financial sector will have to foresee how to engage and collaborate with all crucial stakeholders and Indigenous communities in Ontario (Johns, 2017).

In terms of stakeholder consultation, every resident in Ontario has the right under the Environmental Bill of Rights, 1993, to provide inputs, comments, and recommendations for proposals, decisions, and legislative and regulatory changes pertaining to the environment using the online Environmental Registry of Ontario (Ontario Ministry of Attorney General, 2019). There has been active participation of state and non-state actors in water policy development and water management initiatives, including environmental NGOs, local irrigation advisory committees, and fisheries groups, but leadership and engagement of the private sector have been minimal (Johns, 2017). With the recent federal mandate to institute a new Canada Water Agency, in addition to the reprised federal involvement, a new and collaborative regulatory regime for water management and governance is expected (Brandes et al., 2020).

From the discussion above, it is evident that an understanding of the complex legal and regulatory frameworks for water use, management, discharge as well as governance structures including all levels of the governments, formally instituted local watershed agencies like conservation authorities, non-state actors, and other public groups (NGOs, social justice and action, watershed stewards) are essential for future business and investment decisions. Stakeholder inputs, concerns, and backlash have all contributed to the dynamic water regulations and decisions that can pose an operational and reputational risk to businesses. Thus, watershed-specific water issues and fast-evolving regulatory landscape should capture investor and business interests for assessing risks more accurately rather than maintaining the status quo in Ontario, Canada.

3.4 Reputational Risks, Conflicts, and Legacy Issues

The province of Ontario and specifically the Great Lakes region has had a fair share of conflict amidst competing user groups since the mid-90s over water allocation and use (Kreutzwiser et al., 2004). Ontario has been prone to short-term droughts, low flow conditions, and declining well levels that have created tensions between rural well owners, golf courses, water bottling companies, farmers, aggregate mining and quarry dewatering companies, and suburban water users (Morris et al., 2008; Shifflett, 2014). These conflicts spike in the dry summer months when the municipal, industrial, and agricultural water demand is high. Factors like regional water availability, different types and number of water user groups, economic dependence on water, potential interference with municipal water sources and presence of provincially significant wetlands, areas of natural and scientific interest, and environmentally sensitive areas are considered determinants of conflicts in drought contingency reports (Shifflett, 2014).

In the Canadian context, there is a lack of reliable drinking water infrastructure in Indigenous communities exacerbating existing social inequity issues. More than 600 Indigenous communities with unique histories, cultures, traditions, and close spiritual connection with water are an integral part of Canada. Indigenous peoples (First Nations, Inuit, and Métis) represent <5% of the Canadian population, but they have experienced excessively higher number of drinking water quality issues within on-reserve communities as compared to off-reserve communities (Lam et al., 2017). The jurisdiction of on-reserve Indigenous communities rests with the federal government including the provision of safe drinking water and wastewater treatment. Even with the federal funding support for drinking water and wastewater infrastructure in collaboration with First Nations leadership (Chiefs and Band Council), many First Nations communities are under a long-term or short-term drinking water advisory (White et al., 2012).

Ontario hosts 133 First Nations communities and the largest population of Indigenous peoples in Canada. However, about 400 on-reserve drinking water advisories lasting an average of 294 days were reported between the years 2004–2013 in Ontario due to equipment failure or lack of operator training (Galway, 2016). The largest First Nations reserve "Six Nations of the Grand River" within the Grand River watershed is a mere 100 km away from the megapolis Toronto but has witnessed short-term boil water advisories and contamination from upstream off-reserve activities (Collins et al., 2017). Even in the midst of the COVID-19 pandemic, 51 long-term and 12 short-term drinking water advisories were in effect in October 2020 across Ontario (Indigenous Services Canada, 2020). The underlying social, economic, political, and historical inequities are evident in the lower life expectancies, incomplete education, unemployment rates, lower access to health care, and higher vulnerability to water-borne illness in on-reserve Indigenous communities, invoking sharp criticism regarding environmental discrimination and injustice (Galway, 2016; Lam et al., 2017). Water is central to Indigenous cultural, spiritual, and livelihood traditions; therefore, the lack of reliable drinking water and contamination of source waters resulting in dependence on bottled water is a grave paradox forced upon them (Galway, 2016; White et al., 2012). Lack of access in this context is not a technical or financial lapse but rather a social inequity legacy issue rooted deeply in colonial history (White et al., 2012).

With water quality issues, competing and inefficient water use, environmental degradation, and the growing influence of citizens and social action groups, economic sectors increasingly face significant reputational risks in the region. The public attitudes and perceptions about different aspects of water resource management continue to be captured nationally and regionally within the Great Lakes watershed by surveys and polls (Johns, 2017). While Ontario does offer a competitive advantage to water-dependent businesses and waterways connectivity for trade, existing multi-jurisdictional water management and governance networks do not necessarily favor economic interests over environmental concerns (Heinmiller, 2017).

Environmental action and awareness campaigns, coupled with media coverage on water issues about access, quality, and inequity have shaped public perception and political interest at the grassroots level (Jaffee & Case, 2018). The country-level data on reputational risks is unable to capture these granular and heterogenous issues beyond biophysical metrics, especially when the perception of water security varies considerably in Indigenous communities (Baird et al., 2015). Given the social and contextual nuances involved, undermining these risks in business and investment decisions can have substantial reputational implications on responsible and equitable business and investment practices (Schulte et al., 2012).

4 Opportunities for Sustainable Water Investments

To drive sustainable investment in water and manage portfolios exposed to water risk, a granular assessment of local geographies at the sub-watershed level is necessary. In the case of Ontario, Canada, the interrelated physical, regulatory, and reputational risks at the local level demand corporate and investor attention. Therefore, interdisciplinary investment decision support tools need to be developed using granular data to reflect local hydrological, regulatory, and social conditions. As demonstrated with a case study approach, the deep dive into hydrological risk assessments by local watershed agencies, regulatory developments, past water user conflicts, and legacy of social issues reveal the need for proactive water risk management even in regions like Ontario, perceived to be water secure.

Based on the reviewed literature, the synthesized conceptual water risk assessment framework capturing local sub-watershed risks for Ontario, Canada, is presented in Fig. 2. The framework can be operationalized

Physical Risks

- Statistics Canada freshwater supply and demand assessment
- Environment and Climate Change Canada Water Indicators
- Sub-watershed Source Water Protection Assessment Reports
  - Quaternary watersheds identified for surface water and groundwater quantity risk
  - Regions near municipal water sources at risk for contamination
- Proximity to Great Lakes Areas of Concern and other federal water quality initiatives
- Previous contaminated sites
- Population density and land use
- Annual water extraction by sector and region (PTTW database and Statistics Canada)
- Temporal low flow maps by Ontario Ministry of Natural Resources
- Minimum environmental flow requirements by Province

Regulatory Risks

- Provincial water allocation and extraction charges legislation
- Priority water use groups
- Water consumptive sectors classification by OMECP
- Water quality legislation
- Decentralization of water governance and all state and non-state actors involved
- Transboundary water sharing agreements (conservation, sustainable use, ecological health)
- Existing or past sector-specific moratoriums on water permits
- Developing legislation, regulations around water (allocation, environmental flows, consultation)
- Regulatory actions proposed or undertaken through consultation via Environmental Registry of Ontario (ERO)

Reputational Risks

- Concentration of different water user groups
- Past regional conflicts between water user groups
- Public perception water surveys (RBC Canadian Water Attitudes Survey, Binational Great Lakes Basin Poll)
- EBR comments on water regulations
- Auditor General and Environmental Commissioner of Ontario's reports
- Social action or citizen group campaigns
- Legacy issues
  - Federal database on drinking water advisories (Environment and Climate Change Canada and Indigenous Services Canada)
  - Water contamination events
  - Past or ongoing treaty agreements

Fig. 2 Conceptual water risk assessment framework for Ontario, Canada

into decision support tools using publicly available, reliable, and peer-reviewed data from government databases that allow harmonization of data used for public policies, investments, and corporate water policies. Hence, the confluence of spatial and temporal biophysical and social data provides realistic insights into grassroots water conditions and risk perceptions envisioned to aid proactive decision-making for sustainable water investments and corporate water sustainability. A case study approach can be applied to any geography, and the framework can potentially be adapted for different regions, revealing different risks and available data sources for assessments.

In addition to integrating water risk metrics across investment portfolios, sustainable finance is gaining traction to address the financing gap in sustainable water management initiatives in line with targets of SDG 6 (WWF, 2019b). The financing needs in technical solutions and infrastructure for water security including drinking water treatment, wastewater treatment, water recycling, conservation and efficiency, low impact development, desalination plants, flood control, drought control, and stormwater management are projected as a ripe financial and socially responsible investment opportunity to foster the green economy (Alvarez & Rodriguez, 2015; WWF, 2019b). Even in developed economies like Canada, the federal and provincial governments are unrolling public–private partnerships to finance green technology and infrastructure projects, including sustainable water management that seem like profitable ventures for the financial sector (Infrastructure Canada, 2020).

Canada Infrastructure Bank (CIB) is a crown corporation that has been instituted to invest CAD 5 Billion in revenue-generating green infrastructure projects (including water and wastewater treatment systems) to attract private and institutional investment (Infrastructure Canada, 2020). CIB has already invested CAD 20 million in a water and wastewater project in Mapleton, Ontario (Canada Infrastructure Bank, 2019). The province of Ontario, being a hub for the financial industry, aims to leverage the private sector for pooling sustainable finance (e.g., by issuing green bonds) into future green or cleantech projects (Ontario Ministry of Environment, Conservation & Parks, 2018). Typically, investments in water-related projects are considered to be low risk and of high value in developed economies due to the perception of social and political will pertaining to private sector involvement (Alaerts, 2019). However, a closer assessment reveals many regulatory and reputational risks pertaining to privatization, even in a developed country like Canada. Therefore, water-related risks can be accounted for and assigned appropriate weights using an interdisciplinary granular framework so as to design appropriate communication and engagement strategies within existing and future portfolios to achieve the stipulated financial, social, and environmental outcomes of the project (Vörösmarty et al., 2018).

5 Conclusion

Water risks pertaining to biophysical or hydrological metrics, regulatory trends, social equity, and public perceptions are material for companies and investors. The temporal, spatial, and contextual sensitivities of water make it a more dynamic risk that can have a cascading impact on both individual securities and portfolios (Jorisch et al., 2018). Financial actors like investors, lenders, and insurance companies need to actively account for water risk exposure and response strategies within their current and potential portfolios (Hogeboom et al., 2018). While Task Force on Climate-related Disclosures (TCFD) recommendations for climate risk disclosure are currently being implemented in the financial sector, assessments and scenarios for water-related risks are mostly lacking in investment strategies (WWF, 2019a). Climate change and water management are intricately linked, but unlike globally standardized carbon reduction targets, water-related metrics are more locally defined, as explored in this chapter. Therefore, the corporate and financial sectors need a deeper understanding of interdisciplinary water risks on the local sub-watershed basis to mitigate long-term impacts on their operations and investments (Freyman et al., 2015).

The gaps in water risk accounting and disclosure, as well as the needs of the financial sector to acquire technical and contextual water information to make sustainable investment decisions have been highlighted in academic and practitioner literature. Sustainable investment in water needs to account for social nuances, legal and regulatory environment, existing legacy issues, conflicts amid competing water users, and water intensity trends in local social-economic-environmental context (Jorisch et al., 2018). Moreover, the lack of appropriate risk assessment and management tools is leading to a path dependency where water risks are generalized to certain geographies and sectors without evaluating granular watershed conditions and impacts (Jorisch et al., 2018; Money, 2014a). Contrary to this generalization, this chapter reveals many existing and impending water risks in Ontario, Canada, which is considered a water-rich province.

Assessing water risks and their impacts is crucial to ensuring social well-being, building a strong financial sector, and hence a climate-resilient economy. Given the growing interest of regulators, investors, and businesses to develop robust guidelines to integrate water risks, it becomes crucial to design holistic indicators using granular information (Vörösmarty et al., 2018). The role of the financial sector, regulators, central banks, and credit rating companies in incentivizing sustainable water management and finance is also critical to realizing the objectives of water security (Alaerts, 2019). However, the interdisciplinary understanding of water risks is still developing, and detailed data on a sub-watershed basis is required. Therefore, using a case study approach, this chapter aims to contribute by proposing an interdisciplinary framework that harmonizes biophysical, regulatory, and contextual evidence at the granular watershed level and can be operationalized to assess and rate water risks. Accounting for these multifaceted water risks proactively reveals opportunities for sustainable water management and investments for transitioning into a climate-resilient as well as water-secure economy and society.

References

Alaerts, G. J. (2019). Financing for water-water for financing: A global review of policy and practice. *Sustainability (switzerland), 11*(3), 1–25. https://doi.org/10.3390/su11030821.

Alvarez, M., & Rodriguez, J. (2015). Water-related mutual funds: Investment performance and social role. *Social Responsibility Journal, 11*(3), 502–512. https://doi.org/10.1108/SRJ-08-2013-0104.

Auditor General of Ontario. (2004). Groundwater program. *2004 Annual Report of the Office of Auditor General of Ontario,* 143–169. http://www.auditor.on.ca/en/content/annualreports/arreports/en04/305en04.pdf. Accessed on 15 May 2020.

Auditor General of Ontario. (2015). Management of contaminated sites. *2015 Annual Report of the Office of Auditor General of Ontario,* 400–433. http://www.auditor.on.ca/en/content/annualreports/arreports/en15/3.10en15.pdf. Accessed on 15 May 2020.

Aven, T., & Renn, O. (2009). On risk defined as an event where the outcome is uncertain. *Journal of Risk Research, 12*(1), 1–11. https://doi.org/10.1080/13669870802488883.

Aven, T., & Renn, O. (2010). Risk governance: An overview. In *Risk management and governance: Concepts, guidelines and applications* (Vol. 16, pp. 49–66). Springer Berlin Heidelberg. https://doi.org/10.1007/978-3-642-13926-0_4

Baird, J., Plummer, R., Dupont, D., & Carter, B. (2015). Perceptions of water quality in first nations communities: Exploring the role of context. *Nature and Culture, 10*(2), 225–249. https://doi.org/10.3167/nc.2015.100205.

Bakker, K., & Cook, C. (2011). Water governance in Canada: Innovation and fragmentation. *International Journal of Water Resources Development, 27*(2), 275–289. https://doi.org/10.1080/07900627.2011.564969.

Barton, B. (2010). *Murky waters? Corporate reporting on water risk—A benchmarking study of 100 companies*. Ceres. www.waterfootprint.org/Reports/Barton_2010.pdf. Accessed on 18 May 2020.

Black, K., & McBean, E. (2017). Indigenous water, Indigenous voice—A national water strategy for Canada's Indigenous communities. *Canadian Water Resources Journal, 42*(3), 248–257. https://doi.org/10.1080/07011784.2017.1333044.

Bradford, L. E. A., Ovsenek, N., & Bharadwaj, L. A. (2017). Indigenizing water governance in Canada. In S. Renzetti & D. P. Dupont (Eds.), *Water Policy and Governance in Canada* (pp. 269–298). Springer International Publishing. https://doi.org/10.1007/978-3-319-42806-2_15.

Brandes, O. M., & Curran, D. (2017). Changing currents: A case study in the evolution of water law in Western Canada. In S. Renzetti & D. P. Dupont (Eds.), *Water Policy and Governance in Canada* (pp. 45–67). Springer International Publishing. https://doi.org/10.1007/978-3-319-42806-2_4.

Brandes, O. M., Phare, M. A., & Pomeroy, J. W. (2020). A Canada Water Agency is the first to modernizing water management. In *Policy Options Politiques.* https://policyoptions.irpp.org/magazines/april-2020/a-canada-water-agency-is-the-first-step-to-modernizing-water-management/. Accessed on 5 May 2020.

Bush, E., & Lemmen, D. S. (2019). *Canada's changing climate report*. Government of Canada. https://www.nrcan.gc.ca/sites/www.nrcan.gc.ca/files/energy/Climate-change/pdf/CCCR_FULLREPORT-EN-FINAL.pdf. Accessed on 15 May 2020.

Canada Infrastructure Bank. (2019). *Press release: Canada Infrastructure Bank announces up to $20 million investment commitment in Mapleton water and wastewater project*. https://cib-bic.ca/en/cib-announces-up-to-20-million-investment-commitment-in-mapleton-water-and-wastewater-project/. Accessed on 18 May 2020.

Canadian Securities Administrators. (2019). *CSA staff notice 51–358: Reporting on climate-related risks.* https://www.osc.gov.on.ca/documents/en/Securi

ties-Category5/csa_20190801_51-358_reporting-of-climate-change-related-risks.pdf. Accessed on 15 May 2020.

Christ, K. L., & Burritt, R. L. (2017a). Water management accounting: A framework for corporate practice. *Journal of Cleaner Production, 152*, 379–386. https://doi.org/10.1016/j.jclepro.2017.03.147.

Christ, K. L., & Burritt, R. L. (2017b). What constitutes contemporary corporate water accounting? A review from a management perspective. *Sustainable Development, 25*(2), 138–149. https://doi.org/10.1002/sd.1668.

Christ, K. L., & Burritt, R. L. (2018). The role for transdisciplinarity in water accounting by business: Reflections and opportunities. *Australasian Journal of Environmental Management, 25*(3), 302–320. https://doi.org/10.1080/14486563.2018.1460631.

Cienfuegos Spikin, I. (2013). Risk management theory: The integrated perspective and its application in the public sector. *Revista Estado, Gobierno y Gestión Pública, 21*, 89–126. https://doi.org/10.5354/0717-6759.2013.29402.

Collins, L., McGregor, D., Allen, S., Murray, C., & Metcalfe, C. (2017). Source water protection planning for Ontario First Nations communities: Case studies identifying challenges and outcomes. *Water (switzerland), 9*(7), 1–19. https://doi.org/10.3390/w9070550.

Conservation Ontario. (2018). *Source water protection plans and resources*. http://conservationontario.ca/conservation-authorities/source-water-protection/source-protection-plans-and-resources/. Accessed on 21 May 2020.

Council of Canadians. (2018). *Xinyi proposed floating glass plant: Guelph Eramosa Township (GET)*. https://canadians.org/analysis/xinyi-proposed-floating-glass-plant-guelph-eramosa-township-get. Accessed on 21 May 2020.

Curran, D. (2019). The adaptation potential of water law in Canada: Changing existing water use entitlements. *Water International, 44*(3), 278–291. https://doi.org/10.1080/02508060.2019.1570053.

Disch, J., Kay, P., & Mortsch, L. (2012). A resiliency assessment of Ontario's low-water response mechanism: Implications for addressing management of low-water under potential future climate change. *Canadian Water Resources Journal, 37*(2), 105–123. https://doi.org/10.4296/cwrj3702916.

Environment and Climate Change Canada. (2020). *Great lakes: Areas of concern*. https://www.canada.ca/en/environment-climate-change/services/great-lakes-protection/areas-concern.html. Accessed on 20 May 2020.

Freyman, M., Collins, S., & Barton, B. (2015). *An investor handbook for water risk integration*. Ceres. https://www.ceres.org/sites/default/files/reports/2017-03/Ceres_ESG-WaterRisk_041515_Print.pdf. Accessed on 18 May 2020.

Galway, L. P. (2016). Boiling over: A descriptive analysis of drinking water advisories in First Nations communities in Ontario, Canada. *International Journal of Environmental Research and Public Health, 13*(5), 1–15. https://doi.org/10.3390/ijerph13050505.

Gough, J. (1997). Environmental decision making and risk management for groundwater systems. *RISK: Health, Safety & Environment (1990–2002), 8*(2), 155–172. https://scholars.unh.edu/cgi/viewcontent.cgi?article=1322&context=risk.

Government of Canada. (2020). *About federal contaminated sites*. https://www.canada.ca/en/environment-climate-change/services/federal-contaminated-sites.html. Accessed on 21 May 2020.

Hanemann, W. M. (2006). The economic conception of water. In *Water Crisis: Myth or Reality* (pp. 61–91). Taylor & Francis Group. https://doi.org/10.1201/9781439834275.pt2a.

Heinmiller, B. T. (2017). The politics of water policy development in Canada. In S. Renzetti & D. P. Dupont (Eds.), *Water Policy and Governance in Canada* (pp. 215–229). Springer International Publishing. https://doi.org/10.1007/978-3-319-42806-2_12.

Hoekstra, A. Y. (2014). Water scarcity challenges to business. *Nature Climate Change, 4*(5), 318–320. https://doi.org/10.1038/nclimate2214.

Hogeboom, R. J., Kamphuis, I., & Hoekstra, A. Y. (2018). Water sustainability of investors: Development and application of an assessment framework. *Journal of Cleaner Production, 202*, 642–648. https://doi.org/10.1016/j.jclepro.2018.08.142.

Indigenous Services Canada. (2020). *Ending long-term drinking water advisories*. https://www.sac-isc.gc.ca/eng/1506514143353/1533317130660. Accessed on 12 October 2020.

Infrastructure Canada. (2020). *Canada Infrastructure Bank*. https://www.infrastructure.gc.ca/CIB-BIC/index-eng.html#about. Accessed on 21 May 2020.

Jaffee, D., & Case, R. A. (2018). Draining us dry: Scarcity discourses in contention over bottled water extraction. *Local Environment, 23*(4), 485–501. https://doi.org/10.1080/13549839.2018.1431616.

Johns, C. M. (2017). The Great Lakes, water quality and water policy in Canada. In S. Renzetti & D. P. Dupont (Eds.), *Water Policy and Governance in Canada* (pp. 159–178). Springer International Publishing. https://doi.org/10.1007/978-3-319-42806-2_9.

Jorisch, D., Frank, A., Schneider, M., Stadelmann, M., & Rosenthal, N. (2018). *Water risks and financial market: Overview and analysis*. South Pole. https://www.southpole.com/publications. Accessed on 15 April 2020.

Klinke, A., & Renn, O. (2002). A new approach to risk evaluation and management: Risk-based, precaution-based, and discourse-based strategies. *Risk Analysis, 22*(6), 1071–1094. https://doi.org/10.1111/1539-6924.00274.

Kreutzwiser, R. D., & de Loë, R. C. (2002). Municipal capacity to manage water problems and conflicts: The Ontario experience. *Canadian Water Resources Journal, 27*(1), 63–83. https://doi.org/10.4296/cwrj2701063.

Kreutzwiser, R., de Loë, R. C., Durley, J., & Priddle, C. (2004). Water allocation and the permit to take water program in Ontario: Challenges and opportunities. *Canadian Water Resources Journal, 29*(2), 135–146. https://doi.org/10.4296/cwrj135.

Lam, S., Cunsolo, A., Sawatzky, A., Ford, J., & Harper, S. L. (2017). How does the media portray drinking water security in Indigenous communities in Canada? An analysis of Canadian newspaper coverage from 2000–2015. *BMC Public Health, 17*(1), 1–14. https://doi.org/10.1186/s12889-017-4164-4.

Loucks, D. P. (2000). Sustainable water resources management. *Water International, 25*(1), 3–10. https://doi.org/10.1080/02508060008686793.

Mitchell, B. (2017). The hydrological and policy contexts for water in Canada. In S. Renzetti & D. P. Dupont (Eds.), *Water Policy and Governance in Canada* (pp. 13–28). Springer International Publishing. https://doi.org/10.1007/978-3-319-42806-2_2.

Money, A. (2014a). Corporate water risk: A critique of prevailing best practice. *Journal of Management and Sustainability, 4*(1), 42–59. https://doi.org/10.5539/jms.v4n1p42.

Money, A. (2014b). Corporate water risk: Investor tolerance of the status quo. *Journal of Management and Sustainability, 4*(1), 60–75. https://doi.org/10.5539/jms.v4n1p60.

Morgan, A., Laporte-Bisquit, A., Williams, T., & Maggo, D. (2020). *Right tool for the job: Tools and approaches for companies and investors to assess water risks and shared water challenges.* WWF & WBCSD. https://wwfeu.awsassets.panda.org/downloads/right_tool_for_the_job_1.pdf. Accessed on 20 May, 2020.

Morris, T. J., Mohapatra, S. P., & Mitchell, A. (2008). Conflicts, costs and environmental degradation—Impacts of antiquated ground water allocation policies in the Great Lakes Basin. *Water Policy, 10*(5), 459–479. https://doi.org/10.2166/wp.2008.059.

Morrison, J., Morikawa, M., Murphy, M., & Schulte, P. (2009, February). *Water scarcity & climate change: Growing risks for businesses and investors*. Ceres. http://www2.pacinst.org/wp-content/uploads/2013/02/full_report30.pdf. Accessed on 5 June 2020.

Morrison, J., & Schulte, P. (2010). *Corporate water accounting: An analysis of methods and tools for measuring water use and its impacts*. UNEP, Pacific Institute. https://pacinst.org/wp-content/uploads/2010/04/corporate_water_accounting_analysis.pdf. Accessed on 6 June 2020.

Natural Resources Canada. (2002). *Political map of Ontario. Reference maps—Provincial and territorial*. http://ftp.geogratis.gc.ca/pub/nrcan_rncan/ras

ter/atlas_6_ed/reference/bilingual/ont_new.pdf. Accessed on 10 March 2019.

OECD. (2003). Emerging risks in the 21st century: An agenda for action. *Final Report to the OECD Futures Project*. https://doi.org/10.1787/9789264101227-en.

OECD. (2013). *Water security for better lives*. https://doi.org/10.1787/9789264202405-en

Ontario Ministry of Attorney General. (2019). *Environmental Bill of Rights*. https://www.ontario.ca/page/environmental-bill-rights. Accessed on 21 May 2020.

Ontario Ministry of Environment. (2007). *Water conservation charges proposal* (Report Number PIBS 6134e). http://www.ontla.on.ca/library/repository/mon/16000/272421.pdf. Accessed on 25 May 2020.

Ontario Ministry of Environment, Conservation & Parks. (2017). *Ontario's water conservation and efficiency goals, objectives and programs (in accordance with the Great Lakes-St. Lawrence River Basin Sustainable Water Resources Agreement)*. https://www.ontario.ca/page/ontarios-water-conservation-and-efficiency-goals-objectives-and-programs-accordance-great-lakes-st. Accessed on 25 May 2020.

Ontario Ministry of Environment, Conservation & Parks. (2018). *A made-in-Ontario environment plan: Ontario's new plan to preserve and protect our environment for future generations*. https://www.ontario.ca/page/made-in-ontario-environment-plan. Accessed on 30 May 2020.

Ontario Ministry of Environment, Conservation & Parks. (2020a). *Guide to permit to take water application form*. https://www.ontario.ca/page/guide-permit-take-water-application-form. Accessed on 30 May 2020.

Ontario Ministry of Environment, Conservation & Parks. (2020b). *Updating Ontario's water quantity management framework* (Proposal Paper. ERO number 019-1340). https://prod-environmental-registry.s3.amazonaws.com/2020-06/Proposal%20Paper%20%E2%80%94%20Updating%20Ontario%27s%20Water%20QuantQua%20Management%20Framework.pdf. Accessed on 22 June 2020.

Ontario Ministry of Finance. (2019). *Ontario fact sheet*. Office of economic policy. https://www.fin.gov.on.ca/en/economy/ecupdates/factsheet.pdf. Accessed on 30 May 2020.

Professional Geoscientists Ontario. (2020). *Ontario's bottled water moratorium: A report of a panel of independent experts assembled by Professional Geoscientists Ontario for the Ministry of the Environment Conservation and Parks*. https://ero.ontario.ca/notice/019-1340. Accessed on 22 June 2020.

Renn, O. (1998). The role of risk perception for risk management. *Reliability Engineering and System Safety, 59*(1), 49–62. https://doi.org/10.1016/S0951-8320(97)00119-1.

RobecoSAM. (2015). *Water: The market of the future*. Zurich, Switzerland. https://www.robecosam.com/media/3/2/7/3272659fd94cb56fac2a47a4ec159001_150708-waterstudy-en-Vdef_tcm1011-16007.pdf. Accessed on 15 May 2020.

Russell, M., & Gruber, M. (1987). Risk assessment in environmental policy-making. *Science, 236*(4799), 286–290. https://www.jstor.org/stable/1698638.

Sandhu, G., Wood, M. O., Rus, H. A., & Weber, O. (2020). Bulk water extraction charge calculator: A tool for sustainable water management in Ontario. *Canadian Water Resources Journal / Revue Canadienne Des Ressources Hydriques, 45*(1), 59–76. https://doi.org/10.1080/07011784.2019.1685909.

Sanitation and Water for All. (2020). *Water, sanitation and hygiene: World leaders call for action on COVID-19*. https://www.sanitationandwaterforall.org/about/our-work/world-leaders-call-action-covid-19. Accessed on 15 May 2020.

Schulte, P., Morrison, J., & Gleick, P. H. (2012). Corporate water management. In *The Worlds Water Volume 7* (pp. 23–43). Island Press. https://doi.org/10.5822/978-1-61091-048-4.

Shifflett, S. (2014). *Grand River watershed water management plan-drought contingency plan*. https://www.grandriver.ca/en/our-watershed/resources/Documents/WMP/Water_WMP_Report_Drought.pdf. Accessed on 20 April 2020.

Signori, S., & Bodino, G. A. (2013). Water management and accounting: Remarks and new insights from an accountability perspective. In *Studies in Managerial and Financial Accounting* (Vol. 26, Issue 2013). Emerald Group Publishing Limited. https://doi.org/10.1108/S1479-3512(2013)0000026004.

Slovic, P. (1987). Perception of risk. *Science, 236*(4799), 280–285. https://doi.org/10.1126/science.3563507.

Thistlethwaite, J., Henstra, D., Brown, C., & Scott, D. (2020). Barriers to insurance as a flood risk management tool: Evidence from a survey of property owners. *International Journal of Disaster Risk Science, 11*(3), 263–273. https://doi.org/10.1007/s13753-020-00272-z

United Nations. (2018). *Sustainable development goal 6: Synthesis report 2018 on water and sanitation*. https://www.unwater.org/publication_categories/sdg-6-synthesis-report-2018-on-water-and-sanitation/. Accessed on 25 April 2020.

UN-Water. (2013). *Water security and the global water agenda: A UN-Water analytical brief*. https://www.unwater.org/publications/water-security-global-water-agenda/. Accessed on 25 April 2020.

Von der Porten, S., & de Loë, R. C. (2013). Collaborative approaches to governance for water and Indigenous peoples: A case study from British Columbia, Canada. *Geoforum, 50*, 149–160. https://doi.org/10.1016/j.geoforum.2013.09.001.

Vörösmarty, C. J., McIntyre, P. B., Gessner, M. O., Dudgeon, D., Prusevich, A., Green, P., Glidden, S., Bunn, S. E., Sullivan, C. A., Liermann, C. R., & Davies, P. M. (2010). Global threats to human water security and river biodiversity. *Nature, 467*(7315), 555–561. https://doi.org/10.1038/nature09440.

Vörösmarty, C. J., Osuna, V. R., Koehler, D. A., Klop, P., Spengler, J. D., Buonocore, J. J., Cak, A. D., Tessler, Z. D., Corsi, F., Green, P. A., & Sánchez, R. (2018). Scientifically assess impacts of sustainable investments. *Science, 359*(6375), 523–525. https://doi.org/10.1126/science.aao3895.

WBCSD-IUCN. (2012, August). *Water for business: Initiatives guiding sustainable water management in the private sector*, 1–40. http://www.bcsd.org.tw/sites/default/files/node/domain_tool/678.file.2161.pdf. Accessed on 25 April 2020.

Weber, O., & Feltmate, B. (2016). *Sustainable banking: Managing the social and environmental impact of financial institutions*. University of Toronto Press.

Wellington Advisor. (2018). *Chinese company plans $450-million glass plant in Guelph-Eramosa*. https://www.wellingtonadvertiser.com/chinese-company-plans-450-million-glass-plant-in-guelph-eramosa/. Accessed on 25 May 2020.

White, J. P., Murphy, L., & Spence, N. (2012). Water and indigenous peoples: Canada's paradox. *International Indigenous Policy Journal, 3*(3). https://doi.org/10.18584/iipj.2012.3.3.3.

WHO. (2019). *Factsheet on drinking water*. https://www.who.int/en/news-room/fact-sheets/detail/drinking-water. Accessed on 25 April 2020.

Wolfe, S., & Brooks, D. B. (2003). Water scarcity: An alternative view and its implications for policy and capacity building. *Natural Resources Forum, 27*(2), 99–107. https://doi.org/10.1111/1477-8947.00045.

Wolfe, S. E., & Brooks, D. B. (2017). Mortality awareness and water decisions: A social psychological analysis of supply-management, demand-management and soft-path paradigms. *Water International, 42*(1), 1–17. https://doi.org/10.1080/02508060.2016.1248093.

WWF. (2019a). *Freshwater risks & opportunities: An overview and call to action for the financial sector*. https://wwf.panda.org/our_work/finance/?355750/Linking-water-risk-and-financial-value. Accessed on 10 May 2020.

WWF. (2019b). *Linking water risk and financial value—Part I: Considerations for the financial sector*. https://wwf.panda.org/our_work/finance/?355750/Linking-water-risk-and-financial-value. Accessed on 10 May 2020.

Wyrwoll, P. R., Grafton, R. Q., Daniell, K. A., Chu, H. L., Ringler, C., Lien, L. T. H., Khoi, D. K., Do, T. N., & Tuan, N. D. A. (2018). Decision-making for systemic water risks: Insights From a participatory risk assessment process in Vietnam. *Earth's Future, 6*(3), 543–564. https://doi.org/10.1002/2017EF000777.

Forward Pricing of Embedded Water: A Step Toward Sustainable Development in Agriculture

Paras Mahajan and Hamendra Kumar Porwal

1 Introduction

India has about 140 million hectares of arable land, with 42% of the country's cultivable land situated in drought-prone areas where half of the net sown area is dependent on rain. Poor agricultural productivity due to rain dependency was observed recently in 2015–2016 when drought-like conditions were perpetuated in the subcontinent which resulted in the import of staple crops (Prasanna, 2014). The deficit in food procurement in the early 1950s was mainly due to the shortage in production which prompted the government to initiate the Green Revolution, but today groundwater depletion is a serious cause for food shortages. This has led scientists and observers to reclassify water as a non-renewable resource

P. Mahajan (✉)
Hindu College, University of Delhi, New Delhi, India

H. K. Porwal
Shaheed Sukhdev College of Business Studies,
University of Delhi, New Delhi, India

T. Walker et al. (eds.), *Water Risk and Its Impact on the Financial Markets and Society*, Palgrave Studies in Sustainable Business In Association with Future Earth,
https://doi.org/10.1007/978-3-030-77650-3_9

and treat groundwater and surface water as two different resources so that people use water judiciously (Winter et al., 1998).

India has 18% of the world population, which is likely to surpass China's by 2050 (UN Department of Economic and Social Affairs, 2017), resulting in an increasing demand for water, food, and energy. The agriculture sector accounts for a major chunk of water usage in India which is eighty percent, despite India being home to less than five percent of the world's freshwater resources. The industrial sector is the next big user, consuming around ten percent. Ironically, both the agriculture and industrial sectors are prompted to use groundwater as a source of water despite the fact that farmers are provided with the running cost of surface water in the form of heavy subsidies and that the industrial sector is penalized for using it minimally. India is, therefore, rightly depicted as a recipe for a water scarce country (Dhawan, 2017). India receives an average of 4,000 billion cubic meters of precipitation every year, and less than half of it is used in conservation while only a small percentage of that water is added to the water reserve (Bhat, 2014). The remaining water is not utilized effectively because of a lack of storage and inappropriate water management. On the one hand, the number of people depending on water is significantly higher than the carrying capacity of water; on the other hand, capturing rainwater is uncommon despite water being an essential resource for survival. India's annual rainfall is around 1,183 mm which is, unfortunately, not a year-round phenomenon. Three fourths of the total rainfall is received in just four months from July to September (Ministry of Statistics and Program Implementation, 2012). This skewness in monsoon has impacted agricultural productivity, making a farmer's life full of uncertainties, often leading to a series of choices for survival, such as the crop selection, time of cultivation, irrigation method, and price recovery. India's love for water-intensive crops like rice, wheat, and sugarcane can be seen from the fact that these account for more than ninety percent of the crop production (Srivastava, 2017). These irregularities have led to the ethical questioning of diverting drinking water for irrigation purpose, which could worsen the impending crisis (Bontemps et al., 2001).

We could sum up the water crisis in India as a supply-side problem of water, but here is a thought for introspection:

> imagine you have a hectare of land in one of the driest places in India, say Thar Desert. Only 100 mm of rainwater would fall per year. This is as much

> as one million liters (Ml) of water. Even if you were not able to capture all that water—depending on the rainfall and the type of run-off surface—you could still capture at least 0.5 Ml a year, even with rudimentary technology. (Agarwal, 2000, p. 4)

Assuming that one could capture the 2,000 mm annual rainfall that Eastern India usually receives, one would need only 500 square meters of land to capture one million liters of rainwater. Thus, there is an inverse relationship between the size of the land required to save water and precipitation, given the average rainfall.

Villages becoming 'water-catcher' would lead to a surplus amount for irrigation, highlighting the potential of rainwater harvesting. Agriculture must be sustainable for future generations by conserving rainwater. Yet, how much water should be preserved? Taking this discussion a step further, this chapter argues that we can derive the value of water required to grow crops during an agricultural cycle and save this quantum of water for future use or trade this captured water inter- or intra-regions effectively, making the entire country drought-proof. The government and international community can play a vital role in financing this project through ecological fiscal transfers as being done in countries like India.

2 The Indian Context

2.1 Water Governance: A Miss in India

Over the last century, the burden of water management shifted from individuals and communities to the state, and the inherent inefficiency of the state led to the exploitation of rivers and groundwater through dams and tube wells. Therefore, the problem of water management lies not only in the unscientific extraction of groundwater, but also in the policy paralysis exhibited by the government (Krueger et al., 1991). One such example of policy paralysis is that of providing electricity subsidies disproportionately. Electricity subsidies are widely perceived to impact the economic condition of farmers, and often these are overestimated. Low-price or free electricity offers an incentive to the unchecked lifting of groundwater. It is said that externalities are not unilateral but reciprocal in nature. Hence, the government, by providing subsidy on electricity, and farmers, by pumping out more groundwater, both act in reciprocal ways. Dependence on the state means that the cost of water supply is high. The financial sustainability of agriculture is threatened because of poor cost of recovery and hence, the farmer is not inclined toward water resource management.

One general idea about the state of policy in India about water can be seen by glancing at its constitution. Under Article 246, the Indian Constitution allocates responsibilities of the states and the center into three lists: Union List, State List, and Concurrent List. Water is under Entry 17 of the State List, which reads: "Water, that is to say, water supplies, irrigation and canals, drainage and embankments, water storage and water power," but subjects like "development and linking of inter-state rivers and valley" are in the Union List (Constitution of India, 1949). These policies unintentionally create a water divide among its constituent states and put the policy at the mercy of political intent (Dinar et al., 1998). As most rivers in the country are inter-state, the regulation and development of waters of these rivers become a source of inter-state differences and disputes. This is the major reason for stalled water projects in India.

This has led to an over-reliance on groundwater instead of surface water. The groundwater level has started to decrease heavily as more and more areas were brought under irrigation (Jha & Sinha, 2009). With an increase in areas under irrigation, a shift in the mode of irrigation was observed as farmers started preferring private tube wells rather than using irrigation infrastructure available publicly, which has led to an increase in groundwater extraction; the share of canal irrigation has declined from 39.8% in 1950 to 23.6% in 2013, with a simultaneous increase in the share of groundwater irrigation from 28.7 to 62.4% during the same period, which is an astronomical rise in itself (Ministry of Farmer and Agriculture Welfare, 2014). In general, Indian farmers rely more and more on groundwater irrigation. This is in steep contrast to the global average of forty percent (Siebert et al., 2010). The Central Ground Water Board (CGWB) in its yearbook has categorized 16.2% of the total assessment units as 'over-exploited' and 14% as either at 'critical' or at 'semi-critical' stage, many of which are in the northwest region of the country (CGWB, 2020). This unscientific withdrawal of groundwater without focusing on replenishment has worsened the situation. The depth to water level in 2019 pre-monsoon with its decadal mean indicates that 61% of the aquifers have declined. The government still fails to nudge the farmers in adopting techniques like drip irrigation to outpace water scarcity. This is evident from the fact that a mere three percent of farmers use drip irrigation and more than ninety percent flood their farmlands (Ministry of Water Resources, 2017).

According to the CGWB's 2011 assessment, "India's total annual replenishable groundwater resource is around 433 billion cubic meters (BCM) and net annual ground water availability is 398 BCM of which India withdraws 245 BCM (62%) annually" (Dhawan, 2017, p. 11).

2.2 The Situation of Water Scarcity in India

Water stress is indicated by the per capita water availability in a country. As per the international norms, a country is classified:

- as water stressed if per capita water availability goes below 1700 m^3;
- as water scarce if per capita water availability goes below 1000 m^3.

With 1544 m^3 per capita water availability, India is a water-stressed country and turns progressively into a water-scarce country (Food and Agriculture Organization, 2008). There is an additional pressure on water resources due to population growth and increased irrigation intensity, both per hectare and according to the proportion of irrigated land out of total agricultural land.

As the groundwater level declines, farmers will either deepen their existing wells by buying powerful pumps or reduce the net irrigated area. This will lead to decreased profits either way because farmers require capital to dig deeper wells and because a reduction in the irrigated area will directly reduce the yield. With this thought we concur that, in line with Brouwer et al. (1985), all crops need water to grow and the most important source of water for crop growth is rainfall. When rainfall is sufficient, the yield is good and when the rainfall is insufficient, the farmers are bound to use groundwater for irrigating their fields.

One of the major issues in the eyes of farmers is to understand the irrigation requirement as per the desired crop since excess water leads to crop spoilage. It should neither be too little nor too much. Moreover, the water should be readily available for irrigation as longer spells with insufficient water result in loss of yield or even crop failure. In addition, the irrigation requirement needs to be determined for the proper design of the irrigation system and for the establishment of the irrigation schedules (Jhonson, 2009).

We know that water is not equally distributed across countries and geographies. There are some countries/regions which have surplus in water, and some are scarce areas. Rudimentary trade theory suggests that a country should export commodities that it has in abundance and import those that are scarce. By this logic, an efficient trade in water is achieved when a 'water surplus' region exports it to a 'water deficit' region either through actual means or a redistribution that takes place virtually through an exchange traded product like embedded water derivative, which we discuss in the next section.

Based on these reasons, we need a new outlook which will enable farmers to adopt sustainable agriculture practices and will provide an impetus to policymakers to scrutinize the existing system in favor of inputs required for agriculture in order to advance toward sustainability.

3 Research Methodology

The aim of forward pricing of embedded water is to express the total water consumption per unit value of output as a rough approximation of the opportunity cost of water in agricultural production (Foster & Sekhri, 2008). Thus, it essentially links water resource use to the water footprint of each agricultural commodity. Since the quantum of water remains constant in an ecosystem, it equates total available water with the total water used. The 'water available' is nothing but the amount of rainfall one can capture, and the 'water used' is its corresponding usage. Through forward pricing, one can maintain a balance as it helps to know the amount of water to be saved so that it can be used at a later stage.

3.1 Climate Classification

Since capturing rainwater first involves mapping rainfall, Köppen climate classification has been used (Pidwirny, 2006). It is a land-based classification of climatic zones which divides the earth into five major types, represented with the letters A, B, C, D, and E on the basis of both precipitation and temperature. These five zones are then subdivided into several zones with common characteristics, such as the amount of solar radiation, high and low-pressure zones, topography, and altitude.

A—Tropical moist climates

About 15–25° latitude northwards and southwards of the equator are the tropical moist climates. The annual rainfall in this region is about 1500 mm. Three types of climatic variation may be found within the tropical moist climate region:

- ***Af***, or tropical wet climate, where precipitation occurs year-round.
- ***Am***, or tropical monsoon climate, where precipitation occurs within 7–8 months of a year.
- ***Aw***, or the tropical wet and dry climate, where rain is mainly in the summertime.

B—Dry climates

The regions extend 20–35° latitude northwards and southwards from the equator. There are four subdivisions within this region:

- ***BW***, or dry, arid climate, also known as 'true desert climate.' The letters *h* and *k* are used as suffixes after BW to represent whether the zone is subtropical (h) or mid-latitudinal (k).
- ***BS***, or dry, semi-arid climate. These regions receive more rainfall than their BW counterparts because of mid-latitude cyclones and areas of inter-tropical convergence. Here, the letters *h* and *k* are again suffixed to note the subtropical or mid-latitudinal regions, respectively.

C—Moist mid-latitude climates with mild winters

These zones extend 30–50° latitude northwards and southwards from the equator. The climatic classification is further broken into three types:

- ***Cfa***, or humid sub-tropical climate. Precipitation occurs due to mid-latitude cyclones.
- ***Cfb***, or marine climates with wilder winter. They come with heavy rainfall due to mid-latitude cyclones.
- ***Cs***, or the Mediterranean climate. Precipitation is heaviest during the winters due to mid-latitude cyclones. There is hardly any rainfall during the summer.

D—Moist mid-latitude climates with cold winters

Summers are typically warm but can also be cool, while winters are cold. These regions are typically situated toward the poles. This Köppen climate classification is further divided into three subsections:

- ***Dw***, which denotes dry winters.
- ***Ds***, which denotes dry summers.

- *Df*, for year-round rainfall.

E—Polar climates

Temperatures are usually low all year round in the polar climatic regions. The warmest months see temperatures less than 10° Celsius. These climates are divided into two categories:

- ***ET***, or polar tundra, where soil remains permanently frozen as permafrost and can be hundreds of meters deep.
- ***EF***, or polar ice caps, are the second category, in which the surface is permanently covered in ice or snow (Fig. 1).

The major land area dominated in India is that of humid subtropical climate (Cfa) and tropical savannah climate (Aw). Therefore, the majority of the Indian subcontinent lies between 800 and 1200 mm of rainfall in a year, making it a 'rainfall abundant country.' This fact strengthens the need to harvest rainwater for a sustainable agriculture.

Another interesting concept is that of virtual water. Virtual water is nothing but the amount of water embedded in the product (Allan, 2011). The hidden flow of water refers to the total water employed in the various steps of the assembly chain. This chapter focuses on the impact of embedded water on agricultural commodities only. As these commodities are bought and sold in the market, the water that is used to produce them is also traded (Ray et al., 2018). Thus, water is said to be 'virtual' as one cannot separate the water from the crop. This concept helps us in analyzing the water footprint required to grow agricultural commodities in various regions in India. It is estimated that the country exported about 25 km^3 of virtual water embedded in its agricultural exports in the year 2010. This is equivalent to the demand of nearly 13 million people (Ministry of Finance, 2015). From being a 'net importer' to a 'net exporter' of virtual water, India has the greatest selling story of agricultural prowess. This has led to skewness in groundwater making farming unprofitable in otherwise fertile lands as well. For example, the cost of water extraction for growing wheat in Punjab tends to be on the higher side as compared to that in Madhya Pradesh, when one observes the ratio of water footprint (Mekonnen & Hoekstra, 2011).

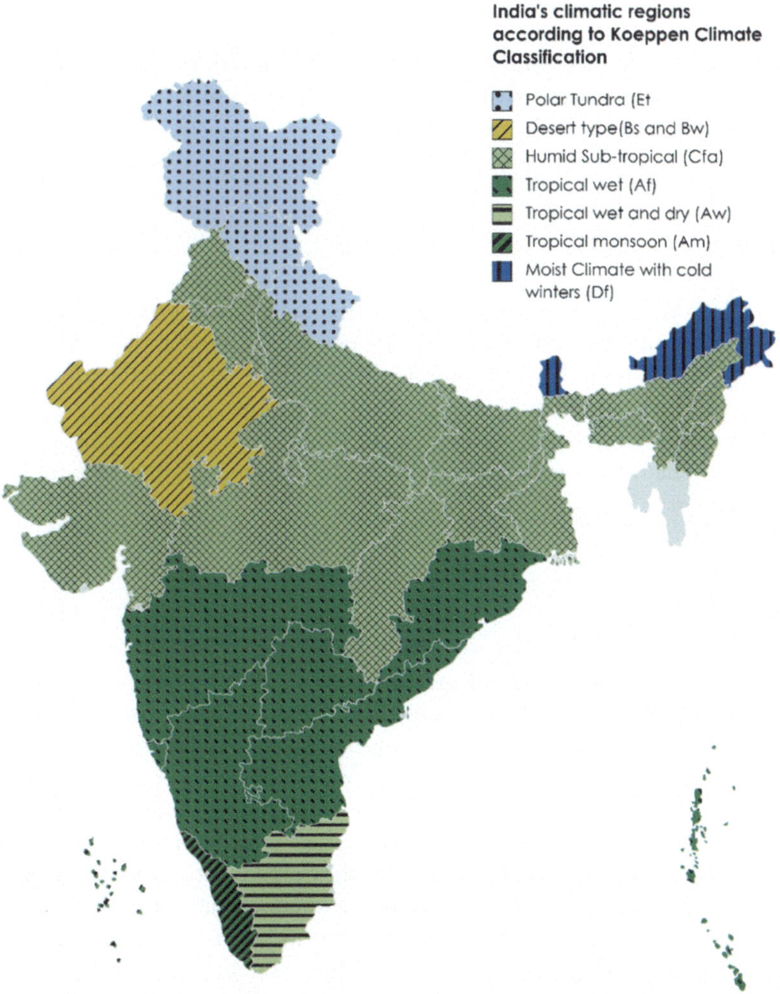

Fig. 1 Köppen climate classification of India (*Source* Own representation [based on mapchart.net])

3.2 *The Model*

Let the current agricultural cycle be n, then the amount of water required to grow the same quantity of crop in the next agricultural cycle $n + 1$ can be denoted with the following formula:

$$F_{n+1} = S_n \times e^{rt}$$

where

F_{n+1}=Water footprint for $n + 1$ or the amount of water required to grow same quantity of crop in the next agriculture cycle.
S_n = The underlying commodity that is the current water footprint.
e = The mathematical irrational constant approximated by 2.7183.
r = The rate of groundwater used for growing crop in cycle n .
t = The duration or the time for which groundwater is extracted in cycle n .

3.3 *The Model Assumptions*

- **Specific assumption**
  - Use of **exponential growth**

    We infer that a population's *per capita* growth rate of water utility stays the same regardless of population size, hence making the population grow faster without a constraint imposed by limited resources in the environment, known as the carrying capacity. Thus, water, as a public good, inhibits the problem associated with the tragedy of commons that is a situation in a shared-resource system where individual users act as per their own self-interest and behave contrary to the common good of all stakeholders by depleting or spoiling the shared resource through their collective action. In relation to water, this behavior signifies that exponential growth produces a reverse J-shaped curve, meaning that initially more water appears at the surface level. However, as the vertical height increases between the surface and aquifer, the water level recedes sharply, keeping the water drawn yield in a time period constant (Fig. 2).

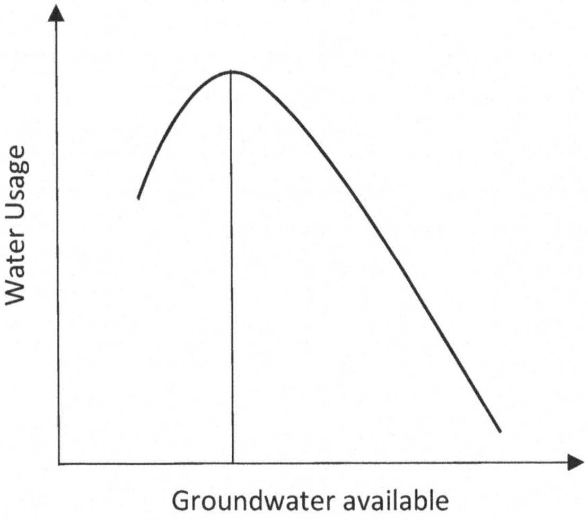

Fig. 2 Inverse J curve graph

- **The rate and time**

  It is assumed that the availability of groundwater in an area is dependent upon the rate of its extraction and the time for which it is being used during an agricultural cycle. When more water is extracted from ground for irrigation, the general water level in an aquifer declines, keeping other variables constant (Fishman et al., 2011). Similarly, there is an inverse relationship between water availability and the time duration.

- **General assumption**
  - The farmer uses water for one and only one commodity. This assumption essentially separates the amount of water used for irrigating two different commodities (wheat and rice). The focus is primarily on the end use of water, which is used to irrigate. This assumption commoditizes water into different products, making water used to irrigate wheat (W_{Wheat}) different from water used to irrigate rice (W_{Rice}). Hence, we must calculate different S-values for both (Mahajan & Porwal, 2018).
  - There is no significant cost other than the one for procuring water. Hence, the focus is only on the procurement of water for

growing agricultural commodities. No other significant costs, like the procurement of seeds, fertilizers, etc., are taken into consideration.

- The average water footprint of a country is taken to calculate the water footprint to standardize the model and to show a better picture. This assumption acts as a neutralizing agent for water-rich and water-deficit areas.
- Water footprint is divided into green (water from precipitation), blue (sourced from surface or groundwater resources), and grey (point-source pollution discharged to a freshwater resource). The total water footprint inclusive of green, blue, and grey information is used for research.
- The r value, i.e., the rate of groundwater extraction, changes as per the geographical area under study. For example, the r value in Uttar Pradesh will be different as the r value in Punjab. The r value is taken as constant in the research for ease of understanding.
- The t value, i.e., the duration or the time for which groundwater is extracted, changes as per the geographical area under study. For example, the t value in Uttar Pradesh will be different as the t value in Punjab. The t value is taken as constant in the research for ease of understanding.
- The rate of groundwater used is calculated as a percentage of total amount of water used in general terms.
- The water footprint or the amount of water used to grow a given quantity of crop is the same throughout the different agricultural cycles. Hence, for agriculture to be sustainable, farmers must aim to save at least S-value derived through the model.
- The total land area is taken to calculate the water capture for an area for the calculation of K-values (carrying capacity).[1]

Hence, the model conclusively takes the form of a forward pricing mechanism where the underlying asset is the water consumption of a particular crop of a given quantity. The spot component is the current water used to grow the given quantity, and the forward component is the

[1] The K-value expresses the volume of rainfall per area.

Table 1 Data variable and source

Data variable	*Source*
Quantity grown that is Q	Department of Agriculture and Cooperation, Ministry of Agriculture, Government of India
Average water footprint that is W	Food and Agriculture Organization, United Nations
S_n or Water footprint for *n* agri-cycle	Calculated from Q and W
F_{n+1} or water footprint for $n + 1$ agri-cycle	Calculated from the formula given above
K-value or water capture	Calculated as total volume of rainfall per area

water to be used in the next agriculture cycle. Since one can derive the quantum of water used and attach a financial value to the use of water, this pricing mechanism will enable a sustainable agricultural practice.

For calculating S_n or the current water footprint for a particular crop in a given area (Table 1),

$$S_n = Q \times W$$

where

Q = Quantity grown.
W = Average water footprint per tonne of crop required in India.

Using this, we derive the S_n-values of wheat for the top 10 wheat-producing states in India, considering that the average water footprint in India for wheat is $W_{\text{Wheat}} = 634\ m^3$ per tonne (Mekonnen & Hoekstra, 2011). For example, in the case of Uttar Pradesh we have:

$$\begin{aligned} S_{\text{Uttar Pradesh}} &= Q_{\text{Uttar Pradesh}} \times W \\ &= 30.25\ \text{MT} \times 634\ \text{m}^3 \text{per tonne} \\ &= 19,178.50\ \text{M m}^3 \end{aligned}$$

Similarly, we derive the S_n-values in Table 2.

Now, we know that for growing 30.25 million tonnes of wheat, a farmer in Uttar Pradesh might use 19,178.50 million m^3(S) of water in

Table 2 Calculation of S-values ($S = Q \times W$)

State	*Wheat* *Quantity in million tonne*	*Given average water footprint in India for wheat per tonne (sourced from FAO) = 634 m^3* *Water footprint for current underlying i.e. S values (mn m^3)*
Uttar Pradesh	30.25	19,178.50
Punjab	17.04	10,803.36
Madhya Pradesh	13.93	8,831.62
Haryana	11.8	7,481.20
Rajasthan	8.92	5,655.28
Bihar	5.08	3,220.72
Gujarat	3.65	2,314.10
Maharashtra	1.6	1,014.40
West Bengal	0.95	602.30
Uttarakhand	0.84	532.56
Total of top 10	94.06	59,634.04

cycle n. Through this, we can also calculate different S-values of variety of crops grown in an area like S_{Rice}, S_{Sugar}.

According to our model, water used for growing wheat in these areas for cycle $n + 1$ (F wheat), considering that $r = 60\%$ (based on our estimates) and $T = 3$ months in a year.

$$\begin{aligned} F &= S \times e^{rt} \\ &= 19,178.50 \times e^{0.6 \times 0.25} \\ &= 19,178.50 \times 1.16 \\ &= 22,282.24 \text{ million m}^3 \end{aligned}$$

Similarly, F-values for various other states are calculated in Table 3.

Therefore, we infer that in year $n + 1$, for growing this quantity of wheat, farmers in Uttar Pradesh must extract 22,282.24 million m^3 of groundwater keeping other variables constant. Through this (F_{wheat}) we can also calculate different F-values of variety of crops grown in an area, such as F_{Rice} and F_{Sugar}.

The results indicate the amount of water to be saved and consumed. Afterward, we calculate the carrying capacity (K), which represents the

Table 3 Calculation of F-values ($F = S \times e^{rt}$)

Given		*r*	*0.60*
		t	0.25
		r x t	**0.15**
State	*Water footprint for current underlying i.e. S values in million m^3*	*Exponential values*	*Water footprint for* n *+ 1 or The amount of water required to grow same quantity of crop in next agriculture cycle i.e. F values in million m^3*
Uttar Pradesh	19,178.50	1.16	22,282.24
Punjab	10,803.36	1.16	12,551.71
Madhya Pradesh	8,831.62	1.16	10,260.88
Haryana	7,481.20	1.16	8,691.91
Rajasthan	5,655.28	1.16	6,570.50
Bihar	3,220.72	1.16	3,741.94
Gujarat	2,314.10	1.16	2,688.60
Maharashtra	1,014.40	1.16	1,178.56
West Bengal	602.30	1.16	699.77
Uttarakhand	532.56	1.16	618.75
Total of top 10	59,634.04		69,284.87

actual amount of rainfall in a given area. This helps in analyzing the surplus or deficit.

For calculating K-values, we equate the land area with annual mean rainfall in that area based on the Köppen classification. In this study, we are using the subtropical climate represented by *Cfa* or Uttar Pradesh.

We know that one hectare = 10,000 m^2; therefore, the area of Uttar Pradesh in m^2 is calculated as per Table 4. Based on average rainfall in meter, we get the K-values or the total rainfall in a given area in m^3 as:

$$K = \text{Area of a given land} \times \text{Average rainfall}$$

With this formula, we get an approximate of the water capture value in billion m^3 (as 1 Billion = 1,000,000,000).

This K-value represents the total volume of rainwater in a given area. Hence, this K-value represents only rainwater used to grow various crops, such as wheat, rice, and soya bean.

Table 4 Calculation of K-values

State	*Area in hectare*	*Area in m^2*	*Lowest rain-fall in m^3*	*Highest rain-fall in m^3*	*Average rain-fall in m^3*	*Water footprint in m^3*	*Waterfootprint in billion m^3*
Uttar Pradesh	24,328,600	243,286,000,000	0.50	1.00	0.75	182,464,500,000	182.46

4 Analysis

4.1 One-Crop Model

This model signifies that the amount of water footprint in cycle n, i.e., '*S*,' will become 'F' in the next agriculture cycle $n + 1$ because of the increasing cost related to extraction. Thus, for the same quantity of crop, farmers require more groundwater which leads to an increase in the cost of growing (Roumasset & Wada, 2010). In order to grow the same quantity of crop year on year, farmers must aim to save at least 'F_{n+1}' m^3 of water so they can use it to grow the crop in $n + 1$ year.

- If '*F*' > '*S*' after $n + 1$, i.e., water used in next cycle is greater than water used in the current cycle, this will be detrimental to water resource in the area, leading to a higher cost per tonne. This result exhibits unsustainable agriculture. Consequently, there will be a gradual increase in the price of an agricultural commodity, which will have to be offset through government subsidy to farmers, otherwise they will have to bear the entire price burden which in turn is passed on to the final consumers. In the long-term scenario, the demand of water will be higher than the supply, still assuming that water used for irrigation is more than the water saved. This will lead to a deficient supply, which will increase the overall prices of agricultural products and lead to supply-side inflation, a difficult condition to handle for developing economies. To provide further hardships, the situation will worsen with an ever-growing demand because of the rise in population. We perceive this situation as being a step toward a worst-case scenario in which there is no more groundwater to extract.

- If '*F*' < '*S*' after $n + 1$, i.e., water used in next cycle is less than water used in the current cycle, this will mean that replenishment is occurring. This is the **ideal scenario** because it will signify that the resource has been stabilized, thus illustrating sustainable agriculture practices. This means that the farmers are extracting less groundwater and using water markets as the demand for water is fulfilled. Consequently, the input cost of extracting water will not be that significant. Although it has no intrinsic value at the inception, over time, the amount of water saved may gain or lose value vis-à-vis its consumption. In every agricultural cycle, there will be surplus water as the amount of water used will be less compared to the amount of water saved. This will give rise to tradable water rights. The farmer will have a choice to either grow crops by using saved water or sell it to other farmers and gain from this transaction.

  Thus, the value of tradable water rights can be calculated as

$$\text{Surplus or Tradable water rights} = \mathrm{F}_{n+1} - \mathrm{S}_n$$

  This practice will be beneficial for small land holders as the farmers can easily conserve water. Thus, in this ideal scenario, excess water, which can be traded intra- or inter-region, can enhance farm produce on one hand and farmers' income on the other.

  This ideal scenario will be the first step to a robust water market in an area. The surplus can then be a part of an ecological fiscal transfer scheme, making it a part of vertical dilution of funds to surplus states in yearly budget. Ideally, this will ensure a standard budgetary allocation and nudge farmers to save more water year after year (Easter & Tsur, 1995).

4.2 *Many-Crop Model*

We calculate S_n-values for one-crop example ($\mathrm{S}_{\mathrm{Wheat}}$) and analyze the F-values needed to save in order to enter a surplus in case of water footprint. In states like Uttar Pradesh, there are many crops grown during an agricultural cycle. Since, the amount of water used for irrigating crops is different, we must calculate S_n-values of each crop grown in an area as $\mathrm{S}_{\mathrm{Wheat}}$ i.e., water used to grow wheat; $\mathrm{S}_{\mathrm{Rice}}$ i.e., water used to grow rice. Hence, we must calculate different S_n-values for as many crops grown in an area as possible as well as the total F_{n+1}-values. Thus, we will get a value for the total amount of water used to grow crops in an area from the total S_n-values of different crops grown. This value is represented as TOT_S.

This variable will enable us to approximate the amount of surplus water collected in an area. To do so, we subtract T_S value from K-value, i.e., water collected in 'total water capture in an area' minus 'total water used in growing crops.'

$$\text{Surplus or tradable water rights} = K - TOT_S$$

Saving water will pave the way to a robust water market, which will lead to a redistribution of water to drought-prone areas and other areas where agricultural productivity is low but rainfall is high. Water collection at a unit level will help us analyze per crop water requirement, and scientific decision-making can be based on water usage. It will have an additional benefit in those states which have water disputes like Karnataka and Telangana, Goa, and Maharashtra. These states can be categorized as upstream states and downstream states. Total water collected in upstream states for growing crops (TOT_S) is known beforehand through forward pricing of embedded water. The surplus water can also be calculated through $K - TOT_S$. This surplus water can be used by downstream states by diverting the said amount of water in river catchment areas as a part of surface run off (Alcamo et al., 2000), negating the need to construct dams if the perennial flow of river is ensured. A river basin authority can easily do that (Winter et al., 1998). Thus, the net amount of water used in the system will always remain in equilibrium, essentially making it a zero-sum usage.

5 Impact of Forward Pricing of Embedded Water

This water pricing strategy of applying the concept of forward contracts to the embedded water is crucial and requires immediate attention for all stakeholders involved in making agriculture sustainable as it enables one to think of water as a commodity and not only as a resource. It will influence the decision to grow only crops that require a minimum of water or that can be grown with water stored for irrigation (*S-values of the crop*). It goes without saying that the core of future farming will rely not only on farmland, but also on adequate water supply. Keeping this in mind, we discuss the impact of forward pricing for various stakeholders in this section.

5.1 For Farmers

- *Quantum of groundwater extraction*

  Groundwater use in irrigation is increasing both in absolute terms and in percentage of total irrigation. Despite the uncertainties associated with statistical data available to track patterns and growth of groundwater use for irrigation, the inventory presented beforehand is a major step toward a more informed assessment of agricultural water use. The quantum of water to be saved is known, and all the work is done accordingly.
- *Cost saving*

  Deepening wells means more investment, high maintenance costs, and an increase of flat electricity tariff (Sayre & Taraz, 2019). There is a strong correlation between the failure of cash cropping due to irrigation, overwhelming debts, and farmer suicides (Natarajan, 2019). The quest to recover investments for drilling provides a reason to grow crops which are not supported by the land. This puts farmers in a debt trap in which they are unable to bear the cost of pumping due to declining groundwater level. Ultimately, this situation affects the well-being of farmers.
- *Sustainable agriculture*

  Groundwater management, through forward pricing of embedded water, will ensure an adequate access to quality water for irrigation in areas where inequity over space and time are acute. We foresee that it will lead to change in the cropping pattern, crop diversification, and to an increase in encouragements toward water use efficiency in agriculture by moving toward food crops from cash crops in areas where cash crops are grown for high profit. These changes will promote sustainability in agriculture, helping its transition toward a net carbon zero activity (Foster & Rosenzweig, 2006).
- *Land management in village*

  Farmers in a bid to earn more often advance toward unscientific crop management. Thus, a land that should have been left fallow so as to gain sufficient nutrient quality might be constantly irrigated, which will reduce its yield. Through forward pricing, villages can earmark land for water management and crop sowing effectively, hence turning them into water-catcher. The adoption of a forward pricing mechanism may lead to coordinated efforts among small-scale landholders for management of groundwater at the local level (Shankar, 1992).

- *Additional income for small farmers*

  Farmers can sell the surplus water to those who need to irrigate their farmlands or for community use. They can charge money for efficient use of water as the water is bought and sold for its own sake and not as an incidental part of the land transfer (Saliba & Bush, 1987). The water can be in turn integrated to an ATM-like vending machine that provides water 24 hours a day. This machine commonly called Water ATM is already operational in various railway stations and cities like Delhi. It is powered by solar energy, integrated with reverse osmosis (RO) and ultra-filtration units with reduced operational costs. For example, Water ATMs at Marathwada provide water at 25 paisa per liter a day.

  Adopting this measure will ensure a steady income even for farmers with small land holdings who do subsistence agriculture, in turn incentivizing water saving behaviors.

- *Step toward traditional techniques*

  Ancient civilizations learnt to harvest rainwater with rudimentary technology. The water markets will be a step toward sustainable methods of saving water as it will achieve a balance between new markets and old techniques like ahars of Bihar and johads of Rajasthan. The practicing of old school agriculture has the added advantage of conserving soil moisture and its nutrients (Ministry of Water Resources, 2005).

5.2 For Government

- *Public availability of groundwater data*

  Making groundwater data available to the public, such as the state of water tables in all aquifers and the availability of surface water in all basins and sub-basins, can facilitate entrepreneurial decision-making. This can act as a warning system for any industrial or agri-food activity in order to stop further expansion in water-stressed regions. It will also facilitate the coordination among users for aquifers and the first step toward water contracts.

- *Cross subsidy*

The intended benefit of many financing schemes formulated by the government does not reach its actual users. In this case, the government can use cross subsidy in water management. Cross subsidization can be done when government purchases at a higher price from a group of farmers indulged in groundwater management or by ensuring sustainable financing schemes related to water use efficiency. Cross subsidy can ensure that existing public irrigation infrastructure is maintained.

- *Comparative advantage*

  Analyzing water resource per unit value of output will accrue benefit in knowing the low, self-sufficient crop (Johansson, 2006). Consequently, importing/growing crops from areas where crops can grow under a higher proportion of green water usage should not be treated as a sign of sectoral weakness but rather as a contribution to securing long-term agricultural competitiveness. Thus, the Indian government should subsidize those crops which are less water intensive by providing appropriate market price, and water-intensive crops should be limited only in those areas which can sustain it. Adopting these measures will create a water-smart policy.
- *Help in quota allotment*

  We know that it is efficient to base prices on the marginal cost of acquiring more water plus its scarcity value (Dinar et al., 1998). However, prices based on marginal costs are unaffordable for low-income farmers. The role of state or community comes into play in these situations through quota allotments, e.g., warabandi system in India, which is a rotational method for equitable distribution of the available water in an irrigation system, in turn fixed according to a predetermined schedule to each irrigator in proportion to the size of his landholding (Venkatachalam, 2008). By allowing quota allotments to be traded, water authorities can address equity concerns while promoting efficient allocations, as it is done in water markets in California and Australia (Howitt, 1994). The trading may lead to the introduction of farm-based measures of water consumption or per farm usage of water.
- *Establishing fully functional water markets*

  It has long been recognized that markets provide a means to allocate water according to its opportunity cost, resulting in efficiency gains (Bjornlund & McKay 2000). Water markets, which rely on market pressures to determine the price for irrigation water, are also more flexible than centrally controlled allocation mechanisms. Market-based mechanisms can address allocation inefficiencies found in traditional irrigation institutions. In fact, water trading is likely to expand the access to water for many small-scale farmers (Rosegrant & Binswanger, 1994). For formal water markets to work, there first needs to be a well-defined, tradable water rights and the appropriate infrastructure and institutions for distributing water.

- *Formalization of an informal sector*

  Informal water markets often develop when water is scarce or when governments fail to respond to rapidly changing water demands. However, given the institutional structure necessary for market-based policy, the sectoral interdependencies, as well as the large, fixed investments costs and uncertain supplies, the prospect of attaining first-best allocations via markets are unlikely yet promising (Brill et al., 1997).
- *Green employment opportunities*

  Often agriculture is mired with disguised employment, i.e., more hands producing less goods, which affects per unit productivity of labor (Datt & Ravallion, 1998). Water resource management will reallocate jobs in an already saturated agriculture market and lead to green employment opportunities in water-rich areas where crops are destroyed due to heavy rainfall. Farmers can capture water and sell it in the water markets in water-deficit areas. Government can also raise special training programs, such as the Namami Gange corps.
- *Sustainable development goals*

  A forward pricing mechanism will help the Indian government in its fulfillment of the Sustainable Development Goals (SDG), as agriculture is an interconnected sector of any economy. It will impact many sectors, from health and nutrition to the economy, and most precisely farm income, ideally leading to sustainable consumption and production. Forward pricing mechanism will cater toward SDG 1 (poverty alleviation), SDG6 (water use), SDG8 (economic growth), and potentially more.
- *Inter-state dispute solving*

  Water markets may lead to systematical use of water for agriculture since the amount of surplus would be 'traded' by water surplus state to water-deficit state. Thus, net water use would remain at the same level. The demand for water across the state would be quantified, and hence better allocation of surface water can take place by following this pricing strategy.

5.3 *The Challenges*

The challenges can be outlined as follows:

- *Natural conditions like drought*
  Since the model is based on the need to capture rainfall, any severe drought or deficit in rainfall will be detrimental, rendering it prone to climate change and weather events. However, when farmers are nudged sufficiently and indulge in water saving techniques (Khanna & Zilberman, 2017), the effects of drought can be reduced considerably by moving away from rainfed agriculture.
- *Lack of pricing mechanism*
  Irrigation water is not priced according to the actual quantity used because volumetric measures are not in place (Bierkens et al., 2019). This contributes to a broader problem related to the widespread use of unregistered wells and inefficient water practices, as well as to the redundant usage of water often leading to an increase in the cost of farming.
- *Focus more on surface water rather than irrigation water*
  The government—and society in general—places a relatively strong emphasis on the management and protection of surface water rather than irrigation. This leads to more financing in the construction of dams rather than in the construction of local infrastructures. Irrigation infrastructures at the local level require small ticket size as compared to dams and other infrastructure projects related to surface run off water (Zaveri et al., 2016).
- *Infrastructure constraint*
  Water markets generally play a second role in government's playbook because of infrastructure constraint. It requires building more capture areas, de-silting them season after season, and maintaining water level in the capture. However, if water markets are developed, they may become an asset whose benefits accrue in the long term (Marino & Kemper, 1999).
- *Rationalization of Minimum Support Price (MSP)*
  India's consumption of irrigation water is not only affected by policies that target irrigation water consumption directly, but also by other agricultural policies that affect farmers' incentives, such as higher MSP to produce water-intensive products. Therefore, the

policies for the pricing of water do not encourage water efficiency or cost recovery (Beaumont, 2000).

- *Potential water price distortion*

  As the water trading market flourishes over time, water price will be determined by supply and demand. The pricing of extracting groundwater could decrease significantly if a large proportion of farmers use water markets. A situation may arise wherein the cost of extracting groundwater will be less than the cost in water markets. This may lead to a distortion of prices in water markets and farmers will be inclined to use groundwater for irrigation.
- *Excess water*

  An increasing number of farmers capturing water will lead to an excess of water. Government policies should aim to make the water trading market robust by enabling the buying and selling of water. Agricultural policies should be more consistently aligned to sustainable long-term management. Water concerns should therefore be reflected extensively in agricultural policies.
- *Regional skewness*

  Groundwater use necessitates demand management and supply augmentation measures for improved water use efficiency in the agriculture sector. There will be regional skewness in case of water markets, such as the Eastern region, where there will be surplus each year (Mukherji, 2004) and the Western region where groundwater utilization is on a large scale. This offers greater scope for harnessing groundwater based on this skewness.

6 Conclusion and Recommendations

There should be a parity between water availability and cropping pattern. Consequently, the current trend of producing water-intensive crops in water scarce areas should be reexamined. Some states which had water surplus like Punjab are becoming water deficit because of the unmonitored extraction of groundwater. India should review its policies associated with the export of water-intensive crops because the ratio of export to import of virtual water is relatively high for India when compared to its neighbors. A forward pricing mechanism of embedded water may help in redistributing water as crops, such as rice, which can be produced at water-rich locations, resulting in transfer of water virtually from water surplus to water-deficit areas. The end goal is to ensure that every farmland in India captures enough water to increase farm revenue, especially during years with a decent amount of rainwater. Once surplus is reached, the villages can store it in tanks for future use or trade it. Such a system will ensure a sufficient stream of finance for continued efforts in water resource management. The focus of the stakeholders should be on making agriculture sustainable with minimum role of the state in water allocation. The bedrock of sustainability can be explained with this quote from Mahatma Gandhi: 'The earth, the land and the water are not an inheritance from our forefathers but on loan from our children. So we have to handover to them at least as it was handed over to us' (Gandhi, n.d.).

This can be applied in international trade as food is the best way to invest in water, that is, to grow food in water-rich areas and transport it for sale in water-poor areas. This method of forward pricing substantiates the need for redistributing water through sustainable water markets, and the subsequent commercialization will ensure that this redistribution is profitable. This method helps to redefine agricultural commodities and to look at it from a new angle.

References

Agarwal, A. (2000). *Drought? Try capturing the rain.* Centre for Science and Environment. https://cdn.downtoearth.org.in/pdf/drought%20-briefing-paper-20160509.pdf

Alcamo, J., Henrich, T., & Rösch, T. (2000). *World water in 2025: Global modeling and scenario analysis for the World Commission on Water for the 21st century.* University of Kassel, Centre for Environmental System Research. http://www.env-edu.gr/Documents/World%20Water%20in%202025.pdf

Allan, T. (2011). *Virtual water: Tackling the threat to our planet's most precious resource.* I.B. Tauris.

Beaumont, P. (2000). The quest for water efficiency – restructuring of water use in the middle east. *Water, Air, and Soil Pollution, 123*, 551–564.

Bhat, T. A. (2014). An analysis of demand and supply of water in India. *Journal of Environment and Earth Science*, *4*(11), 67–72.

Bierkens, M. F. P., Reinhard, S., de Bruijn, J. A., Veninga, W., & Wada, Y. (2019). The shadow price of irrigation water in major groundwater-depleting countries depleting countries. *Water Resources Research, 55*(5), 4266–4287. https://agupubs.onlinelibrary.wiley.com/doi/full/10.1029/2018WR023086

Bjornlund, H., & McKay, J. (2000). Do water markets promote a socially equitable reallocation of water? A case study of rural water markets in Victoria Australia. *Rivers, 7*(2), 139–152.

Bontemps, C., Couture, S., & Favard, P. (2001). Is the irrigation water demand really convex? *SSRN Electronic Journal.* https://doi.org/10.2139/ssrn.291940

Brill, E., Hochman, E., & Zilberman, D. (1997). Allocation and pricing at the water district level. *American Journal of Agricultural Economics, 79*(3), 952–963. https://doi.org/10.2307/1244435

Brouwer, C., Goffeau, A., & Heibloem, M. (1985). *Irrigation water management: Training manual No. 1 - Introduction to irrigation.* Chapter four. Food and Agriculture Organization of the United Nations (FAO). http://www.fao.org/3/r4082e/r4082e00.htm#Contents

CGWB—Central Ground Water Board. (2020). *Ground water year book - India 2019-20.* Government of India. http://cgwb.gov.in/GW-Year-Book-State.html

Constitution of India. (1949). *Seventh schedule.* https://www.india.gov.in/sites/upload_files/npi/files/coi-eng-schedules_1-12.pdf

Datt, G., & Ravallion, M. (1998). Farm productivity and rural poverty in India. *Journal of Development Studies, 34*(4), 62–85. https://doi.org/10.1080/00220389808422529

Dhawan, V. (2017). *Water and agriculture in India - Background paper for the South Asia expert panel during the Global Forum for Food and Agriculture* (GFFA). https://www.oav.de/fileadmin/user_upload/5_Publikationen/5_Studien/170118_Study_Water_Agriculture_India.pdf

Dinar, A., Balakrishnan, T., & Wambia, J. (1998). *Political economy and political risks of institutional reform in the water sector.* http://documents.worldbank.org/curated/en/568881468761346683/Political-economy-and-political-risks-of-institutional-reform-in-the-water-sector

Easter, K. W., & Tsur, Y. (1995). The design of institutional arrangements for water allocation. In A. Dinar & E. T. Loehman (Eds.). *Water quantity/quality management and conflict resolution.* Praeger.

Fishman, R. M., Siegfried, T., Raj, P., Modi, V., & Lall, U. (2011). Over-extraction from shallow bedrock versus deep alluvial aquifers: Reliability versus sustainability considerations for India's groundwater irrigation. *Water Resources Research, 47*(6). https://doi.org/10.1029/2011WR010617

Food and Agriculture Organization. (2008). *Coping with water scarcity* (Water Report 38). Rome, Italy: Food and Agriculture Organization of the United Nations (FAO).

Foster, A., & Rosenzweig, M. (2006). *Inequality and sustainability of agricultural growth: Groundwater and green revolution in rural India.* https://eml.berkeley.edu/~webfac/emiguel/e271_s07/foster.pdf

Foster, A., & Sekhri, S. (2008). *Can expansion of markets for groundwater decelerate the depletion of groundwater resource in rural India?* http://afosterri.org/papers/sekhri2.pdf

Gandhi, M. (n.d.). Quote. In Chemtech International (2020). *6 quotes about water pollution to inspire sustainable industrial practices* (June 4, 2020). https://chemtech-us.com/6-quotes-about-water-pollution-to-inspire-sustainable-industrial-practices/

Howitt, R. E. (1994). Empirical analysis of water market institutions: The 1991 California water market. *Resource and Energy Economics, 16*(4), 357–371. https://doi.org/10.1016/0928-7655(94)90026-4

Jha, B. M., & Sinha, S. K. (2009). *Towards better management of ground water resources in India.* http://cgwb.gov.in/documents/papers/incidpapers/Paper%201-B.M.Jha.pdf

Jhonson, C. (2009). *Biology of social science.* Oxford Book Company.

Johansson, R. C. (2006). *Micro and macro-level approaches for assessing the value of irrigation water.* https://openknowledge.worldbank.org/handle/10986/8471

Khanna, M., & Zilberman, D. (2017). Theme overview: Inducing water conservation in agriculture: Institutional and behavioral drivers. *Choices: The Magazine of Food, Farm, and Resource Issues, Agricultural and Applied Economics Association, 32*(4), 1–3. https://doi.org/10.22004/AG.ECON.264597

Krueger, A. O., Schiff, M., & Valdes, A. (Eds.). (1991). *The political economy of agricultural pricing policy.* Johns Hopkins University Press.

Mahajan, P., & Porwal, H. K. (2018). Embedded water derivatives. *International Journal of Management Studies, 3*(3), 53–61. https://doi.org/10.18843/ijms/v5i3(3)/08

Marino, M., & Kemper, K. E. (1999). *Institutional frameworks in successful water markets* (World Bank Technical Papers). https://elibrary.worldbank.org/doi/abs/10.1596/0-8213-4459-5

Mekonnen, M. M., & Hoekstra, A. Y. (2011). The green, blue and grey water footprint of crops and derived crop products. *Hydrology and Earth System Sciences, 15*(5), 1577–1600. https://doi.org/10.5194/hess-15-1577-2011

Ministry of Farmer and Agriculture Welfare. (2014). *Agricultural-statistics-at-glance.* Government of India. https://eands.dacnet.nic.in/PDF/Agricultural-Statistics-At-Glance2014.pdf

Ministry of Finance. (2015). *Economic survey 2015-16: Chapter 4.* Government of India. https://www.indiabudget.gov.in/budget2016-2017/es2015-16/echapter-vol1.pdf

Ministry of Statistics and Program Implementation. (2012). *Statistical year book of India 2012.* Government of India. http://mospi.nic.in/statistical-year-book-india/2012/

Ministry of Water Resources. (2005). *General guidelines for water audit and water conservation*. Government of India. https://www.iitr.ac.in/wfw/web_ua_water_for_welfare/water/General_Guidelines_for_Water_Audit_and_Water_Conservation.pdf

Ministry of Water Resources. (2017). *Fifth minor irrigation census*. Government of India. http://mowr.gov.in/report-5th-minor-irrigation-census-has-been-released

Mukherji, A. (2004). Groundwater markets in ganga-meghna-brahmaputra basin: Theory and evidence. *Economic and Political Weekly, 39*, 3514–3520.

Natarajan, N. (2019). *Changing dynamics of tobacco production and exchange in south India: Towards a socio-ecological analysis* (Doctoral Dissertation, SOAS University of London). https://eprints.soas.ac.uk/32309/1/4750_Natarajan.pdf

Pidwirny, M. (2006). Climate classification and climatic regions of the World. *Fundamentals of physical geography* (2nd ed.). http://www.physicalgeography.net/fundamentals/7v.html

Prasanna, V. (2014). Impact of monsoon rainfall on the total foodgrain yield over India. *Journal of Earth System Science, 123*(5), 1129–1145. https://doi.org/10.1007/s12040-014-0444-x

Ray, C., McInnes, D., & Sanderson, M. (2018). Virtual water: Its implications on agriculture and trade. *Water International, 43*(6), 717–730. https://doi.org/10.1080/02508060.2018.1515564

Rosegrant, M. W., & Binswanger, H. P. (1994). Markets in tradable water rights: Potential for efficiency gains in developing country water resource allocation. *World Development, 22(11)*, 1613–1625. https://www.sciencedirect.com/science/article/abs/pii/0305750X94000751

Roumasset, J. A., & Wada, C. A. (2010). Optimal and sustainable groundwater extraction. *Sustainability, 2*(8), 2676–2685. https://doi.org/10.3390/su2082676

Saliba, B. C., & Bush, D. B. (1987). *Water markets in theory and practice*. Westview Press.

Sayre, S. S., & Taraz, V. (2019). Groundwater depletion in India: Social losses from costly well deepening. *Journal of Environmental Economics and Management, 93*, 85–100. https://doi.org/10.1016/j.jeem.2018.11.002

Shankar, K. (1992). Water markets in eastern UP. *Economic and Political Weekly, 27*(18), 931–933.

Siebert, S., Burke, J., Faures, J. M., Frenken, K., Hoogeveen, J., Döll, P., & Portmann, F. (2010). Groundwater use for irrigation – a global inventory. *Hydrology and Earth System Sciences, 14*, 1863– 1880. https://doi.org/10.5194/hess-14-1863-2010.

Srivastava, P. K. (2017). Satellite soil moisture: Review of theory and applications in water resources. *Water Resources Management, 31,* 3161–3176. https://doi.org/10.1007/s11269-017-1722-6
UN Department of Economic and Social Affairs. (2017). *World population prospects 2017.* https://population.un.org/wpp/Publications/Files/WPP2017_KeyFindings.pdf
Venkatachalam, L. (2008). *Market-based instruments for water allocation in India: Issues and the way forward.* http://freshwater.issuelab.org/resources/23458/23458.pdf
Winter, T. C., Harvey, J. W., Lehn Franke, O., & Alley, M. W. (1998). *Ground water and surface water: A single resource.* https://doi.org/10.3133/cir1139
Zaveri, E., Grogan, D. S., Fisher-Vanden, K., Frolking, S., Lammers, R. B., Wrenn, D. H., Prusevich, A., & Nicholas, R. E. (2016). Invisible water, visible impact: Groundwater use and Indian agriculture under climate change. *Environmental Research Letters*, *11*(8), 84005. https://doi.org/10.1088/1748-9326/11/8/084005

Misbehaving Drinking Water Systems: Risk and the Complex Nature of Failure

Bryan Karney and *John Gibson*

1 Introduction

Most engineered systems probably never behave exactly as designers had hoped, and water distribution pipe networks are no exception. For example, despite much careful thought, water supply systems often leak, sometimes burst, waste energy, and occasionally risk contamination. For example, an overview article (Colombo et al., 2009) compiles a range of fault detection by transients while a more recent exploration of a related fault-detection topic is provided by Wang et al. (2020). These are samples of a rather vast and growing literature on leak and fault detection but generally represent a literature that primarily looks at faults

B. Karney (✉)
Department of Civil and Mineral Engineering,
University of Toronto, Toronto, ON, Canada
e-mail: bryan.karney@utoronto.ca

J. Gibson
Department of Chemical Engineering and Applied Chemistry,
University of Toronto, Toronto, ON, Canada
e-mail: jh.gibson@utoronto.ca

T. Walker et al. (eds.), *Water Risk and Its Impact on the Financial Markets and Society*, Palgrave Studies in Sustainable Business In Association with Future Earth,
https://doi.org/10.1007/978-3-030-77650-3_10

from the rather limited perspective of the utility in charge of water delivery. Certainly, such systems can sometimes fail to adequately deliver the product (e.g., discharging insufficient flow or too low a pressure) or deliver an inferior product (e.g., biological, physical, or aesthetic attributes compromised).

Less frequently considered is the fact that the costs of these failures extend well beyond the water utilities. Additional costs can include disrupted businesses or transportation routes, and can extend to acute illness costs, long-term health consequences, property damage, an increased risk of traffic collisions, and even death. What makes an evaluation of the cost of these failures especially difficult is that they are often borne by different entities, ranging from householders, businesses, insurance companies, to banks and municipal taxpayers.

Given this complexity, it is not surprising that estimates of the total costs associated with water systems failures have seldom been attempted. The absence of accurate cost of failure information represents a significant gap in our knowledge. This paper argues that it is time to take better account of these many failure-related costs.

2 Features of Water Distribution Networks

A large urban water distribution network contains thousands of pipes and valves, many pumps, and several water towers and reservoirs. They are large, complicated machines. A sample of a water distribution network appears in Fig. 1. Typically, the layout of the pipes follows the streets.

In addition to containing thousands of components, many networks were built in the middle of the last century and thus many pipes have remained buried and in constant service for at least fifty years. Many cities in North America and Europe have pipes more than a hundred years old still in service. This makes failures of these components almost inevitable. Even though this problem has been around for a long time, almost everyone involved with water utilities, whether engineers, workers, or planners, including the associated municipal officials and residential taxpayers, only ever see a part of the system and a portion of its challenges. Indeed, even specialists are often organized into distinct departments with specific responsibilities, so that, say, transportation issues, property insurance matters, public health concerns, firefighting considerations, and water supply preoccupations almost inevitably address only that portion of the system that specifically touches a more limited domain. Thus, overall,

Fig. 1 A small portion of a typical water distribution network

it remains deeply challenging to understand or estimate the true costs of pipe failure.

3 Failure Types

To better understand and quantify failures, it may be helpful to define them. The proposed failure categories are dramatic, significant, chronic, and benign, and are described below.

> **Dramatic failures** make the news. For example, the contamination of drinking water by lead in Flint, MI, represents a dramatic failure of drinking water infrastructure (Masten et al., 2016; Pieper et al., 2017). Unfortunately, recent evidence suggests that lead in drinking water may be a source of concern in other parts of the USA as well, thus making the Flint situation part of a larger national conversation (Grant et al., 2020). Events in Flint have added new vigor to the US Environmental Protection Agency's (EPA) mandatory lead sampling requirements, and thousands of lead service lines are being removed in cities across North America (Masten et al., 2019). These failures tend to be dealt with quickly, and adequate resources to

correct them are made available. Other examples of newsworthy failures include sink holes caused by pipe failures, disease outbreaks, and major road closures. However, like many of the failures described here, few resources are devoted to their prevention, until a crisis occurs.

Significant failures are serious, but not widely reported. These failures are often the results of cascading steps. For example, a transformer flooded by rain can result in the loss of power in homes. Loss of power can allow pipes to freeze. When they freeze, the pipes can rupture, and since the system is continuously pressurized, the home is flooded. This can result in loss of access to the home by the homeowner and a costly insurance claim. Other significant failures include short-term outages of water supply, a pump station being out of service, or a host of similar events. This sequence of consequences and interdependencies has been dramatically illustrated in February 2021 by the widely reported events in North America, but particularly in Texas, in which a severe winter storm induced widespread water, energy, and infrastructure failures with profound human and economic costs (Black & Chiglinsky, 2021). In such cases, the consequences, though most intense locally, are typically short-term, geographically confined, and economically limited to the water supply company or system owner.

Chronic failures are persistent inefficiencies and might include leakage, pumps operating at their sub-optimal level, system velocities that are too low so that flushing is routinely used to improve water quality, aesthetic complaints, marginal low pressures that might increase local insurance rates to homeowners, and a long list of related issues. One example of a chronic failure arises from overdesign—if pipes are too large or too strong, if pumps are too big, if redundant pipes are too common, then an ongoing cost occurs that does not result in dramatic events but results in a system that is more expensive to build, operate, and repair. Though these failures can be relatively small, their long-term consequences can be substantial. A water distribution component often has a service life longer than almost anything we use, sometimes more than a hundred years. There is ample time for the effects of chronic failures to accumulate.

Benign failures are ones that have few negative complications. Small rates of leakage below the level that economically warrants repair is one example. But most examples generated, such as supplying

> even slightly more pressure than needed to a zone, have some consequences that are worth considering. Perhaps the one profound example of a benign failure is a system that is "fail-safe" by design, i.e., one that is thought through to be "safe in failure," similar to the way a rupture disk might be for high pressures or a circuit breaker protects an electrical system from fire and short circuit.

In the minds of the public, only the dramatic failures are well known. For researchers and practitioners, the costs of these dramatic failures are widely reported and available for scrutiny. Other failure types, significant, chronic, and benign, are largely unknown to the public. Some may be noticed by the utility company and be scheduled for improvements, but many are not. Benign failures may persist for decades without being considered.

4 Propagating Failures

As neighborhoods are developed, it is convenient to bury water pipes along other services, such as water, gas, and electric lines, under the streets. This efficiency comes at a cost, however. Due to their proximity to other infrastructure, pipe failures can propagate through the urban landscape.

Drinking water distribution networks not only provide drinking water, but also supply water for firefighting, and fire hydrants are connected to the same network. In case of fire, earthquake, or natural disaster, large water flows must be immediately available. Typically, the water in the network is continuously pressurized to about 60 pounds per square inch (psi) by the weight of water in elevated water towers. This pressurization is helpful in case of fire, but it tends to waste energy, an example of the many trade-offs that must be made in network design (Ghorbanian et al., 2017). Pipes are sized to guarantee large flows in case of fire, often around 1,500 gallons per minute (gpm). When pipes fail, there is a sudden release of water, like opening a fire extinguisher. This can result in the formation of sinkholes and urban geysers that block roads. Because utilities are buried together under the streets, other infrastructure including gas, electricity, phone lines, and roads are also affected (Fig. 2). Pipe failures propagate to other urban users in ways that are difficult to quantify.

Figures 2 shows just some of the ways that pipe breaks can have localized effects. But these effects tend to cascade outward. A simple broken

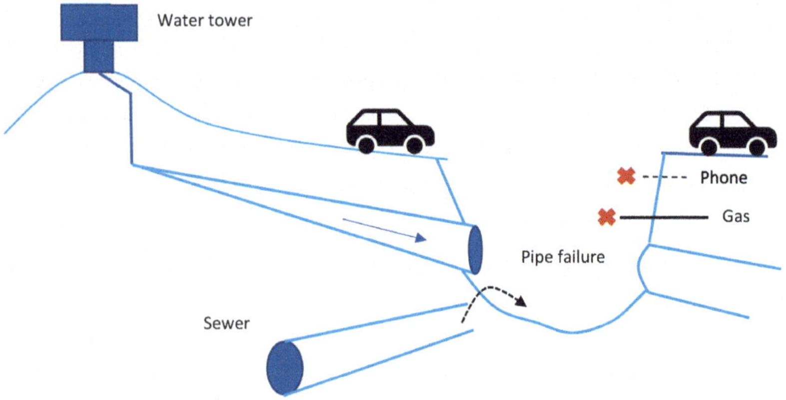

Fig. 2 Water main breaks and their effects on other types of urban infrastructure

phone line can stop a business in its tracks. A blocked road can stop a delivery of vital components, or a fire truck, or an ambulance. As the effects of these failures cascade, they get more difficult to quantify. Figure 3 shows some pathways in which losses can occur from pipe breaks.

The utility company is usually responsible for the direct costs of repairing pipe breaks. Much research has been devoted to characterizing and predicting water main breaks (Wilson et al., 2015; Zamenian et al., 2017), investigating breaks (Grigg, 2017), reliability assessment (Tee et al., 2014), water hammer effects (Fernandes & Karney, 2004; Wang et al., 2014), and life cycle costs (Francisque et al., 2017). High repair costs are not restricted to larger water mains, with the smaller pipes used to connect to the homes resulting in an annual global cost of US$5 billion (Ward et al., 2017).

However, relatively little effort has been devoted to how these costs tend to cascade to the larger society. These costs are borne by citizens, insurance companies, business, and others. Blocked roads, flooding, and loss of services affect the larger society. There is evidence that sandy soils, which tend to wash out when the pipe breaks, are especially vulnerable. The location of the water main is important as well, with serious cascading failures occurring when in close proximity to major highways, gas lines, or other important services (Farewell et al., 2018).

There has been some recent research into the total societal costs of main breaks. In their study of large diameter water main breaks, Piratla

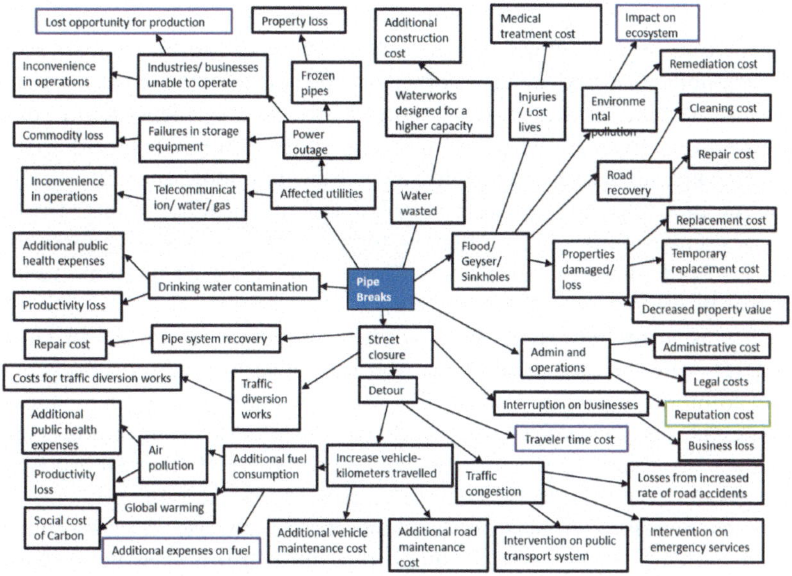

Fig. 3 Potential loss pathways from pipe breaks

et al. (2015) consider the costs of lost product, pipe repair and return to service (e.g., disinfection of new pipes), travel delay, supply outage and substitution, health risk, and property damage. They find that the direct costs of repairing the pipe and putting it back into service was the largest single contributor to the cost, but this only represented 34% of the total cost. About 66% of the cost of the failure were indirect costs borne by the larger society. Put another way, for each dollar spent by the utility company to repair a broken pipe, society paid two dollars.

The above work closely follows an earlier attempt to quantify the total cost failure conducted by the American Water Works Association Research Foundation (Gaewski et al., 2007). This work also focuses on large diameter mains and includes estimates related to the loss of firefighting capacity during main break. The average estimated total direct and societal cost of water main failure is \$1.7 million, with approximately 50% being direct costs. The largest societal cost is related to flooding damage (57%), which some utility companies will pay for, while some will not.

The above work represents significant progress in our understanding of the true cost of pipe failure, but there are more questions to be answered. For example, Gaewski et al. (2007) do not include the costs of road repair or vehicle damage in their study. Increased accidents from road rage resulting from traffic delays were also not considered. Health costs are very difficult to quantify in dollars as well. For some amount of time after the pipe fails, and before it is isolated, drinking water is mixed with the local environment, which may include seepage from sewer pipes (see Fig. 2). Epidemiologists and doctors may be able to tell us more about the real health impacts of these breaks.

Perhaps most important, both studies only consider breaks in large diameter water mains. Smaller distribution pipes in neighborhoods are more common than large water mains, and typically represent the majority of pipe failures (Pietrucha-Urbanik, 2015). Smaller distribution pipes are in closer proximity to drinking water consumers, and their failure may have greater health impacts. Therefore, many questions remain.

All these poorly understood and potentially adverse effects can be traced back to the failure of a single pipe in a distribution network of thousands of pipes. Pipe failures propagate throughout the urban landscape in ways that are rarely considered and difficult to quantify. The failure of a single pipe can be compared to the loss of a single nail in the classic proverb below, while the rider can be compared to an ambulance held up due to a water main break, thus illustrating the cascading effects pipe failure may have at the societal level.

> For want of a nail the shoe was lost.
> For want of a shoe the horse was lost.
> For want of a horse the rider was lost.
> For want of a rider the message was lost.
> For want of a message the battle was lost.
> For want of a battle the kingdom was lost.
> And all for the want of a horseshoe nail.

5 Competing Modes of Failure: Hydraulics and Water Quality

So far, we have talked mainly about a failure in which the failure mode is unambiguous: the pipe break. However, there are other cases where a failure in one aspect is chosen to have success in another. The competing requirements of hydraulics and water quality is an example of this, as typified in the tension between water age and firefighting capacity.

To clearly illustrate this point, consider that two key purposes of modern water supply systems are to provide water that is safe to drink and the high flows needed to fight fires. Logically, failure in either of these objectives can result in the loss of property, illness, or even death. Yet, achieving both of these objectives simultaneously requires compromises. In simple terms, the largest diameter pipe has the lowest average velocity and thus the lowest hydraulic losses for a given flow, and consequently larger diameter pipes have greater ability to supply a lot of water during firefighting conditions. However, these same large diameter pipes can have an adverse effect on water quality, since the low velocity leads to long residence times or even to water stagnation. Thus, pipes must be sized to balance these two competing objectives (Fig. 4).

Fig. 4 Long-term deterioration can reduce hydraulic capacity, including the ability to supply fire flows, and can clearly have significant water quality implications (Picture taken by the authors)

Some of the effects of water stagnation are well known, and others are just emerging. Stagnation often results in the loss of disinfectant residual. Disinfectants are used to maintain the microbial integrity of the water. As water travels through pipes, it can be contaminated by low pressure water intrusions, for example (Besner et al., 2011). Disinfectant residuals can help mitigate this problem (Propato & Uber, 2004). Some disinfectants can form harmful by-products that tend to increase in concentration over time. Both these phenomena are adversely affected by stagnation. Long periods of water stagnation have been associated with taste and odor problems, bacterial growth, temperature rise, and loss of residual (AWWA, 2002).

Historically, chlorine was used as a disinfectant, but it tends to decay quickly and can form harmful by-products. Many utility companies today use chloramine, a mixture of chlorine and ammonia, instead, despite the lack of research and regulations on this product. Figure 5 shows how chloramine usually decays much slower than chlorine. Disinfectant decay can be especially bad in dead-end pipes, where water tends to stagnate (Abokifa et al., 2016).

A new water quality issue that we are just beginning to understand is the risk of legionellosis. It is estimated that 18,000 people in the USA are hospitalized annually with legionellosis, a collection of lung infections caused by *Legionella* and related bacteria. The severity of legionellosis varies from mild to the often-fatal form of pneumonia called Legionnaire's disease. *Legionella* bacteria are commonly found in soil, but also in the biofilms in drinking water distribution systems (Percival & Williams, 2014). Currently, the World Health Organization recommends regular

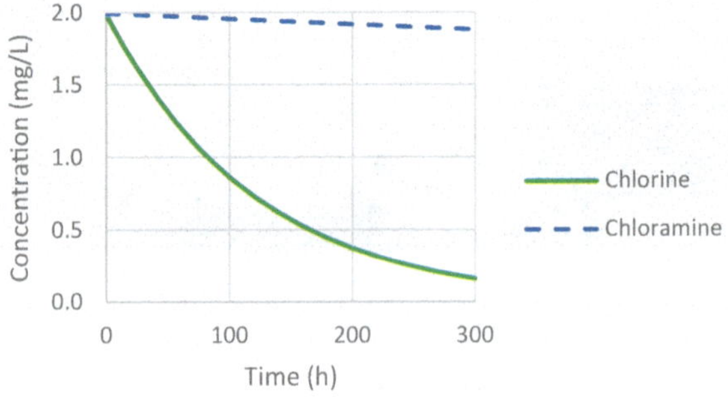

Fig. 5 Disinfectant decay of chlorine and chloramine

surveillance for *Legionella* in drinking water supply systems (Bartram, 2007).

The precise biology of legionellosis remains poorly understood. *Legionella* bacteria are known to replicate in water distribution networks (Declerck et al., 2009; Douterelo et al., 2017) and can become detached under certain flow conditions (Horn et al., 2003; Shen et al., 2015). These bacteria can be both amplified, or inhibited, in the water heater, depending on the temperature. A minimum temperature of 60 °C is often recommended to prevent *Legionella* growth, but this can increase scalding risks in hospitals where the risks of legionellosis are more acute (Percival & Williams, 2014). Premise plumbing likely also plays a role, with the small diameter pipes having a large surface area on which bacteria can grow. Figure 6 shows some of the risk factors associated with Legionnaire's disease.

Strategies to control *Legionella* bacteria include maintaining disinfectant residuals and limiting biofilm growth (Gray, 2014). Unfortunately, disinfectants decay over time in a distribution system. Over-designed networks can result in a loss of disinfectant residuals. Nor is the answer to simply add more chlorine, since high chlorine concentrations can often produce a set of harmful disinfection by-products. *Legionella* and biofilm development should be minimized within our distribution systems; however, aging networks, episodes of low flow, and reactant pipe

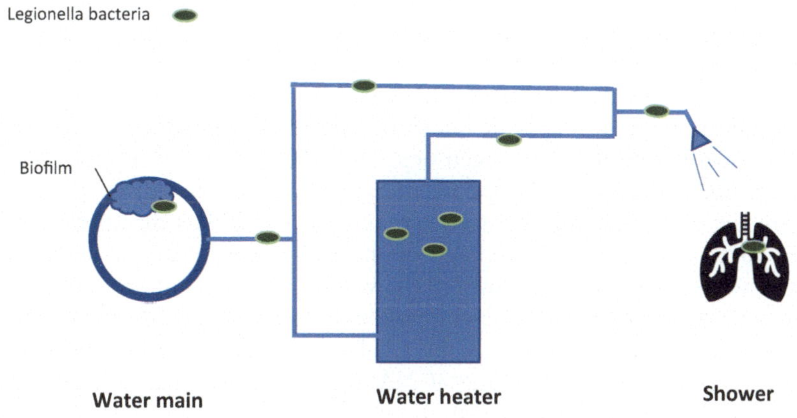

Fig. 6 Suspected risk factors associated with Legionnaire's disease

walls make this a significant challenge. Moreover, system upsets from sudden large demands, such as fire-fighting flows or pipe bursts, can induce low pressures and high flow velocities. Low pressure events can draw a wide range of pathogens, including *Legionella* into the distribution system from leaks, and high flow velocities can cause *Legionella* to be released by biofilm detachment (Horn et al., 2003; Shen et al., 2015).

Thus, a clear need in this area is to assess hydraulic events that have the potential to increase waterborne diseases and to seek effective long-term solutions. To this end, a comprehensive inventory of hydraulic threats needs to be undertaken, including the pressure transients and high velocities caused by firefighting, main breaks, and pump failures. Low disinfectant residuals arising from long residence times, over-sized pipes, deteriorated system materials, and uncontrolled biofilm growth all need to be assessed.

6 Benefits of Failure

It is perhaps too easy to be starkly negative about failure. Ironically, this is perhaps another form of failure, or at least an occasional mistake. Pumps in water distribution networks use a lot of electricity. In a case of chronic failure, a pump can be operated far from its point of maximum efficiency for years or decades. The electricity bills come in each month, and since they are not increasing, complacency sits in, and they get paid. Meanwhile, money and energy are wasted. It is only when the pump starts to groan and finally overheat that its operation is closely examined. In this sense, a significant pump failure can be a gift to the utility. The costs of low efficiency for decades can easily exceed the cost of a new pump.

Failures, especially dramatic ones, are a wonderful opportunity to draw attention to areas that have been neglected. This creates an opportunity to galvanize the political support needed to make change, or to seek a greater understanding. Historical examples of engineering opportunities have arisen from the footage of the suspension bridge at Tacoma Narrows bouncing up and down in the wind—something almost all engineers are forced to watch—which provides a potent reminder of what can go wrong. Another example is the Challenger Space Shuttle explosion, which resulted in new thinking at NASA and changed the field of risk analysis, as discussed in some length by Vick (2002). If thinking is mired in complacency, or if faults are accumulating and gradually becoming serious, if understanding is conventional or static, what is needed is sometime precisely the kind of event that challenges the status quo. Dramatic failures can do this.

Failures are chances to improve. They can galvanize political or senior management support for better ways of doing things based on lessons learned. We believe that Samuel Beckett captures the essence of the value of failure very well in one of his most famous quotations, from his prose piece "Worstward Ho" (1983):

> Ever tried. Ever failed. No matter.
> Try again. Fail again. Fail better.

7 Need for Risk Analysis

Most of the training that engineers receive is related to design. This is important since cities tend to grow over time and new neighborhoods are continuously being developed. New designs are needed to meet this need. However, most of the water distribution network has been in the ground since about the middle of the last century. Maximizing the performance of existing networks requires thinking beyond design.

Thinking beyond the design stage to improve how a structure, or device, will perform when it is in service is a common problem in the practice of engineering. Practicing civil engineers tend to think a lot about this issue, since infrastructure usually lasts decades, and many problems can happen in this time. The relationship between an infrastructure and those responsible can last decades and is not unlike parenting. Vick (2002, p. 124) describes the relationship and the need for risk analysis as follows:

> When a structure is conceived, it is a gleam in the designer's eye reflecting the hopes that the design assumptions and all other good intentions will come to pass. But once born, it develops a personality with idiosyncrasies all its own, and as it grows older the creaks and groans of age will settle in...engineering brings its fair share of problem children into the world, some who will become cranky and cantankerous. It is for identifying and learning how best to cope with these offspring that risk analysis finds its most successful and useful application.

Risk analysis is a process for identifying what is likely to go wrong and estimating how serious the effects will be. Good advice for any parent. Risk assessment is the product of two estimates: the probability of failure

(p_f) and the consequences of failure (C_f).

$$\text{Risk} = p_f \, x \, C_f$$

One of the main arguments of this chapter is that pipe breaks are relatively common (i.e., they have a high probability), but the consequences of these failures are often underestimated, so that the overall risk is underestimated. Thus, for a given level of risk, a higher probability of failure is tolerated than would be the case if the costs of failure were more comprehensively included. The utility pays to fix the pipes, but society pays for the traffic delays, flooding, water damage, and interruption of other services like phone, gas, and electricity. The true risks associated with pipe failure are much larger than people and utility companies realize. In contrast, terrorist attacks, such as poisoning the water supply, get a great deal of attention and study (Larsen et al., 2017; Ostfeld et al., 2008). However, it has been suggested that the chance of successfully executing such an attack is exceedingly low, even if someone wanted to do so, making the overall risk very low in comparison with chronic pipe failures (Bloetscher, 2009).

Some risks, like *Legionella*, are emerging topics for study. *Legionella* bacteria are often found in water distribution networks, so they have a high probability of being observed. However, the consequences of this remain unclear. For example, even if *Legionella* bacteria were found in a water system, the bacterial dose might not be enough to result in infection. Moreover, it is difficult to account for differences in susceptibility in healthy and compromised individuals.

One of the things that makes risk analysis so challenging in water distribution is that it involves so many specialties. Typically, the network is designed by someone with expertise in hydraulics. But when a pipe breaks, and the road is blocked, expertise in traffic or city planning is required. If a fire truck cannot reach its destination, then expertise in fire protection is required. If a transformer is flooded by the sudden release of water, then an expert in power distribution is required. Understanding health risks associated with *Legionella*, or other pathogens, typically requires training in epidemiology or medicine. The fact that these failures cascade to so many parts of urban infrastructure means that expertise in many areas is required to assess the risk. A first step to better understanding these failures is the use of multidisciplinary teams, including experts in traffic, fire protection, and electricity supply.

Furthermore, assessing these risks often requires subjective estimates of probability, something that usually requires expertise to provide. This makes risk assessment of water distribution networks a considerable challenge and may explain why it has rarely been attempted in the published literature. There is a need for a better understanding of the consequences of pipe breaks, among others. For example, the probability of infection from *Legionella* in the distribution system is currently poorly understood, illustrating one case where multidisciplinary work is required to not only understand its health risk, but also its financial and environmental impacts.

8 Summary and Conclusions

Water distribution networks often do not perform as engineers wish. Dramatic failures, like a burst pipe next to a major highway, tend to get a lot of attention and get fixed quickly. These dramatic failures can focus attention and free up resources. However, there are many smaller, yet significant, types of failure. A main break that results in the flooding of a single home or blocks a single residential street has diverse consequences that can often propagate far beyond a single failed pipe. The overall risks associated with the lowly pipe break are often under appreciated.

Water distribution networks are linked to other urban infrastructure including gas, electricity, and roads. Water distribution networks are in service for many decades, and the effects of chronic failures can accumulate over time. The failure modes of water distribution networks are complex. One reason for this is that the costs of these failures are often borne by those outside the utility company, such as homeowners, taxpayers, and insurance companies. A more comprehensive use of risk analysis, preferably with multidisciplinary teams, will increase our understanding and ability to mitigate these risks.

Pipe breaks are usually bad for everyone, but there are other kinds of failure which are more in a gray area. There is a natural tension between the hydraulic capacity of the network, important for fighting fires, and water quality that is provided, important to consumers. Essentially, we are asking the network to do two different jobs. In North America, many regulations used to design water distribution networks were shaped by the Great Chicago Fire of 1871 (Linder, 1991). Modern building codes and firefighting practices have improved, but the requirements on the water supply system have changed little since the nineteenth century. Emerging contaminants, such as *Legionella* bacteria, may change how we think about stagnation in water distribution networks. Balancing the needs for fire protection and water quality is a challenging task.

There is clearly a need for additional research. How do firefighting requirements affect water quality? Are there health risks here? What are the true societal costs of pipe failures and what can be done to prevent them? One of the greatest obstacles to progress here is the multidisciplinary nature of the problem. Firefighting, flood control, electrical supply, and public health are all affected by water distribution networks. Teams of researchers with expertise in these areas are needed to answer these questions.

The fact that water distribution networks so frequently misbehave does create an opportunity. With the appropriate resources, there is a chance to study these failures and better understand them. With understanding comes the opportunity to mitigate their effects.

Water distribution networks are large, complex machines. We rely on them every day to provide high-quality drinking water and protect our homes in case of fire. Designing, maintaining, operating, and studying these networks can be a challenge. Water can be our friend when it does what we want, but also our enemy in failure. Enzo Levi (1995, p. xiii) writes in the *The Science of Water: The Foundation of Modern Hydraulics*:

> City dwellers who day after day see water flowing from faucets, docile to their needs, have no idea of its idiosyncrasy. They cannot imagine how much patience and cleverness are needed to handle our great friend-enemy; how much insight must be gained in understanding its arrogant nature in order to tame and subjugate it; how water must be 'enticed' to agree to our will, respecting its own at the same time.

References

Abokifa, A. A., Yang, Y. J., Lo, C. S., & Biswas, P. (2016). Water quality modeling in the dead end sections of drinking water distribution networks. *Water Research, 89*, 107–117. https://doi.org/10.1016/j.watres.2015.11.025.

AWWA. (2002). *Effects of water age on distribution system water quality*. U.S. Environmental Protection Agency Office of Ground Water and Drinking Water Standards and Risk Management Division, Washington, DC.

Bartram, J. (Ed.). (2007). *Legionella and the prevention of legionellosis*. World Health Organization.

Black, T., & Chiglinsky, K. (2021, February, 19). Water pipes burst across Texas, setting off a hunt for plumbers. *Bloomberg*. https://www.bloomberg.com/news/articles/2021-02-19/texas-thaw-sets-off-frantic-hunt-for-plumbers-ready-to-wrench.

Beckett, S. (1983). *Worstward Ho*. https://genius.com/Samuel-beckett-worstward-ho-annotated.

Besner, M.-C., Prévost, M., & Regli, S. (2011). Assessing the public health risk of microbial intrusion events in distribution systems: Conceptual model, available data, and challenges. *Water Research, 45*, 961–979. https://doi.org/10.1016/j.watres.2010.10.035.

Bloetscher, F. A. (2009). *Water basics for decision makers: Local officials' guide to Water & Wastewater Systems*. American Water Works Association.

Colombo, A., Lee, P., & Karney, B. (2009). A selective literature review of transient-based leak detection methods. *Journal of Hydro-Environment Research, 2*(4), 212–227. https://doi.org/10.1016/j.jher.2009.02.003.

Declerck, P., Behets, J., Margineanu, A., van Hoefa, V., De Keersmaeckera, B., & Olleviera, F. (2009). Replication of Legionella pneumophila in biofilms of water distribution pipes. *Microbiological Research, 164*, 593–603. https://doi.org/10.1016/j.micres.2007.06.001.

Douterelo, I., Jackson, M., Solomon, C., & Boxall, J. (2017). Spatial and temporal analogies in microbial communities in natural drinking water biofilms. *Science of the Total Environment, 581–582*, 277–288. https://doi.org/10.1016/j.scitotenv.2016.12.118.

Farewell, T. S., Jude, S., & Pritchard, O. (2018). How the impacts of burst water mains are influenced by soil sand content. *Natural Hazards and Earth Systems Sciences, 18*, 2951–2968. https://doi.org/10.5194/nhess-18-2951-2018.

Fernandes, C., & Karney, B. (2004). Modelling the advection equation under water hammer conditions. *Urban Water Journal, 1*, 97–112. https://doi.org/10.1080/15730620412331290038.

Francisque, A., Tesfamariam, S., Kabir, G., & Haider, H. (2017). Water mains renewal planning framework for small to medium sized water utilities: A life cycle cost analysis approach. *Urban Water Journal, 14*, 493–501. https://doi.org/10.1080/1573062X.2016.1223321.

Gaewski, P. E., Tata and Howard Inc., Blaha, F. J., & AwwaRF. (2007). *Analysis of the Total Cost of Large Diameter Pipe Failures*. http://infra-tect.com/wp-content/uploads/2014/11/Analysis-of-Total-Cost-of-Large-Diameter-Pipe-Failures.pdf..

Ghorbanian, V., Karney, B., & Guo, Y. (2017). Intrinsic relationship between energy consumption, pressure, and leakage in water distribution systems. *Urban Water Journal, 14*, 515–521. https://doi.org/10.1080/1573062X.2016.1223325.

Grant, A., Scherer, M. M., Land, D., Cwiertny, D. M., Edwards, M. A. Mount, J., & Latta, D. E. (2020). Estimating consumers at risk from drinking elevated lead concentrations: An Iowa case study. *Environmental Science & Technology Letters*. acs.estlett.0c00753. https://doi.org/10.1021/acs.estlett.0c00753.

Gray, N. F. (2014). Pathogen control in drinking water. In *Microbiology of waterborne diseases* (pp. 537–569). Elsevier.

Grigg, N. S. (2017). Water supply pipeline failures: Investigative procedures and data management. *Journal of Performance of Constructed Facilities, 31*, 06017004. https://doi.org/10.1061/(ASCE)CF.1943-5509.0001113.

Horn, H., Reiff, H., & Morgenroth, E. (2003). Simulation of growth and detachment in biofilm systems under defined hydrodynamic conditions. *Biotechnology and Bioengineering, 81*, 607–617. https://doi.org/10.1002/bit.10503.

Larsen, S. L., Christensen, S. C. B., Albrechtsen, H.-J., & Rygaard, M. (2017). GISMOWA: Geospatial risk-based analysis identifying water quality monitoring sites in distribution systems. *Journal of Water Resources Planning and Management, 143*, 04017018.

Levi, E. (1995). *The science of water: The foundations of modern hydraulic*. ASCE Press.

Linder, K. (1991). *Fire Protection Handbook* (17th ed.). National Fire Protection Association.

Masten, S. J., Davies, S. H., Haider, S. W., & McPherson, C. R. (2019). Reflections on the lead and copper rule and lead levels in Flint's water: Reflections on the lead and copper rule and lead levels in Flint's water. *Journal—American Water Works Association, 111*, 42–55. https://doi.org/10.1002/awwa.1360.

Masten, S. J., Davies, S. H., & Mcelmurry, S. P. (2016). Flint water crisis: what happened and why? *Journal-American Water Works Association, 108*(12), 22–34.

Ostfeld, A., Uber, J. G., Salomons, E., Berry, J. W., Hart, W. E., Phillips, C. A., Watson, J.-P., Dorini, G., Jonkergouw, P., Kapelan, Z., di Pierro, F., Khu, S.-T., Savic, D., Eliades, D., Polycarpou, M., Ghimire, S. R., Barkdoll, B. D., Gueli, R., Huang, J. J., ... Walski, T. (2008). The battle of the water sensor networks (BWSN): A design challenge for engineers and algorithms. *Journal of Water Resources Planning and Management, 134*, 556–568.

Percival, S. L., & Williams, D. W. (2014). Legionella. In *Microbiology of Waterborne Diseases* (pp. 155–175). Elsevier.

Pieper, K. J., Tang, M., & Edwards, M. A. (2017). Flint water crisis caused by interrupted corrosion control: Investigating "ground zero" home. *Environmental Science and Technology, 51*, 2007–2014. https://doi.org/10.1021/acs.est.6b04034.

Pietrucha-Urbanik, K. (2015). Failure analysis and assessment on the exemplary water supply network. *Engineering Failure Analysis, 57*, 137–142. https://doi.org/10.1016/j.engfailanal.2015.07.036.

Piratla, K. R., Yerri, S. R., Yazdekhasti, S., et al. (2015). Empirical analysis of water-main failure consequences. *Procedia Engineering, 118*, 727–734. https://doi.org/10.1016/j.proeng.2015.08.507.

Propato, M., & Uber, J. G. (2004). Vulnerability of water distribution systems to pathogen intrusion: How effective is a disinfectant residual? *Environmental Science and Technology, 38*, 3713–3722. https://doi.org/10.1021/es035271z.

Shen, Y., Monroy, G. L., Derlon, N., Janjaroen, D., Huang, C., Morgenroth, E., Boppart, S. A., Ashbolt, N. J., Liu, W.-T., & Nguyen, T. H. (2015). Role of biofilm roughness and hydrodynamic conditions in *Legionella pneumophila* adhesion to and detachment from simulated drinking water biofilms. *Environmental Science and Technology, 49*, 4274–4282. https://doi.org/10.1021/es505842v.

Tee, K. F., Khan, L. R., Chen, H. P., & Alani, A. M. (2014). Reliability based life cycle cost optimization for underground pipeline networks. *Tunnelling and Underground Space Technology, 43*, 32–40. https://doi.org/10.1016/j.tust.2014.04.007.

Vick, S. (2002). *Degree of belief: Subjective probability and judgement*. American Society of Civil Engineers.

Wang, R., Wang, Z., Wang, X., & Yang, H. (2014). Pipe burst risk state assessment and classification based on water hammer analysis for water supply networks. *Journal of Water Resources Planning and Management, 140*. https://doi.org/10.1061/(ASCE)WR.1943-5452.0000404.

Wang, X., Waqar, M., Yan, H.-C., Louati, M., Ghidaoui, M. S., Lee, P. J., Meniconi, S., Brunone, B., & Karney, B. (2020). Pipeline leak localization using matched-field processing incorporating prior information of modeling error. *Mechanical Systems and Signal Processing, 143*(11). https://doi.org/10.1016/j.ymssp.2020.106849.

Ward, B., Smith, D., Savic, D., Roebuck, J., & Collingbourne, J. (2017). Optimized investment planning for high-volume low-value buried infrastructure assets. *Journal of Pipeline Systems Engineering and Practice, 8*, 0. https://doi.org/10.1061/(ASCE)PS.1949-1204.0000266.

Wilson, D., Filion, Y. R., & Moore, I. D. (2015). Identifying factors that influence the factor of safety and probability of failure of large-diameter, cast iron water mains with a mechanistic, stochastic model: A case study in the city of hamilton. *Procedia Engineering, 119*, 130–138. https://doi.org/10.1016/j.proeng.2015.08.863.

Zamenian, H., Faust, K. M., Mannering, F. L., Abraham, D., & Iseley, T. (2017). Empirical assessment of unobserved heterogeneity and polyvinyl chloride pipe failures in water distribution systems. *Journal of Performance of Constructed Facilities, 31*(5). https://doi.org/10.1061/(ASCE)CF.1943-5509.0001067.

Multi-dimensional and Interacting Water and Climate Risks and Pricing Them in the Industry Context

Jason Siegel, Paulina Concha Larrauri, Luc Bonnafous, and Upmanu Lall

1 Introduction

Historically, water risk concerns have primarily concentrated around water scarcity and the competition for water resources among stakeholders. Given water's importance as a critical resource for both industry and society, investment in water-related infrastructure and resource planning has focused on mitigating the impact of water-related shortages and ensuring that there is sufficient water to support foreseeable requirements.

The findings, interpretations, and conclusions expressed in this chapter are entirely those of the authors. They do not necessarily represent the view of the World Bank where one of the authors is currently employed.

J. Siegel
Columbia University, New York, NY, USA
e-mail: Jason.Siegel@columbia.edu

P. Concha Larrauri
Columbia Water Center, New York, NY, USA
e-mail: pc2521@columbia.edu

T. Walker et al. (eds.), *Water Risk and Its Impact on the Financial Markets and Society*, Palgrave Studies in Sustainable Business In Association with Future Earth,
https://doi.org/10.1007/978 3 030 77650 3_11

However, extreme events related to water are potentially as important and can leave a lasting impact on financial markets and our society. These include flooding due to extreme rainfall; persistent drought; and pollution, either due to the industrial operations or due to the failure of infrastructure. The global impact of the COVID-19 pandemic has made what are perceived as “tail risks” particularly topical. Unprecedented data from the epidemic on what happens when society and the economy come to an outright halt is useful for inferences and to quantify the damage of these low-probability, high-impact water-induced extreme events. Given that climate is the causal factor for the frequency and consequence of many of these observable incidents, it is important to highlight the pervasiveness of this risk across industries, the likely change in the occurrence rate of water-related extreme events, the geographic locations most exposed, and what if anything can be done to anticipate, disclose, or mitigate such risks.

The mining industry represents a small subset of the industrial activity impacted by water-related extreme events. However, the methodology we apply to assess risk in the mining sector can be viewed as a beginning of a roadmap for quantifying water exposure in other industries. Mining can be viewed as a microcosm for essential business activity globally affected by water risk in that the impact of shutdowns to mining operations is pervasive. The sector produces essential commodities that the world relies on for its necessities. In many regions, the industry is a major source of employment and tax revenue, and companies vary in terms of management styles, risk tolerance, size, geographic diversity, and financial capacity. As climate change and a focus on ESG become important to investors and stakeholders, mitigating water risk is near the top of their mind. While the results of water scarcity are clearly important to consider, the impact of climatic extremes must be considered as part of

L. Bonnafous (✉)
World Bank, Washington, DC, USA
e-mail: lbonnafous@worldbank.org

U. Lall
International Research Institute for Climate and Society, Columbia Water Center, Columbia University, New York, NY, USA
e-mail: ula2@columbia.edu

the wider equation. While several products have been developed to assess carbon budgets and emissions as an ESG measure, the consideration of extremes and specifically of stochastic events requires a separate and more focused treatment. We provide the structure of such a measure, including the minimal disclosure requirements for companies so that the materiality of the water risks can be assessed in a standard value at risk framework.

1.1 Climate-Induced Tail Risk

Tail events are those that have a "very low probability" of occurring (i.e., they are at the "tails" of a probability distribution), commonly below a 5 or 1% annual probability threshold. Tail events can produce large disturbances. The COVID-19 pandemic is an example of a tail event, and while pandemics occur infrequently, when they do, the consequences are devastating and widespread. Similarly, the impacts of extreme weather events such as intense floods and persistent droughts can be experienced in a wide range of spatial scales, directly or indirectly affecting people's livelihoods, the economy, and the environment. In the context of this chapter, climate risks are defined by the probability of occurrence of an extreme rainfall event of a given magnitude (e.g., the 1-in-100-year flood) and the ensuing financial impact in the mining sector.

Mining companies are usually well-equipped to deal with typical fluctuations in water availability but may not be prepared to respond to tail events, despite the existence of contingency plans at the asset level. When extreme weather events occur, the financial impact to a company can cascade down to an entire region. As an example, the 2015 El Niño-related drought in Papua New Guinea had devastating economic, social, and environmental consequences for the country (Jorgensen, 2016) and significantly affected its biggest copper miner, Ok Tedi Mining. The mining company was forced to shut down for five months because of the drought (Field, 2015), laying off a sizable portion of its workforce and closing schools and other services it supported (Jorgensen, 2016). This brought significant losses to the government as Ok Tedi's taxes account for around 20% of its total revenue. The El Niño event of 1997 had a similar effect in Papua New Guinea (Field, 2015), yet no risk mitigation actions were implemented. Examples such as Ok Tedi Mining are not uncommon in the mining sector (Kunz, 2020) and illustrate how diverse stakeholders are affected by unassessed tail risks. Risk can for instance stem from a lack of proper data to assess climate risk (data not going far enough back in time, leaving all engineering design consideration to short time record and safety coefficients), or legacy issues.

Further, even if a company has contingency plans at the asset level, (e.g., financial provisions and planning for infrastructure damage), such procedures are generally designed separately for each asset. Thus, they do not account for potentially correlated risks across the portfolio of physical assets. This creates the potential for fat-tail risk at the portfolio level, as climate extremes are not independent from each other due to the underlying structure of the climate system (driven, in part, by climate oscillations regimes of different frequencies). Correlated shocks in the same year across multiple assets of a company with a significant market share can translate into market supply and price disruptions at a global scale.

1.2 Importance of Incorporating Climatic Extremes into Risk Models

Extreme climate events can be incorporated into risk models to prevent and mitigate detrimental financial and societal impacts. The El Niño phenomenon (a tail event whose occurrence can be predicted with some accuracy and its impacts on global weather extremes have been widely studied) is a case of climate's cyclical variability, occurring every 3–7 years with different levels of intensity. The event intensity itself is modulated by regime like or oscillatory phenomena with much longer time scales of recurrence. Other oscillations have lower frequencies such as the Pacific Decadal Oscillation (PDO) (Mantua & Hare, 2002), and the Atlantic Multidecadal Oscillation (AMO). Combinations of the phase of these climate oscillations can lead to extreme weather events in different regions and countries (Frankcombe et al., 2010; Mantua & Hare, 2002) that can often be confused with climate change given their unprecedented impacts in places with short historical records, and their long recurrence intervals. Considering that climate cycles can span decades and that facilities are often designed using short climate time series (in the mining sector, the remote locations where mines are typically built often lack long-term climate data), and physical and insurance protections against floods and droughts may be insufficient.

Furthermore, extreme weather events can be temporally and spatially correlated (e.g., an intense flood might happen in Australia while a prolonged drought is hitting Peru). El Niño and la Niña events are known to cause simultaneous weather extremes in multiple regions across the globe. Therefore, in the context of portfolio management, the global space-time clustering of climate risks needs to be considered (Bonnafous & Lall, 2020). This is particularly relevant for companies and investors in the mining sector, where the concentration of minerals and metals in specific parts of the world makes climate risk management at the portfolio level more important.

The question we aim to answer is how can stakeholders identify and price the exposure to the extreme weather events with the limited information that companies disclose related to their climate risk protections? Quantifying the financial impact of extreme climate events requires consideration of the multidimensional consequences across stakeholders and time horizons described as probability distributions (Fig. 1). The immediate direct ramifications of extreme events in terms of foregone economic profit may be measurable; however, the indirect and lasting impacts of individual occurrences may be more challenging to quantify. Because decision-makers may have differing objectives in terms of their incentive structures surrounding externalities and timescales, perceived financial impact may be very different across stakeholder groups.

In the following sections, we provide a basic framework to quantify *relative* risk (in terms of potential losses and damages) from extreme climate events with a methodology for assessing both the temporary impacts of more minor incidents and the enduring damage of permanent asset closure resulting from catastrophic loss, focusing specifically on the

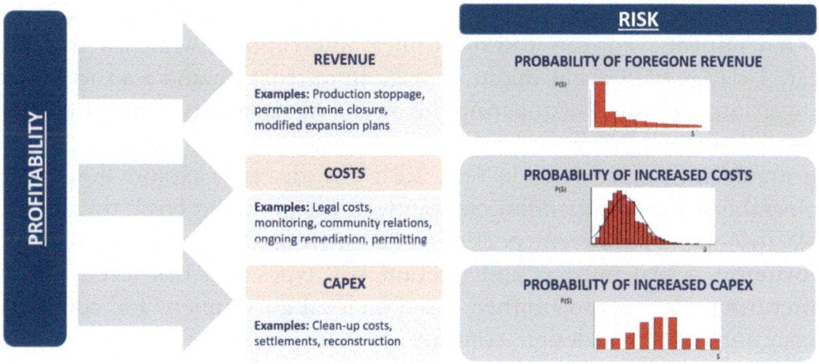

Fig. 1 Quantifying financial impact

direct financial impacts on the underlying asset owners. We use the mining industry as an example of the application of this risk indexing framework to real companies and assets using an approach previously developed by Bonnafous et al. (2017a, 2017b). We highlight how companies with more geographically concentrated assets carry increased risk related to climate, which should be considered in stakeholder decision-making.

2 Likelihood of Extreme Climate Events

2.1 Choosing Climate Data to Quantify Impacts Across Space and Time

In recent years, a significant body of research on extreme climate events was developed. Focus areas range from theoretical definitions of climate extremes with a statistical, physical, or meteorological emphasis, to investigations of the effects on the ground. Here, we focus on a flexible framework to explore climate risk for portfolios of physical assets previously developed (Bonnafous et al. 2017a, 2017b). The approach enables users to choose from different climate datasets publicly available, allowing to empirically explore such datasets and the spatiotemporal relationships in the climate structure to uncover potentially correlated risks, and thus fat-tailed risk at the portfolio level (Bonnafous & Lall, 2020).

The practical application highlighted is the comparison of the exposure of different portfolios of assets to the same type of extremes, with the view of preparing contingency plans or insurance tools. Risk levels of different orders of magnitude are considered as representing different types of consequences. For instance, for a given asset, a 1-in-10-year, 30-day rainfall event can lead to business interruption, while a 1-in-100-year, 1-day rainfall event might damage infrastructure and lead to much longer interruption, endangering the asset's sustainability without proper government help, or even to the complete wipe-out of the asset (this is illustrated with more detail in Sect. 2.2). The idea is to explore how these vulnerabilities might manifest conjointly at the portfolio level, thus being able to compare different portfolios (Bonnafous et al., 2017a, 2017b). Obviously, actual impacts and relevant risk types and risk levels might differ from one asset to another, based on local government and company policy, infrastructure design, contingency plans, etc. However, an investor might not have access to data granular enough to differentiate between assets, and if they do, it is easy to generalize the approach presented here to different risk types and risk levels. In any case, the method can be used to compare relative risk exposures of portfolios of assets, highlighting how the geographic structure of portfolios makes them more or less vulnerable to correlated climate risk stemming from the spatiotemporal organization of the climate system. The hypothesis and generalizations of the base case can be used by an investor with basic knowledge as a way to explore risk and initiate a discussion with companies. This approach can also be extended to explore how different risk levels evolve conjointly.

Climate extremes are by definition rare, and often modeled using extreme value theory assuming that the annual maximum events considered for return period analysis conform to independent and identically distributed random processes (referred here as iid). However, these assumptions require relatively long time series to assess and to reliably model the risks. We note that the presence of regimes and oscillations in the data may actually invalidate the assumptions in typical return period analysis, and that the presence of these features will typically lead to fatter tails of the probability distributions than what is assumed in traditional iid analyses.

We use global climate datasets because we are interested in portfolios of assets of global industries, and their relation to the spatial structure of climate. Extreme rainfall events generally manifest over periods lasting from a few minutes to several days, and daily rainfall generally captures such events, even when they happen on a shorter time scale. Thus, using datasets with a daily or sub-daily temporal resolution are advised for the particular example developed. If the hazard of interest is drought for instance, such fine temporal granularity is not needed (Bonnafous et al., 2017b).

Climate model outputs for the Intergovernmental Panel on Climate Change (IPCC) still fail to capture many of the oscillations and spatial patterns in the twentieth-century data and exhibit significant biases especially in even the basic statistics (e.g., monthly mean, standard deviation and skew, daily maxima, dry period durations, wet spells) of precipitation over most of the land areas of the planet (Woldemeskel et al., 2012). As a result, investigators apply point-by-point bias corrections (typically multiplicative factors of 3 to 5, or quantile by quantile adjustments) to the statistics and extrapolate them to the future. In our view, this can lead to an unreliable approach to risk assessment, since a model that performed poorly enough to not even reproduce historical statistics cannot be expected to extrapolate well to future conditions with previously unseen carbon forcing conditions. Thus these model outputs, while useful to understand uncertainty and potential for higher volatility and intense conditions, are not ready to be used to provide estimates of exposure into the future. Of course, this concern is amplified with the large uncertainty in the future carbon-emission scenarios, and the inability of these models to provide the space-time structure that translates into persistent and spatially correlated fat-tail risk. Further, since the companies have typically not even scored the baseline portfolio risk based on the historical record, it is imperative that at least an initial attempt uses a reliable global historical data set.

Observation-based global datasets are generally constructed with ground and satellite data (global products that are exclusively ground-observation based tend to not propose daily data). As a result, they often do not go far back in time, which can be problematic for the above-mentioned reason. Reanalysis datasets are modeled data constrained by daily global observations. They effectively provide a physics-based interpolation of such observations and preserve information from climate oscillations. The models used in reanalysis datasets limit the propagation of uncertainty and biases since their values are constrained to follow the historical observations available at the planetary scale on each day of the historical record, while they still suffer from such issues. Thus, they represent the best-case scenario from a model run in that sense, which is important for us. Examples of such products include (1) the CERA-20C reanalysis from the European Center for Medium Range Weather Forecasting (ECMWF) (ECMWF, 2021a) covers the period from 1900 to 2010 with a spatial resolution of 125 km and a daily temporal resolution; (2) the twentieth-century reanalysis V.3 available from the National Center for Atmospheric Research (NCAR) covering the period 1836–2015, with a 75 km grid resolution (NOAA, 2021b), (3) ERA5 from the ECMWF covering 1950 to the present at 25 km grid resolution (ECMWF, 2021b), and (4) MERRA-2 from the National Aeronautical and Space Administration (NASA) covering 1980 to present at a 50 to 65 km resolution (NASA, 2021). These products differ in the time periods covered, in the daily observations that they assimilate, and to an extent in the underlying model physics and the algorithm used for data assimilation. However, over the same time period, the precipitation fields tend to be well but not perfectly correlated. Given the differing record lengths and the varying spatial coverage of observations over time, the precision and the fidelity of the "observations" vary in space and time, and some characterization of these variations and inter-comparison in the context of the analysis of extremes and their recurrence intervals is recommended prior to the use of one or more of these products for analysis. We note that a variety of global station and gridded data on precipitation are also available, including some that go back to 1901 (NOAA, 2021a). Their coverage, especially for daily values, can be sparse over much of the world.

2.2 *Defining an Extreme Event*

The nature of impactful climate events and the degree of impact depends on a wide variety of factors, including the type of asset of interest, the design specifications and data used, the location, the existence of contingency plans, etc. Here we take the perspective of a user who can get informed on a set of assets using publicly available data and potentially get some inside knowledge on some assets (e.g., engineering design), and link this to current practices in climate science to define extremes.

In the climate sciences, extreme events are defined either based on the physical or meteorological phenomenon they represent, on a nominal level for a physical variable (e.g., diurnal temperature above 40 °C), or on their probability of occurrence given their intensity (Karl et al., 1999). For our examples here, we use the annual exceedance probability of occurrence and corresponding return levels, since these can then be mapped directly to a value at risk (VaR) framework.

In the examples presented, the event of interest is a precipitation event over *k* days, with a return-level r, i.e., an annual probability of exceedance $1/r$. The methodology can also be applied to other types of risks, as briefly touched in the conclusion of this paragraph.

2.3 *Calculating Climate Exposure—Basic Algorithm*

The basic algorithm to compute climate exposure at a portfolio level follows simple steps:

- Compute the yearly time series of maximum k-day precipitation for each site of interest s, thus obtaining time series $\text{Precip}_{k,s}^{\max}$.
- Compute the empirical distribution function as Rank $(\text{Precip}_{k,s}^{\max}/\text{NY})$, where NY is the number of years of data, or

fit a Generalized Extreme Value distribution (GEV) to $\text{Precip}_{k,s}^{\text{max}}$ (details on how to proceed for this particular step are provided in Annex I).

- For each site, compute the threshold corresponding to the risk level of interest.
- Then, in each year, count the number of times this threshold was exceeded (we call each exceedance a "hit") in separate non-overlapping occurrences, to obtain time series of "hits".

For instance, using data on mine locations from 2015, Bonnafous et al. (2017a) apply this method to the portfolio of 40 mines in operation for Rio Tinto, using a century-long climate dataset, and a 30-day, 10-year precipitation event. The time series of the number of hits can be observed on Fig. 2 (allowing for several hits per year as long as they have no days overlapping).

In 1981, 36 hits were counted for 22 different sites, while in theory, four ($40 \times 0.1 = 4$) of such exceedances would be expected under assumption of iid events, indicating that there is significant fat-tail risk across the portfolio relative to what would be expected by chance. The probability that one would experience a number of hits by chance even once

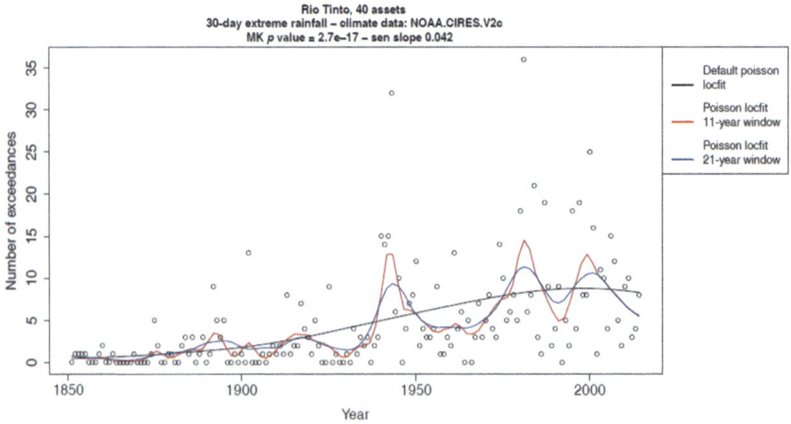

Fig. 2 Time series of the yearly number of 30-day extreme rainfall events exceeding the 10-year return level for the Rio Tinto portfolio computed using the 20CR dataset (Reproduced from Bonnafous et al. [2017a, p. 2085])

in the 165 years of data across 40 locations is essentially zero. As can be seen from Fig. 2, this point is amplified by the observation that there are 21 years with more than 10 hits each year, which also has zero probability of happening by chance and underscores the significant amplification of portfolio risk from climate extremes once their spatial coincidence in the same year is accounted for. If an insurance provider was to underwrite a policy for the company and was cognizant of the spatially correlated risk, the insurance premium would be dramatically amplified by this observation. Conversely, if the company was self-insured, they would likely not be prepared for the magnitude of the loss of business use and restoration funds needed to address these events. It is also worth noting that the frequency and severity of these portfolio impacts have an increasing trend and also has bi-decadal variability (note the 11- and 21-year smoothing averages presented in Fig. 2). The increasing trend could be related to climate change or to better data quality post-Second World War. The implication of bi-decadal variability is that at a site with short records in which a risk assessment is conducted during the asset commissioning stage, the company may be in for a rude surprise in the subsequent decade, as a previously dry period may be followed by an extremely wet period or vice versa, depending on which regime was experienced in the pre-completion phase relative to the post-completion phase. Clearly, this problem is amplified when portfolio risk for this company is considered and will likely be misconstrued as an anthropogenic climate change problem rather than a natural climate regime shift problem. Recall our concerns with the biases and uncertainties in the future climate projections, especially of oscillatory phenomena, and hence of the misclassification of climate risk if one solely relied on those projections. It is important that the baseline risk and its time variation be well understood and addressed before potential future changes are speculated on. To translate these portfolio level events into financial risk measures:

- Each "hit" at each asset location can be weighted by appropriate loss functions at the asset level. This will be discussed in more detail in Sect. 3.
- The weighted sum across the assets can then provide a measure of the value at risk, since each "hit" corresponded to a specific probability of an at site threshold being exceeded.

The potential losses at each site can be used for weighting and translating risk into a potential financial cost. It provides a basis for climate/water risk disclosure by the company; such a disclosure would force the company to critically examine the physical and operational risks that are residual to existing risk mitigation measures, or to their potential failure. For instance, in the mining context, tailing dams are infrastructure designed to retain wastes such that they can accommodate runoff from extreme rainfall events up to some threshold which has a certain annual exceedance probability. Losses that could occur if that infrastructure overtops and fails can be significant as demonstrated in the recent cases of dam failures in Brazil (Concha Larrauri & Lall, 2017). Production losses for a year or more, and compensation for environmental restoration losses, loss of life and property, as well as brand reputation can impact long-term and short-term share values and future insurability (Concha Larrauri & Lall, 2018). At present, companies may disclose that they have such infrastructure and that it is designed to meet engineering and risk standards appropriate for the country, but they do not necessarily disclose the potential loss if such an infrastructure were to fail. A key part of our framework would be to motivate the company to identify the key thresholds of concern, their exceedance probabilities, and the resulting losses for a proper climate risk disclosure. This could be done and reported at the asset or portfolio level properly accounting for the spatially correlated risk.

3 Quantifying Financial Impact of Extreme Climate Events (Application to the Mining Industry)

3.1 Metrics to Quantify Loss/Damage

3.1.1 Overview

Defining the financial impact of climate events requires consideration of the short-term and long-term impacts on cash flow and valuation of the affected assets (Lall, 2017). Extreme climate events must be characterized by their relative severity to take into consideration whether the impact of such incidences is likely to be temporary or more prolonged (Bonnafous et al., 2017a, 2017b). The direct financial impact can be measured in terms of foregone revenue, cost and capital expenditures with the severity of the extreme event defining whether the asset has any residual value and

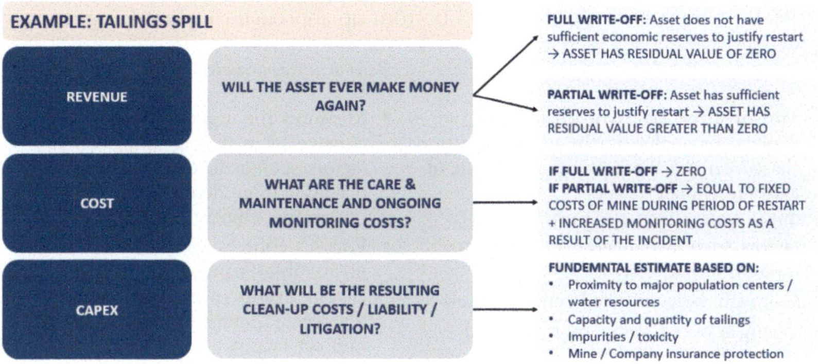

Fig. 3 Comparing full write-off vs. partial write-off for extreme events

whether an eventual re-start is infeasible, in which case a full "write-off" would occur (Fonseca et al., 2016). This is illustrated in Fig. 3 with an example of a tailings dam failure.

The resulting damage from extreme climate events can be assessed in two primary ways (Blanchet et al., 2015):

- Top-down approach: measures the *perceived* damage by the market to be lost as a result of extreme climate events (e.g., the immediate destruction to the enterprise value of publicly traded company following the announcement of an extreme climate event); and
- Bottom-up approach: measures the *expected* damage based on fundamental analysis of the company's underlying assets (e.g., a calculation of different components of loss resulting from an extreme event including foregone revenue, cost of maintenance and monitoring as well as the capital investment required for restart and clean-up).

A more detailed description of these approaches is presented in Table 1.

We focus on the bottom-up approach to quantify damage, acknowledging that oftentimes there are reputational and other non-financial factors that may concern stakeholders (Fig. 4). While these non-financial factors cannot easily be incorporated into a risk framework, these incremental indirect costs are important to highlight as they often play a critical role in decision-making.

Table 1 Comparing top-down and bottom-up approach to quantifying damage

Top-down approach	*Bottom-up approach*
• Measures the *perceived* quantifiable damage by analyzing the market reaction (based on the share price) as a result of extreme climate events • Takes into consideration non-financial factors such as damage to an operator's reputation and fear • Can only be completed retrospectively • Changes over time as more detailed information is released	• Measures the *expected* quantifiable damage by analyzing fundamental factors such as foregone revenue and net asset value destruction as a result of extreme climate events • Focuses entirely on the direct measurable financial impact of extreme climate events (permanent or temporary), ignoring behavioral and indirect factors • Can be completed prospectively and incorporated into risk analyses • Compares relative risk across companies at a static point in time based on consistent data

Fig. 4 Factors to incorporate when quantifying financial loss and damage

The bottom-up methodology put forth in this chapter provides a technique to quantify short- and long-term financial exposure and compare *relative* risk across portfolios. Using indexing methodology enables us to highlight whether certain companies are more exposed to extreme climate events than others based on the composition of their portfolios (Bonnafous et al., 2017a, 2017b). The following sub-sections describe the methodologies which can be used to rank relative risk by company, by quantifying temporary and permanent impacts of extreme climate events.

3.1.2 *Temporary Impact*

The temporary impact of extreme climate events refers to the immediately measurable lost cash flow resulting from an asset temporarily being shut down. For example, if as a precautionary safety measure, a company considered it prudent to halt production for a predetermined period, we could measure the foregone revenue that otherwise could have been generated had the asset continued operating as a means of quantifying loss.

In the example, we use "foregone revenue" as a means for quantifying temporary damage. More specifically, we weight each asset owned by each company by a factor equal to the production of each commodity at each asset multiplied by the dollar commodity price (the "Asset Weighting Factor", W_{tempi}) as shown in Eq. 1.

$$W_{tempi} = \sum_{c}^{i} P_c Q_{i,c} \quad (1)$$

where, for each asset i, P_c is the average unit price of commodity c, and $Q_{i,c}$ is the quantity of this commodity produced by mine i in the year of analysis (to produce a proxy for foregone revenue).

The result is a portfolio of assets where the relative significance of a temporary shutdown at each asset is characterized by the amount of foregone revenue that would be lost for a predetermined period, which may correspond to our exceedance and severity threshold. The logic is that extreme climate events disrupting production at those assets which generate the largest amount of revenue will naturally be more meaningful relative to those which are not producing (Bonnafous et al., 2017a).

A flaw in this analysis is that it does not take into consideration the fact that assets which have no production (construction-stage or development assets) may still be as or more disrupted by these extreme climate events as those in production. In addition, the impact of temporarily halting an asset that is less profitable may be significantly less meaningful than halting an asset that produces at a much lower cost. The next sub-section discusses how to incorporate "net asset value" into the analysis, taking into consideration factors such as cost and future profitability (beyond just foregone revenue).

3.1.3 *Permanent Impact*

The permanent impact of an extreme climate event represents the resulting damage from an event of sufficient severity to result in the permanent shutdown of an asset. For the purpose of this analysis, we assume that the maximum loss or liability from a catastrophic event is equal to the *net asset value* ("NAV") of the underlying asset itself. However, in extreme circumstances, whereby a company has been deemed to be grossly negligent or committed some form of misconduct, there have been cases where this assumption would not hold true (Varvastian & Kalunga, 2020).

The advantage of quantifying loss in terms of NAV rather than the temporary metrics discussed in the previous section is that this metric more holistically quantifies the loss. For example, NAV considers variables such as future production levels, asset life, cost of production, and sovereign risk that analyzing a point-estimate of an asset's revenue would not take into account. We re-run the same analysis outlined in the previous section; however, we weight each asset by NAV instead of than foregone revenue to account for the permanent impact (resulting from a full write-off of the underlying asset) by using asset weighting factor $Wper_i$ (Eq. 2). For companies where a large portion of their value may be derived from assets not currently in production, their risk of catastrophic loss from more severe climate events may be very different than their risk of temporary loss from less severe but more frequent events (that would simply cause a short-term production disruption).

$$W_{\mathrm{per}\,i} = \mathrm{NAV}_i \tag{2}$$

where, for each asset *i*, NAV is the net asset value determined from research analyst valuations of each asset on a consistent date using consistent commodity price and other valuation assumptions.

This analysis yields a result whereby the relative significance of an extreme event on each asset is based on its NAV. Said differently, the damage resulting from an extreme climate event of sufficient significance will be equal to the maximum quantifiable damage derived from our analysis, which is a full "wipe-out" of the value of the impacted asset. Alternatively, the relative damage from an extreme event of sufficient magnitude will proportionately impact an asset based on its NAV as a function of the aggregate value of company (Bonnafous et al., 2017a).

In continuing to apply our running example to the mining industry, we consider events such as major (irreparable) tailings breaches to be permanent impact, given the likely result that the asset will perpetually be unable to restart (as exemplified in Fig. 3). Shorter-term impacts such as drought or precautionary shutdowns are more akin to the temporary damage described in the previous section (Bonnafous et al., 2017a). In calibrating risk models, both temporary and permanent impacts must be considered to make prudent decisions.

3.2 Quantifying the Portfolio Impact

Companies will often own assets in specific geographic regions where they have developed skills, relationships, and local expertise. While the objective is to maintain an efficient organization and corporate focus, this can also lead to increased exposure to extreme climate events at the portfolio level. In the mining industry in particular, production tends to occur in those countries which are endowed with the highest quality resources. About 60% of global copper is produced in the top 5 copper producing countries, with over 80% of copper coming from the top 15 copper producing countries (USGS, 2021). Furthermore, within the mining industry, individual companies tend to own mines that are even more geographically concentrated as investors often prefer companies to remain focused on regions where they have direct operating experience. These regions might be impacted by the same climate oscillations, which can lead to a very different set of risks and exposure to extreme climate events across companies.

As described previously, we can use different weighting methodologies to compare the potential temporary and permanent exposure across companies to extreme climate events. We can explore different stress levels by varying the duration of the climate extreme and its exceedance probability. Each combination may reveal a different type of exposure and by varying the exceedance probability of the threshold, we can build a loss probability distribution for the asset and portfolio.

As an example, Fig. 4 shows a comparison of Barrick Gold to both a 10-year, 30-day rainfall event and a 100-year, 1-day rainfall event. The 10-year event exemplifies an extreme climate event of sufficient magnitude to potentially catalyze temporary shutdowns while the 100-year event exemplifies a sudden extreme climate event that could potentially result in catastrophic loss. We therefore quantify impact of the 10-year event (temporary) by foregone revenue and the 100-year event (permanent) by NAV (values obtained from TD Securities). From Fig. 5 (taken from Bonnafous et al., 2017a), 30% of Barrick's NAV is exposed to the 100-year event with a 1% per year probability while 80% of Barrick's revenue is exposed to the 10-year event with a 1% per year probability. At the 5% per year level, 7% of Barrick's NAV is exposed to catastrophic loss (resulting from the 100-year event) while 45% of Barrick's revenue is exposed to a more temporary production stoppage (resulting from the 10-year event).

This methodology allows us to assess the financial exposure of a company based on an event of pre-defined severity. For a given set of design parameters (for tailings dams for example which may be well understood), this technique also allows to consider design parameters and makes the determination of what constitutes an "extreme event" that would result in temporary versus permanent damage (Concha Larrauri & Lall, 2017).

Using the above methodology, we are able to calculate simple risk metrics such as "value at risk" (VaR) and "conditional value at risk" (CVAR) at the portfolio level for each company or a group of companies in an investor's portfolio (Bonnafous et al., 2017a; Dolan et al., 2018). Calculating VaR and CVAR are important as they highlight to key stakeholders the materiality of the risk from such extreme events relative to other risk-informed decisions considered in the ordinary course whereby the impacts may be more frequent but less severe, in both dollar terms and as a function of a company's aggregate value (adjusted for likelihood). The exposure component of the analysis is particularly relevant for low-probability, high-impact events to ensure that appropriate risk mitigants are considered proactively to the extent damage can be contained.

3.3 Relative Risk and Indexing Across Portfolios

Once we quantify the exposure of each company's portfolios to extreme events, we can use an indexing method to rank those companies most exposed to extreme climate events (Bonnafous et al., 2017a, 2017b). This is particularly useful for stakeholders that may be concerned about evaluating the most extreme tail events and determining exposure at the portfolio level.

If we pick a level of tail exposure (defined by percentile), we can compare the resulting impact across companies, assuming that the damage

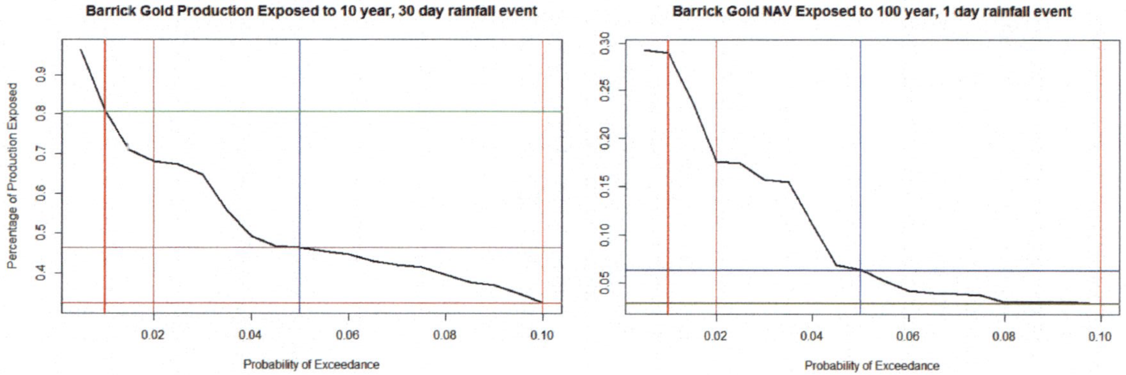

Fig. 5 Comparison of Barrick temporary and permanent exposure to extreme climate events of varying severities (Reproduced from: Bonnafous et al. [2017a, p. 2098])

(permanent or temporary) caused by any such extreme event is proportionate to the weighting method applied. We can sensitize the severity of the tail event to (1) hypothesize what quantile and duration of rainfall or drought are necessary to produce such an extreme climate event or (2) see which companies are most exposed to more frequent, less extreme climate events, and which are more exposed to permanent catastrophic loss.

Table 2 ranks 15 mining companies by their NAV exposure to a 12-month drought event (with data from Bonnafous et al., 2017b). The logic is that such an event may be sufficiently severe to cause significant social problems surrounding the asset (resulting from competition to water), potentially rendering the asset permanently unable to operate.

As shown in Table 2, exposure varies from 31% of NAV to 82% of NAV. Companies such as Capstone that have few geographically concentrated assets are more exposed to such events while companies such as Franco Nevada and Barrick that have large, diversified exposures are less exposed. Of course, there are potential mitigants related to the design parameters of the individual assets (such as desalinization) that could render a company less exposed than presented. However, it is still important to understand relative risk across companies. In addition, investors may have a portfolio of companies which should help provide an additional degree

Table 2 Ranked exposure to 12-month drought event (Modified from Bonnafous et al. [2017b])

Company	*Exposure to event (%)*	*Rank*
Agnico Eagle	44	9
B2Gold	33	13
Barrick Gold	31	14
Capstone Mining	82	1
Eldorado	54	4
First Quantum Minerals	44	8
Franco Nevada	28	15
Goldcorp	47	6
Hudbay	51	5
Iamgold	45	7
Kinross	34	11
Lundin Mining	35	10
New Gold	56	3
Newmont	34	12
Teck Resources	64	2

of geographic diversification. Nonetheless, the potential impact of climate extremes still needs to be considered in the context of overall risk and portfolio construction.

3.4 Parallels with the COVID-19 Crisis

The impact of the COVID-19 crisis was widespread across jurisdictions and industries. Disruption has made its way into almost every sector. The impact of such an extreme event on global supply chains has never been more evident as companies scrambled to produce essential goods and provide much needed services. Globalization has made it such that consumers of essential supplies and raw materials are dependent on producers abroad. As borders closed, procurement strategies needed to adjust to the reality that the impact of the pandemic made it such that certain materials were unsafe to produce in certain places.

This was seen at the most basic level as governments struggled to source masks and vaccines but had a particular impact on mining as countries and companies scrambled to prioritize the health and safety of their mine-site employees. Assets were shut down indefinitely until management determined that employees could do their on-site jobs safely. This had both a macroeconomic impact on the prices of certain commodities as well as a microeconomic impact on individual assets and companies.

While precautions were pervasive across the industry, the duration and impact of prolonged shutdowns varied by region based on both severity of the outbreak and public policy response. In work performed in 2020, Bank of Nova Scotia research analysts ranked jurisdictions according to how their response to the crisis affected mining operations classifying them in four categories (Bank of Nova Scotia, 2020). They then presented how potential production, cash flow, reserves, and NAV might be exposed, thus providing an example of how an extreme event, with uncertain outcomes, might impact global mining portfolios across different jurisdictions as the severity and public policy responses were still being assessed.

More specifically, the analysis compared company exposure by several metrics and whether mining in each jurisdiction was deemed essential or shut down altogether. For example, the report concluded that Agnico Eagle (assets predominantly in Canada) and Buenaventura (assets predominantly in Peru) had the highest percentage of their production located in jurisdictions where mining was halted, while Barrick and

Kinross (who have more diversified portfolios) had the lowest percentages respectively.

It remains to be seen how the impacts will manifest on companies in the long term, as the crisis impacted demand and prices, and support for companies varied widely. Yet, this crisis is a *realization* of a tail event. The methodology outlined can help simulate such realizations for climate extremes in a way that enables portfolios to be compared. In particular, COVID-19 presented a real-world backward-looking example of what happens when a tail event occurs that is widespread but also with differing in severities spatially distributed.

When we are finally able to look back on the financial impact of COVID-19, we will have the benefit of a real-world non-simulated extreme event as a further basis for quantifying the financial impact on companies. Companies were forced to shut down assets, in some cases as a precautionary measure and in other cases were mandated to do so by federal and local governments. Impacts varied from company to company and across assets of companies with global portfolios.

Such a response could be imagined in reaction to an extreme climate event of differing severities. Mining assets usually run "around-the-clock," so it is seldom that we get data regarding the cost of shutting down and restarting assets as well as actual care and maintenance expenses resulting from sustained unanticipated production stoppages. The direct impacts of the pandemic on production and cash flow as well as the permanent impact on net asset value in certain circumstances could serve to further inform our analysis on quantifying damage and risk resulting from extreme climate events. Once we have the final data, the analog of the pandemic and its spatial and temporal impacts on individual assets and companies could be a useful tool to quantify and price the risk of an extreme climate event.

4 Summary

The key points conveyed are

1. The potential disruptions from spatially correlated risks related to weather extremes need to be considered in portfolio or company-level risk assessments. The aggregate risk is much greater than the sum of asset-level risks and needs to be evaluated as such.

2. Disclosure of climate risk (in this chapter defined as the physical risk from rainfall extremes but could apply to other stressors), needs a bottom-up process. The process considers specific thresholds, their probability of exceedance, the loss when that threshold is crossed, and the proper aggregation of these risks across assets due to the high likelihood that many assets around the world may be impacted in the same year. These could be flood as well as drought impacts. Consideration of both types of extremes enhances the risk of failure. Companies often have measures for at site risk mitigation through insurance and/or structural solutions but may not evaluate the potential loss when these measures fail or the conditions in which they fail. These elements are relevant to assess and disclose.
3. Losses that result from extreme weather events may be short-term, operational losses, including those that propagate through the economy, suppliers, and purchasers, and/or long-term losses that reflect an asset write-off or stranding. Different measures need to be applied to both types of losses recognizing that the event duration and probability of exceedance for the two types of losses are likely to be very different. We provide examples for these cases and discuss how reported losses from COVID-19 could be used as a proxy for establishing ways to assess some of the short-term loss companies incurred, once the full data becomes available.

Annex 1: Extreme Value Theory Applied to Extreme Rainfall

For simplicity, let's call X_i the random variable of cumulative *k*-day rainfall at a given site starting on day *i* of the year. If we assume that the X_i are independent and identically distributed (iid), extreme value theory tells us that if $M_n = \max_n(X_i)$, if there exist sequences (this existence is here assumed) a_n and b_n such that $M_n^* = \frac{M_n - a_n}{b_n}$ converges to a non-degenerate cumulative distribution function, this distribution is of the form:

$$G(z; \mu, \sigma, \xi) = \exp\left[-\left(\frac{\xi(z-\mu)}{\sigma}\right)^{\frac{-1}{\xi}}\right]$$

μ is called the location parameter, σ is called the scale parameter, and ξ is called the shape parameter.

For our purpose, we fit the above-mentioned parameters to the $\text{Precip}_{k,s}^{\max}$ time series. Careful assessment of the validity and usability of the model has to be conducted, as not all assumptions enunciated here are validated. In particular, the iid hypothesis does not hold for us, but academic work (Katz, 2013) shows that adjustment to the location and scale parameters can account for this.

References

Bank of Nova Scotia. (2020). *Introducing the COVID 19 response tracker for mining countries.*

Blanchet, J., Dolan, C., Iyengar, G., & Lall, U. (2015). *Towards a robust top-down model for valuation of mining assets.* http://water.columbia.edu/files/2015/11/mine_6_Sept_2015.pdf

Bonnafous, L., & Lall, U. (2020, March). Space-time clustering of climate extremes amplify global climate impacts, leading to fat-tailed risk. *Natural Hazards and Earth System Sciences*, 1–19. https://doi.org/10.5194/nhess-2019-405

Bonnafous, L., Lall, U., & Siegel, J. (2017a). A water risk index for portfolio exposure to climatic extremes: Conceptualization and an application to the mining industry. *Hydrology and Earth System Sciences, 21*(4), 2075–2106. https://doi.org/10.5194/hess-21-2075-2017

Bonnafous, L., Lall, U., & Siegel, J. M. (2017b). An index for drought induced financial risk in the mining industry. *Water Resources Research, 53*(2), 1509–1524. https://doi.org/10.1002/2016WR019866

Concha Larrauri, P., & Lall, U. (2017). *Assessing risks of mine tailing dam failures.* http://water.columbia.edu/files/2019/01/TSF-paper-CWCrev2.pdf

Concha Larrauri, P., & Lall, U. (2018). *The value of water.* https://drive.google.com/file/d/1aKtIgDdUoV6l9iZaOPit9HJUm_uxu-r5/view

Dolan, C., Blanchet, J., Iyengar, G., & Lall, U. (2018). A model robust real options valuation methodology incorporating climate risk. *Resources Policy, 57,* 81–87.

ECMWF. (2021a). *CERA-20C dataset.* https://www.ecmwf.int/en/forecasts/datasets/reanalysis-datasets/cera-20c

ECMWF. (2021b). *ERA5 dataset.* https://www.ecmwf.int/en/forecasts/datasets/reanalysis-datasets/era5

Field, M. (2015). *Government finances threatened as drought shuts Ok Tedi mine.* Nikkei Asia. https://asia.nikkei.com/Business/Government-finances-threatened-as-drought-shuts-Ok-Tedi-mine

Fonseca, F., Bonnafous, L., & Lall, U. (2016). *Asset-level analysis and modeling of water risks associated with mining*. http://water.columbia.edu/files/2015/11/Internal-Report-Asset-Level-analysis-and-modeling_ul_lp.pdf

Frankcombe, L. M., von der Heydt, A., & Dijkstra, H. A. (2010). North Atlantic multidecadal climate variability: An investigation of dominant time scales and processes. *Journal of Climate, 23*(13), 3626–3638. https://doi.org/10.1175/2010JCLI3471.1

Jorgensen, D. (2016). The garden and beyond: The dry season, the Ok Tedi shutdown, and the footprint of the 2015 El Niño drought. *Oceania, 86*(1), 25–39. https://doi.org/10.1002/ocea.5121

Karl, T. R., Nicholls, N., & Ghazi, A. (1999). CLIVAR/GCOS/WMO workshop on indices and indicators for climate extremes: Workshop summary. *Climatic Change, 42*, 3–7.

Katz, R. W. (2013). Statistical methods for nonstationary extremes. In A. AghaKouchak, D. Easterling, K. HsuSiegfried, & S. Sorooshian (Eds.), *Extremes in a changing climate*. Springer.

Kunz, N. C. (2020). Towards a broadened view of water security in mining regions. *Water Security, 11*(November). https://doi.org/10.1016/j.wasec.2020.100079

Lall, U. (2017). *Mining and water risk: Diagnosis, benchmarking, and quantitative analysis of financial impacts*. http://water.columbia.edu/files/2015/05/NBIM-Synthesis-Chapter-FInal-4.11.18.pdf

Mantua, N. J., & Hare, S. R. (2002). The Pacific decadal oscillation. *Journal of Oceanography, 58*(1), 35–44. https://doi.org/10.1023/A:1015820616384

NASA. (2021). *MERRA-2 dataset*. https://gmao.gsfc.nasa.gov/reanalysis/MERRA-2/

NOAA. (2021a). *Gridded climate datasets: Precipitation*. https://psl.noaa.gov/data/gridded/tables/precipitation.html

NOAA. (2021b). *The twentieth century reanalysis project dataset V3*. https://psl.noaa.gov/data/20thC_Rean/

USGS. (2021, January). *Mineral commodity summaries*. https://pubs.usgs.gov/periodicals/mcs2021/mcs2021-copper.pdf

Varvastian, S., & Kalunga, F. (2020). Transnational corporate liability for environmental damage and climate change: Reassessing access to justice after Vedanta v. Lungowe. *Transnational Environmental Law, 9*(2), 323–345. https://doi.org/10.1017/S2047102520000138

Woldemeskel, F. M., Sharma, A., Sivakumar, B., & Mehrotra, R. (2012). An error estimation method for precipitation and temperature projections for future climates. *Journal of Geophysical Research: Atmospheres, 117*(D22). https://doi.org/10.1029/2012JD018062

Water Risk: An Overview and Inspiration for Future Work

Kalima Vico and Thomas Walker

As discussed throughout this book, water risk affects every part of our society and economy differently: An overabundance of water related, e.g., to floods can be just as destructive as a shortage stemming from droughts. Similarly, water pollution can have devastating effects on our flora, fauna, and our own (human) lives. While water risk affects us on a global scale, certain locations, industries, and even segments of our population are more affected than others. For instance, a given company may only face water risk exposure in some of its operations or may be exposed in whole (Craig et al., chapter "How Do Investors Assess Water Risk?"). Similarly, the "actual impacts and relevant risk types and risk levels might differ from one asset to another, based on local government and company policy, infrastructure design, contingency

K. Vico · T. Walker (✉)
Emerging Risks Information Centre (ERIC), John Molson School of Business, Concordia University, Montréal, QC, Canada
e-mail: thomas.walker@concordia.ca

K. Vico
e-mail: k_vicoge@live.concordia.ca

T. Walker et al. (eds.), *Water Risk and Its Impact on the Financial Markets and Society*, Palgrave Studies in Sustainable Business In Association with Future Earth,
https://doi.org/10.1007/978-3-030-77650-3_12

plans, etc." (Siegel et al., chapter "Multi-dimensional and Interacting Water and Climate Risks and Pricing them in the Industry Context"). We know, for example, that the global food and agribusiness sector is highly exposed to water risk, that consumer goods and other industry sectors that depend on raw inputs are exposed not only to upstream but also to downstream risks, and that mining and the high technology industry (which depends on mined materials) are highly dependent on water for their operations and carry their own unique risks with respect to water quality. Finally, many utilities are heavily dependent on a steady and reliable source of water and cannot generate electricity without it. As we have learned from Zhou et al. (chapter "Financial Implications of Parched Power: Insights from an Analysis of Indian Thermal Power Companies") "water shortage risks cause occasional significant financial impacts for thermal power companies in India and if drought conditions worsen, water shortage risks could become more significant in the future, causing more severe outages and greater financial impacts."

Other industries might not directly depend on water in their operations, but are nevertheless subject to water risks. For instance, the real estate sector and the tourism and leisure industry can be affected by flooding, coastal erosion, water scarcity, and other physical water risks. These risks tend to directly affect the market price of a property and, at the same time, can affect the accessibility to certain areas such as hotels and tourist attractions. "Even emerging sectors such as cannabis production and software development have their own complex water risks" some of which me might still be unaware of (Craig et al., chapter "How Do Investors Assess Water Risk?"). We must also remember that water risks are complex and diverse. While they can materialize in the form of physical, reputational, financial, and regulatory threats to a given entity, they often do so in more than one category.

When governments and legislators try to address and combat these risks through water usage policies and related legislation, they may create inequalities (e.g., by favoring businesses over local farmers) and exacerbate social inequity issues. This is demonstrated, e.g., by Sandhu et al. (chapter "Water Risks, Conflicts and Socially Responsible Investment: A Case Study of Ontario, Canada") who highlight governmental failures to cater to indigenous tribes in Ontario.

In an era when many of us are worried about a rise in political extremism, mass migration, and economic uncertainty—not to mention a deadly pandemic—longer-term issues such as climate change and water uncertainty are easily ignored in light of seemingly more pressing issues. Although both air and water pollution have declined in many parts of the world as a result of business closures, and a mere halt of tourism and business travel during the pandemic, it can be fully expected that once life returns to its pre-pandemic level, so will man-made pollution. We must also realize that when this happens, the issues that plagued us will resume, that the contamination and depreciation of our natural resources may exacerbate the risk of pandemics and that as pollution affects our health, it will also affect our ability to fight future disease outbreaks. Despite the literal breather our environment has been granted since the beginning of the pandemic, this book attempts to remind us that water risk is a highly complex issue that continues to be severe and increasingly urgent. It cannot be lessened in the long term unless we meaningfully address it now. At the same time, the COVID-19 pandemic can give us a glimpse on what would happen to our services if an extreme event were to shut them down for a certain period of time. In this context, Siegel et al. (chapter "Multi-dimensional and Interacting Water and Climate Risks and Pricing them in the Industry Context") note that "the direct impacts of the pandemic on production and cash flow as well as the permanent impact on net asset value in certain circumstances could serve to further inform our analysis on quantifying damage and risk resulting from extreme climate events."

In recent decades, industries, governments, news outlets, and investors have seemingly paid more attention to (and have given a higher priority to) carbon transition and more generic climate change risks than to water risk; yet, assessing climate risks too broadly will provide an incomplete picture of the risk landscape. Only evaluating one set of risks may lead to wrong investment and regulatory decisions, which in turn may increase the level of overlooked risks and perpetuate bad decisions. This is highlighted by Mirando and Tan (chapter "Chronic Coastal Threats Warrant a Valuation Re-Think") who point out that "central banks will drive chronic risks to be factored in, or else all valuations will be off and the negative finance feedback loop will be perpetuated." As mentioned by Craig et al. (chapter "How Do Investors Assess Water Risk?") "the concept of green and low carbon portfolio construction can be extrapolated to include 'blue' [i.e., water-related] components. There are several investment opportunities in water as a commodity, ranging from individual stocks of firms in water-related businesses, indexes like the Dow

Jones U.S. Water Index or the S&P Global Water Index, water exchange traded funds (ETFs), mutual funds, and investment trusts."

Dramatic water-related events can focus the public's attention on water risk and can provide resources for affected communities. For example, the Flint, Michigan, water pollution crisis garnered heavy media attention and led to widespread conversations about water quality around the United States. However, even in the absence of such large-scale events, one has to be aware that there are many small or less popularized, yet significant water problems throughout the world. Chronic water stresses, for instance, are often underappreciated even though they are more prevalent, costly, and damaging in their own right, such as the "financial impacts of water being material and less acute in the short term but negatively affecting revenue, profit margins, and earnings in the long term" (Craig et al., chapter "How Do Investors Assess Water Risk?"). Similarly, the fact that improperly maintained water supply systems can lead to "traffic delays, flooding, water damage, interruption of services, health and even political complications" often goes unrecognized (Karney and Gibson, chapter "Misbehaving Drinking Water Systems: Risk and the Complex Nature of Failure"). If these indirect risk exposures are to be properly modeled and better understood, we must consider the whole spectrum of water risks, from extreme weather events that "can be temporally and spatially correlated (e.g., an intense flood might happen in Australia while a prolonged drought is hitting Peru)" (Siegel et al., chapter "Multi-dimensional and Interacting Water and Climate Risks and Pricing them in the Industry Context", and also discussed by Rayer et al., chapter "Water Insecurity and Climate Risk: Investment Impact of Floods and Droughts") to a pipe breaking and causing localized havoc (Karney and Gibson, chapter "Misbehaving Drinking Water Systems: Risk and the Complex Nature of Failure").

The chapters featured in this book point out several related issues:

1. *We are not evaluating water risks correctly*: Josset and Concha Larrauri (chapter "Data for Water Risks: Current Trends in Reporting Frameworks, Shortcomings, and the Way Forward") indicate that water risk-related costs are often treated as operating expenses through the costs of infrastructure, distribution, and treatment; therefore, when evaluating water stresses, we are missing some of the most important risks and expenses related thereto (e.g., water scarcity, floods, and pollution). These specific

risks represent blind spots in many evaluation models and are often missing in risk quantification. Disconcertingly, this suggests that water will not be integrated in investment decisions until its externalities can be quantified through improved methodologies, data collection, regulations, and transparency. Similarly, Xu and Tan (chapter "The Developing Field of Water Risk Valuation for the Financial Industry") find that even when water risks are specifically analyzed and quantified, water pollution is frequently ignored. Relatedly, Rayer et al. (chapter "Water Insecurity and Climate Risk: Investment Impact of Floods and Droughts") indicate that the potential liabilities stemming from extreme floods and droughts (events intrinsically linked to global warming) may not be reflected in share prices, which can lead to over- or undervaluing stocks. Even "irrigation water is not priced according to the actual quantity used because volumetric measures are not in place" (Mahajan and Porwal, chapter "Forward Pricing of Embedded Water—A Step Towards Sustainable Development in Agriculture"). Ultimately, "the lack of appropriate risk assessment and management tools is leading to a path dependency where water risks are generalized to certain geographies and sectors without evaluating granular watershed conditions and impacts" (Sandhu et al., chapter "Water Risks, Conflicts and Socially Responsible Investment: A Case Study of Ontario, Canada").

2. *We are missing data*: "Water use in itself is not informant of water-related risks if the local context is not taken into account. The water footprint at each step of the supply chain would be a better informant yet that would require considerable information and an often arbitrary estimate of what can be attributed to a product, a corporation, or a sector play (Josset and Concha Larrauri, chapter "Data for Water Risks: Current Trends in Reporting Frameworks, Shortcomings, and the Way Forward"). Ideally, as pointed out by Xu and Tan (chapter "The Developing Field of Water Risk Valuation for the Financial Industry") in the context of utilities, "to analyze water risk exposure, knowledge of each power plant's water use, specific location, and cooling type is required. Water risk is location specific, yet companies rarely disclose facility level water data. There is a long way ahead before this level of information is disclosed and the limited accuracy of underlying data may lead to under/over-estimating the

degree of water availability challenges." Zhou et al. (chapter "Financial Implications of Parched Power: Insights from an Analysis of Indian Thermal Power Companies") run into similar problems when analyzing the Indian electrical sector because the available climate simulation and projection data only includes physical changes in water supply, but does not include localized socioeconomic scenarios and local water demand. In addition, the data available is difficult to download and not in a format that is familiar for investors. They highlight that while climate models project an unambiguous increase in temperatures across the world in the coming decades, there is a lack of data and methodologies to incorporate temperature risks into analyses, including potential efficiency losses and impacts from temperature regulations on cooling water discharges. Even climate models provided by the Intergovernmental Panel on Climate Change (IPCC) are biased and fail to capture many of the oscillations we experience, while observation-based global datasets tend to be outdated (Siegel et al., chapter "Multi-dimensional and Interacting Water and Climate Risks and Pricing them in the Industry Context"). Nevertheless, even though "data problems continue to be a challenge, [...] we must start assessing these risks today so that the data can be improved as waiting for the perfect set may be leaving it until it is too late" (Mirando and Tan, chapter "Chronic Coastal Threats Warrant a Valuation Re-Think").

3. *Water risk and its drivers are highly interconnected.* Risks to water security are a "multi-dimensional construct of quantity, quality, access, and public perception that needs to be tethered together in policies and investments for a sustainable future. Given the wide social, economic, and ecological implications, it becomes necessary to conceptualize, assess, and manage interdisciplinary water risks as a means to ensure water security" (Sandhu et al., chapter "Water Risks, Conflicts and Socially Responsible Investment: A Case Study of Ontario, Canada"). Water issues may result from deep systemic risks such as climate change, and the mismanagement or misuse of water. Yet, even local manifestations of water risk can have a global impact. Equally, water is a "multi-impact investment since it is intrinsically linked to several other PRI impact themes like food supply, industrial production, health, productivity, and the environment" (Craig et al., chapter "How Do Investors Assess Water Risk?"). Most chapters point out that this water interconnectedness is one of the

reasons that evaluating the risks is complicated, either because multi-sector expertise is needed (such as hydraulics, traffic or city planning, fire protection, power distribution, epidemiology or medicine, in order to properly manage and quantify water distribution related risks as discussed by Karney and Gibson, chapter "Misbehaving Drinking Water Systems: Risk and the Complex Nature of Failure"), or because the tools used to quantify water risks are too complex. For instance, as explained by Xu and Tan (chapter "The Developing Field of Water Risk Valuation for the Financial Industry"), shadow pricing tools are still limited to a very small group of professionals—both because of their complexity, and because the risks affect multiple localities and sectors.

In addition to highlighting important problems and inconsistencies, this book also discusses opportunities for the development of innovative and practical solutions. For example, Karney and Gibson (chapter "Misbehaving Drinking Water Systems: Risk and the Complex Nature of Failure") indicate that since water distribution networks misbehave so frequently, with the appropriate resources there is a chance to study these failures and better understand them. They argue that with this understanding comes the opportunity to mitigate their effects. Similarly, Craig et al. (chapter "How Do Investors Assess Water Risk?") confirm that "addressing water pollution can also drive business opportunities." For example, they estimate that "if the negative environmental and societal externalities associated with the apparel and textile industry were addressed, the benefit to the global economy is [...] upwards of US$190 billion per year."

Sandhu et al. (chapter "Water Risks, Conflicts and Socially Responsible Investment: A Case Study of Ontario, Canada") offer us a framework to quantify water risks, which can be operationalized into decision support tools using publicly available, reliable, and peer-reviewed data from government databases to allow the harmonization of data used for public policies, investments, and corporate water policies. This can aid decision making for investments and corporate water sustainability. While their case study is focused on Ontario, it could be applied to any geography. Similarly, Zhou et al. (chapter "Financial Implications of Parched Power: Insights from an Analysis of Indian Thermal Power Companies") provide a "robust way for investors and companies to analyze climate-related risks and conduct scenario analysis based on the Task Force on Climate-related Financial Disclosures (TCFD's) recommendations," adaptable to any geography, when the proper data is available. In the same context, Siegel et al. (chapter "Multi-dimensional and Interacting Water and Climate Risks and Pricing them in the Industry Context") use the mining industry as an example, and provide "a basic framework to quantify relative risk (in terms of potential losses and damages) from extreme climate events" which includes the ability to asses both the temporary impacts of minor incidents and enduring damages to permanent assets. Equally, they offer recommendations to limit the propagation of uncertainty and biases, suggesting to use reanalysis datasets to extrapolate the proper climate data and quantify impacts across space and time.

Xu and Tan (chapter "The Developing Field of Water Risk Valuation for the Financial Industry") estimate the shadow price of water in order to use it as a proxy that reflects exposure to water risk, with the shadow price incorporating different factors such as population, water use competition, and the characteristics of the local economy. Once the shadow price of water has been quantified, it can be incorporated in a company's income statement in the form of, e.g., operating expenses. They also propose the use of baseline water stress maps to assess and compare exposure to water stress across companies, sectors, or commodities. Similarly, Craig et al. (chapter "How Do Investors Assess Water Risk?") discuss different tools investors can use to evaluate water risks and identify opportunities, indicating that "several of the risks and impacts caused by water-related issues could be better mitigated through smart water management, including increasing agricultural efficiencies as well as improving coastal water management and preparedness in case of extreme physical water risks, like flooding or droughts." In addition, they recommend adapting circular economy concepts to the way we think about and manage water.

Mahajan and Porwal ("Forward Pricing of Embedded Water—A Step Towards Sustainable Development in Agriculture") suggest that we look at agricultural commodities from a new angle. They indicate that to help with allocation inefficiencies and discrepancies currently found in our water management systems we should use market-based mechanisms to trade water as a commodity, allowing the market to "allocate water according to its opportunity cost," noting that "water trading is likely to expand the access to water for many small-scale farmers." They argue that this type of water resource management will "reallocate jobs in an already saturated agriculture market and lead to green employment opportunities in water rich areas where crops are destroyed due to heavy rainfall." They further recommend using a forward pricing method and to focus on growing food in water-rich areas and transporting it for sale in water-poor areas.

In order to fully take advantage of these opportunities, certain things need to be implemented:

1. We need more comprehensive risk analysis approaches, preferably involving multidisciplinary teams; this will increase our understanding and ability to mitigate water risks (Karney and Gibson, chapter "Misbehaving Drinking Water Systems: Risk and the Complex Nature of Failure"). Similarly, a multi-scale approach is needed to allow for the integration of relevant datasets from multiple disciplines. This will allow researchers and practitioners to embed and merge the unique insights provided by each database, navigate from a local to a global scale, and perform transparent and robust assessments of the link between water risks and financial risks. As Sandhu et al. (chapter "Water Risks, Conflicts and Socially Responsible Investment: A Case Study of Ontario, Canada") point out: to make better informed investment choices, "interdisciplinary investment decision support tools need to be developed using granular data to reflect local hydrological, regulatory, and social conditions." These risks must be assigned proper weights using a granular framework. This will allow investors to fully capture the intricacies of water risk and "to design appropriate communication and engagement strategies within existing and future portfolios to achieve the stipulated financial, social, and environmental outcomes of the project."
2. We need to assess all water-related risks in depth, with full transparency, and with quantifiable data; to do so, we need a comprehensive assessment and management framework and regulatory/industry standards that control how water risks should be disclosed, estimated, and modeled. The disclosure standards must include: "human intervention on the water systems (e.g., land use, withdrawals, built infrastructure, return flows, management patterns and practices) [...] to understand what climate patterns lead to water scarcity risks, exacerbate floods, and indirectly impact pollution events" (Craig et al., chapter "How Do Investors Assess

Water Risk?"). Corporations' water and land use and their related management practices must be included for any assessment to be meaningful due to the pivotal role that the industrial, energy, agricultural, mining, and food and beverage sectors play (Josset and Concha Larrauri, chapter "Data for Water Risks: Current Trends in Reporting Frameworks, Shortcomings, and the Way Forward"). Disclosure of climate risk needs to be a bottom-up process. "The process needs to consider specific thresholds, their probability of exceedance, the loss when that threshold is crossed, and the proper aggregation of these risks across assets due to the high likelihood that many assets around the world may be impacted in the same year" (Siegel et al., chapter "Multi-dimensional and Interacting Water and Climate Risks and Pricing them in the Industry Context"). To avoid that data is misleading or subject to misinterpretation (such as false confidence and underestimation) we need to use contextual information and confidence intervals (Josset and Concha Larrauri, chapter "Data for Water Risks: Current Trends in Reporting Frameworks, Shortcomings, and the Way Forward"). Data previously ignored, such as the effect of legal cases, should also be considered as any movement toward an active liability regime could risk shares in at-risk companies becoming stranded assets (Rayer et al., chapter "Water Insecurity and Climate Risk: Investment Impact of Floods and Droughts"). Similarly, extreme climate events need to be incorporated into risk models in order to "prevent and mitigate detrimental financial and societal impacts" (Siegel et al., chapter "Multi-dimensional and Interacting Water and Climate Risks and Pricing them in the Industry Context"). Therefore, "the potential disruptions from spatially correlated risks related to weather extremes need to be considered in portfolio or company level risk assessments," taking into consideration that "the aggregate risk is much greater than the sum of asset-level risks and needs to be evaluated as such" (Xu and Tan, chapter "The Developing Field of Water Risk Valuation for the Financial Industry"). We must also give consideration to how data should be disclosed so that it is relevant and represents an adequate and robust overview of the data collected (Josset and Concha Larrauri, chapter "Data for Water Risks: Current Trends in Reporting Frameworks, Shortcomings, and the Way Forward").

3. We need to evaluate risks locally and globally: While risk interpretation starts at the asset level, it is passed on through an industry's supply chain and gets compounded across a company's assets; therefore, consideration of local plant-level water risks is essential to accurately assess the potential water-related losses/opportunities for companies. Relatedly, to attract sustainable, long-term investments in water and better manage portfolios exposed to water risk, a detailed assessment of local geographies at the sub-watershed level is necessary (Mirando and Tan, chapter "Chronic Coastal Threats Warrant a Valuation Re-Think"). When looking at water risk exposure from an asset perspective, local water availability matters. Thus, it is important to understand local water stresses, including, e.g., local regulatory risks which can be disruptive in the short term but could provide long-term water stress alleviation (Xu and Tan, chapter "The Developing Field of Water Risk Valuation for the Financial Industry"). Such assessments are necessary even for areas or sectors that seem water secure, such as Canada, as there are other risks involved such as physical, regulatory, and reputational risks (Sandhu et al., chapter "Water Risks, Conflicts and Socially Responsible Investment: A Case Study of Ontario, Canada").
4. Investors need to have access to readily available data and an established information chain to assess the water risk of an investment. The greatest challenge facing investors who are trying to assess water risks is the sourcing of detailed and reliable information from their portfolio companies. To address this issue, organizations need to employ specialists who can develop and apply the necessary frameworks to assess water risks, so that the organization can then properly disclose this information (Craig et al., chapter "How Do Investors Assess Water Risk?"). In the same vein, investors must be take into account "social nuances, the legal and regulatory environment, existing legacy issues, conflicts among competing water users, and water intensity trends in the local social-economic-environmental context" in order to be able to fully attend to all the risks (Sandhu et al., chapter "Water Risks, Conflicts and Socially Responsible Investment: A Case Study of Ontario, Canada").

We must remember that the complexity of these risks and their related opportunities and solutions means that we must not look at them through a focal lens, or assign responsibility to only one party. To identify, mitigate, and manage such complex risks requires that we assume collective responsibility, combine collective intelligence, and collaborate. Otherwise—due to failures by one party—we may end up suffering collective failures.

In one way or another, we all contribute to and are responsible for today's water problems: indirectly by contributing to climate change or directly through our use of water. After all, we have affected the environment in more ways than any other species. We are responsible for the eight million tons in annual plastic waste that pollute the oceans and make up 80% of all marine debris in both the oceans' surface water and deep-sea sediments; and even though we represent only 0.01% of all living things by weight, humanity has caused the loss of 83% of all wild mammals and half of all plants (Craig et al., chapter "How Do Investors Assess Water Risk?").

Our ecological and water footprint can be mitigated through a careful and conscientious use of water in our daily life, e.g., in the bathroom, kitchen, garden, etc. We can also attempt to choose products or investments based on a company's water policies. For instance, instead of eliminating coffee from our diet or investment choices, we might want to choose a coffee producer that exhibits more responsible water usage. This would require relevant information disclosures by companies. Therefore, while we can all contribute to alleviating local and global water stress, some of the first steps in that direction may have to be taken by governments and businesses.

Firms in the fossil fuel sector and their customers carry most of the responsibility for climate change, which has a direct connection to water risks; therefore, the onus falls on them to correct the damage done by decades of neglecting how harmful their products truly are. "If major carbon producers wished to make more than a minimal positive contribution (i.e., do more than only reduce the harm caused by their products), their distinctive political power, wealth, and expertise qualify them to lead the transition to an energy regime which would be safe for future generations to rely on" (Rayer et al., chapter "Water Insecurity and Climate Risk: Investment Impact of Floods and Droughts"). In this context, industrial suppliers carry the responsibility to provide efficiency gains and alternative technologies that enable transport that minimizes both a company's carbon and water footprint.

Industries do have their own direct effect on water issues. For instance, 19% of global water withdrawals stem from industrial operations such as textile, energy, and chemical production (Craig et al., chapter "How Do Investors Assess Water Risk?"). At the same time, industries release 300 to 400 megatons of waste into water per year (Xu and Tan, chapter "The Developing Field of Water Risk Valuation for the Financial Industry"). Other issues affecting water quality and stress such as pollution, increased water acidification, and agricultural runoff (containing contaminants, residues of fertilizers or pesticides, or other chemicals) are also directly linked to industrial operations (Craig et al., chapter "How Do Investors Assess Water Risk?").

Businesses play a crucial role in designing proactive sustainable water management practices that incorporate the various types of water risk and related response strategies (Mirando and Tan, chapter "Chronic Coastal Threats Warrant a Valuation Re-Think"). If firms manage their water risks, they can reduce operational interruptions and equipment losses, and may take advantage of otherwise unrecognized opportunities and efficiencies. For example, "food and beverage companies can improve the resilience of their supply chains by supporting sustainable farming practices in the watersheds where they source their most significant commodity inputs" (Craig et al., chapter "How Do Investors Assess Water Risk?").

In the same vein, Craig et al. point out that "for companies to have an effective water stewardship process in place, they need to build their knowledge of global water challenges and establish which ones are relevant for their operations, supply chains, customers, and communities." These companies need to better understand their dependence on water and where that dependence is highest within the value chain. Ultimately, firms must understand that water may affect them in two possible ways: It may impact company operations internally through their direct effect on corporate operations and the supply chain, and may affect local water basin levels and the upstream and downstream water supply.

Although companies increasingly engage in at-site risk mitigation through insurance and/or structural solutions, they may not evaluate the potential losses they are exposed to if these measures fail or the conditions under which they fail. However, as Siegel et al. (chapter "Multi-dimensional and Interacting Water and Climate Risks and Pricing them in the Industry Context") point out, these elements are equally relevant to assess and disclose.

Even if companies remain inactive in the face of increasing water risk, investors can encourage a water dialogue by requesting improved disclosure of relevant and material information. "Investors who take an active approach in engaging with their portfolio companies on water risks can raise important issues and instigate change of corporate practices and policies. By doing so, investors can improve investment returns, create long-term value, and mitigate water-related and other risks" (Craig et al., chapter "How Do Investors Assess Water Risk?"). In the same vein, Sandhu et al. (chapter "Water Risks, Conflicts and Socially Responsible Investment: A Case Study of Ontario, Canada") point out that "the role of the financial sector, regulators, central banks, and credit rating companies in incentivizing sustainable water management and finance is also critical to realizing the objectives of water security."

Government is another major player due to its privileged position to access and request data. Arguably, given their positioning, it should be government's duty to disclose and host the data and tools that are necessary for risk assessments and interpretation. In this context, Josset and Concha Larrauri (chapter "Data for Water Risks: Current Trends in Reporting Frameworks, Shortcomings, and the Way Forward") argue that government involvement is needed to ensure transparency, accuracy, and compliance with standards. If information such as groundwater data is made available to the public, it can "facilitate entrepreneurial decision-making, facilitate coordination among users of aquifers, and be the first step towards water contracts" (Mahajan and Porwal, "Forward Pricing of Embedded Water—A Step Towards Sustainable Development in Agriculture"). If trading markets for water are to originate, government action is equally needed in order to formalize the markets, as there would need to be "well-defined, tradable water rights and the appropriate infrastructure and institutions for distributing water."

The provision and setting of subsidies, quotas, and allocations to encourage efficient water use are also a government responsibility. When properly designed and implemented, cross subsidization strategies can ensure sustainable financing schemes related to water use efficiency, or allow quota allotments to be traded, in order to address equity concerns while promoting efficient allocations (Mahajan and Porwal, "Forward Pricing of Embedded Water—A Step Towards Sustainable Development in Agriculture").

Because government actions around the globe tend to be inconsistent and unique, one of the challenges of assessing risks such as coastal threats lies in understanding what proxy indicators governments use in their risk assessment procedures. Therefore, a proper identification of these threats by non-governmental entities must include an analysis of related government actions (Mirando and Tan, chapter "Chronic Coastal Threats Warrant a Valuation Re-Think"). Governments can take direct action to diminish climate change and related water stresses such as: (1) pushing for a reduction in carbon emissions in order to avoid the worst risks (mitigation), and (2) investing in protecting people and assets from the expected physical threats arising from climate change (adaptation). A government's adaptation efforts can directly reduce the risks faced by its citizens and the economy. In the long term, this will help sovereigns to be financially resilient, and will ultimately lower their costs (Mirando and Tan, chapter "Chronic Coastal Threats Warrant a Valuation Re-Think").

Academic research is urgently needed to both enhance our understanding of water risk and its related challenges and to improve our practices. As Xu and Tan (chapter "The Developing Field of Water Risk Valuation for the Financial Industry") point out, further research and development on the various tools will only help improve investment decision making and risk management strategies in the future. In this context, Josset and Concha Larrauri (chapter "Data for Water Risks: Current Trends in Reporting Frameworks, Shortcomings, and the Way Forward") raise the following central but unanswered questions: "What metrics, methods, and tools [can be] useful to achieve relevant, standardized, meaningful, robust, and investment-related quantification of water risks? And what are the feedback mechanisms that would help reach societal objectives?" Additional research is also needed on such questions as: How can we gather better data that allow us to assess water risk, are there alternative ways to quantifying water risk, how can water risk models be stress tested, how does water risk affect company performance (and possibly survival), and how can countries and financial institutions better understand the interaction between water risk and other emerging risks? Other research venues that were not fully explored in our book include: wastewater reuse, water recycling, rainwater and stormwater harvesting, and the desalination of salt water (among many others). All of these have important implications for our society and financial markets and will hopefully be addressed in future research.

Ultimately, we can all help address the water problems our world is facing today: Academics can contribute by conducting both theoretical and practice-oriented research and by providing collective intelligence in areas non-academics may have disregarded or questions they were unable to answer; citizens can increase their water consciousness, both with respect to their personal water usage and by becoming better informed about the water footprint of the products they consume; investors can use the tools described in this book to make better informed investment decisions which should reduce their exposure to at-risk companies and punish companies that disregard the importance of water as a precious limited resource; and, finally, companies can use the insights provided in this book to improve their understanding of how water risks can be properly assessed and managed, and how they can reduce their exposure to water risks—thereby mitigating potential revenue losses and avoiding production stoppages, while at the same time protecting the communities in which they operate. Ideally, in this scenario, governments provide assistance and guidance to make sure that all parties have access to reliable information and that they employ suitable measurement and management techniques. In doing so, all these parties can improve the well-being of our society and can mitigate many of the direct and indirect consequences of water risk. In fact, biodiversity losses, poverty, malnourishment, disease, mass migration, political extremism, civil unrest, and armed conflicts all have been linked to water risks, highlighting the importance of the basic water molecule for both humankind and our broader planet. It will take a multipronged approach and dedication from all of us to reduce water risks and the potential issues that will come with them. We cannot address water risk separately, we need to unite: We call the world to action to protect water as our arguably most precious commodity and to enhance our efforts in managing it for the good of our environment, our economy, and our own well-being. We hope this book has inspired our readers to take on this challenge.

Index

T. Walker et al. (eds.), *Water Risk and Its Impact on the Financial Markets and Society*, Palgrave Studies in Sustainable Business In Association with Future Earth,
https://doi.org/10.1007/978-3-030-77650-3

The manufacturer's authorised representative in the EU is Springer Nature Customer Service Centre GmbH, Europaplatz 3, 69115 Heidelberg, Germany. If you have any concerns regarding our products, please contact ProductSafety@springernature.com

Printed and bound by CPI Group (UK) Ltd, Croydon, CR0 4YY

30/09/2026

02225700-0010